AF556953

PIONEERING V·E·N·U·S

A Planet Unveiled

The Pioneer Project and the Exploration of the Planet Venus

Library of Congress Cataloging-in-Publication Data

Fimmel, Richard O.
Pioneering venus : a planet unveiled / Richard O. Fimmel, Lawrence Colin, Eric Burgess ; prepared at NASA Ames Research Center.
p. cm.
Includes bibliographical references and index.
ISBN 0-9645537-0-8 (hardcover). --ISBN 0-9645537-1-6 (pbk.)
1. Venus probes. 2. Pioneer (Space probes) I. Colin, Lawrence. II. Burgess, Eric. III. Title.
QB621.F55 1995
523.4'2--dc20 95-16551
CIP

For Sale by the Superintendent of Documents
U.S. Government Printing Office, Washington, D.C. 20402

PIONEERING VENUS

A Planet Unveiled

Richard O. Fimmel
Manager, Pioneer Missions
Ames Research Center

Lawrence Colin
Project Scientist

Eric Burgess
Science Writer

Prepared at NASA Ames Research Center

CONTENTS

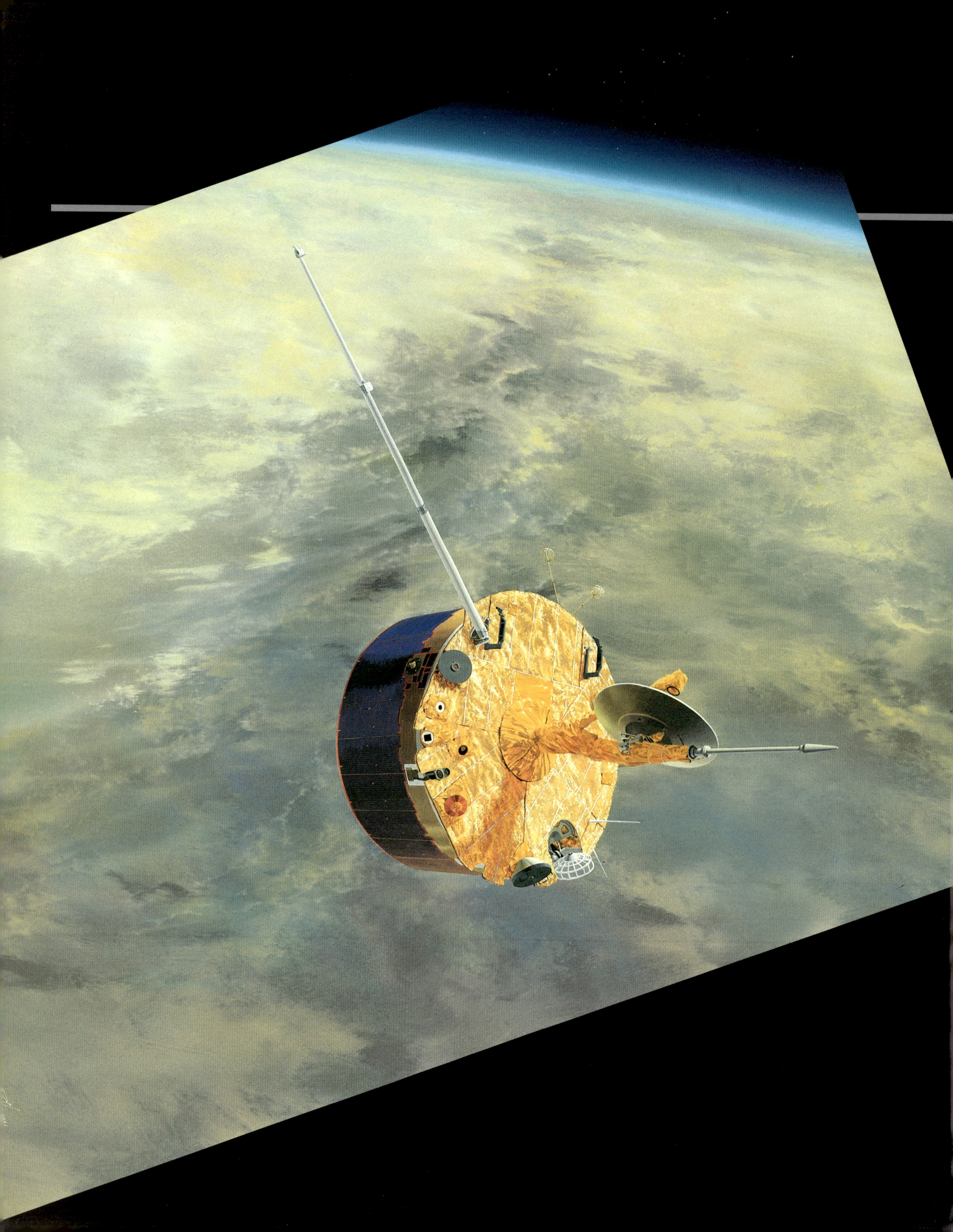

FOREWORD

Since their first launches in 1958, pioneering spacecraft have made many trail-blazing discoveries in exploring the Solar System. Initial attempts of our nation to probe interplanetary space were with spacecraft named Pioneer. The first of several small spacecraft was launched successfully by the Air Force Ballistic Missile Division in November 1958. One of these spacecraft escaped completely from Earth's gravity and went into solar orbit. It was the first interplanetary spacecraft.

Subsequently, NASA's Pioneers 6 through 9 made major discoveries about interplanetary particles and fields and the solar wind. Pioneers 10 and 11 explored the asteroid belt and the magnetospheres and physical natures of the giant outer planets Jupiter and Saturn. They discovered a new ring and new satellites of Saturn and sent to Earth the first images from spacecraft of the Galilean satellites. Pioneer 10 showed that spacecraft could safely travel through the asteroid belt and survive passage through the intense radiation environment of Jupiter, paving the way for the Pioneer 11 mission to Saturn. After their planetary encounters, Pioneers 10 and 11 headed out from the Solar System to the distant stars.

Pioneer Venus carried on the pioneering tradition. The two spacecraft of the mission, an Orbiter and a Multiprobe Bus, were launched from Kennedy Space Center in 1978. Four probes and the Multiprobe Bus penetrated the atmosphere of Venus and gathered important new data from the exosphere to the planet's hot surface. This probe gave scientists new insights not only about the Venusian atmosphere, but also about planetary atmospheres in general.

The 14-year Orbiter mission was equally successful. Its payload of advanced science instruments gathered a wealth of data about the atmosphere and ionosphere of Venus and their interactions with the solar wind. Additionally, the instruments penetrated the dense clouds of Venus for the first time and revealed global details of the planet's intriguing surface.

Pioneer Venus yielded a high scientific return for a relatively low cost. It produced valuable scientific data from 1978 through 1992—over more than one complete solar activity cycle. For 14 years the spacecraft continued in excellent working order. Its conservative design maintained all functions with only modest reductions from their original performance at the beginning of the mission. No complete failure of any critical component occurred.

The Pioneer Venus program was remarkable in the way it successfully pursued investigations over a broad range of planetary sciences. These included information gathered by the probes, radar and gravity mapping of the surface of Venus, investigations of the atmosphere from the surface through the clouds and the ionosphere to the exosphere, and the interaction of Venus with solar wind.

Also of great importance was the way the Pioneer Venus program created a sense of collegial scientific investigators cooperating with engineers and other mission personnel as a highly effective team for nearly 20 years of planning and mission operations.

Although the Orbiter's final entry into the Venusian atmosphere in October 1992 ended Pioneer Venus operations, the mission continues as scientists access its extensive archives of data about Venus. The mission also provided important groundwork for NASA's highly successful Magellan mission to Venus. Undoubtedly, the experience gained from Pioneer Venus will continue to be of great value to planners of future missions to the strange twin of Earth.

Ken K. Munechika
Director, NASA Ames Research Center

PREFACE

Pioneer Venus Orbiter completed an unprecedented 14 years in orbit about Venus, from December 1978 to October 1992. In this NASA Special Publication we describe for a wide readership the scientific discoveries not only of the Orbiter but also of the four probes and the Multiprobe Bus that entered the atmosphere of Venus and made many scientific measurements within that atmosphere.

The great excitement of any age has been created by pioneers—those who sought out new lands, new ideas, new social systems, new forms of governance and new goals for humankind. In our time we have been privileged to witness and be part of an outstanding human achievment of pioneers probing a great new frontier, space. Space pioneering has been a team effort of many people; dreamers, planners, technicians, engineers, scientists, and managers. The objective has been to broaden human knowledge about the wider environment beyond Earth and how this environment affects our own planet. To this end NASA sought information about the other planets of the Solar System through a series of interplanetary missions. Of these pioneering missions, the one described in this book targeted cloud-shrouded Venus which in several ways seemed to be a twin of Earth, but in others quite different from our planet. Scientists wanted to know how and why it differed.

The Pioneer Venus mission studied practically all aspects of the environment of Venus. Scientific investigations covered surface geology and electrical properties, gravity field, intrinsic and induced magnetic fields, neutral atmosphere composition and temperature structure, cloud structure and microphysics, atmospheric electrical discharges, ionospheric composition and temperature structure, and the complicated physics of the interaction of the solar wind with the planet over more than one solar activity cycle.

Many space scientists devoted a major part of their professional careers to this mission. They published a wealth of scientific papers; well over 1000 in a wide range of science journals. Of these Pioneer Venus scientists, 45% were from institutions in academia, 47% from federal laboratories, and 8% from industrial laboratories. Thirty-four colleges and universities, 14 federal laboratories, and 15 industrial laboratories were involved, and ten countries outside the U.S. were represented.

The mission demonstrated how a large amount of scientifically important information can be obtained in an extremely efficient and cost effective manner. A small number of management and spacecraft operations personnel supported the Pioneer Venus mission, relative to the large number of benefitting scientists. The mission was one of the most scientifically beneficial, low-cost programs conducted by NASA. It benefitted from a "lean and mean" highly professional project management and operations organization. The average annual funding to operate the mission over 13 years was $5 million, of which 60% was spent on science and 40% on management and operations. These laudable results were obtained by a team of dedicated, hardworking engineers and scientists who never underestimated the value of what they were doing.

In preparing this final report about the pioneering mission to our neighbor planet, we set out to make the presentation of information suitable for a wide readership including current and future students. Toward this end we appreciated the work of John Boeschen, our editor, who helped us simplify much of the involved science and technical material.

We are grateful to R. Z. Sagdeev, V. I. Moroz, and T. Breus who supplied the material for Chapter 7 about Soviet missions to Venus. Also, we thank T. M. Donahue for contributing the important final chapter in which he points out the relevance of studies of Venus to improving our understanding of Earth's evolution.

Richard O. Fimmel
Lawrence Colin
Eric Burgess — July 1994

DEDICATION

To the Memory of:

ROBERT BOESE
.... Original Principal Investigator,
Large Probe Infrared Radiometer.

HAL MASURSKY
.... Interdisciplinary Scientist,
Radar team.

FRED SCARF
.... Original Principal Investigator,
Orbiter Electric Field Detector.

JOHN H. WOLFE
.... Original Principal Investigator,
Solar Wind Plasma Experiment.

Pioneer Venus team members and colleagues.

CHAPTER 1

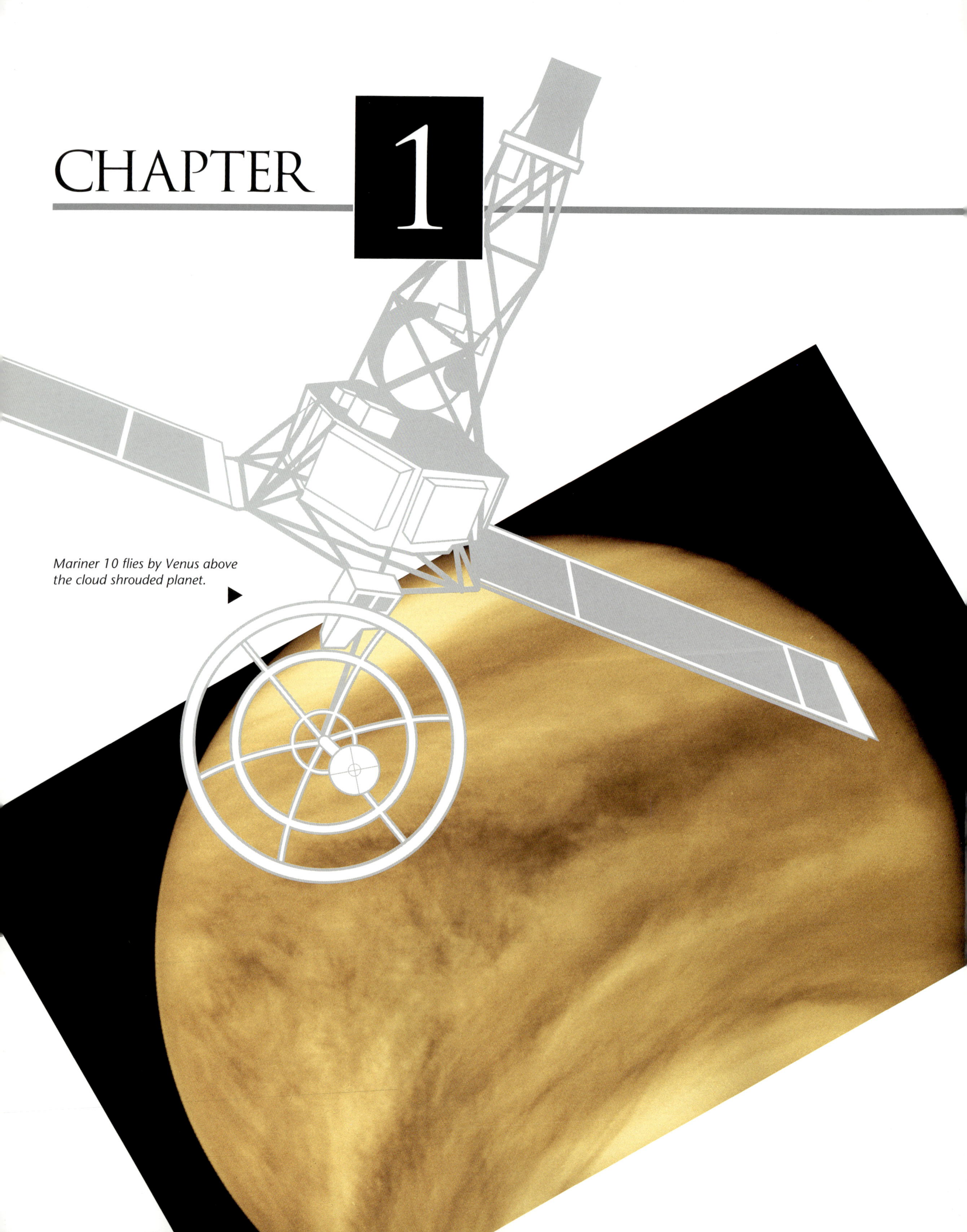

Mariner 10 flies by Venus above the cloud shrouded planet.

VENUS BEFORE PIONEER

This special publication presents the exciting story of Pioneer Venus, a National Aeronautics and Space Administration (NASA) program. In the following pages, you track the mission from its start through its highly successful operations and conclusion. This chapter flips back the calendar to the late 1960s when initial planning for an in-depth exploration of Venus began.

You might wonder what it was about Venus that rallied thousands of scientists behind Pioneer Venus. To understand their enthusiasm for Earth's sister planet, a review of what we knew about Venus before Pioneer is important. *Pioneer Venus* gives you this review. It also describes intriguing new knowledge that earlier U.S. and U.S.S.R. spacecraft missions brought us and why those missions emphasized the need for a Pioneer mission. Later chapters describe the mission's spacecraft, experiments, results, and their implications, and then provide background information about related Soviet/Russian missions.

Pre-Space-Age Knowledge

The brilliant planet Venus has intrigued humans since ancient times. The highly reflecting, cloud-shrouded planet is clearly visible from Earth, and shines brighter than all other objects in the sky except the Sun and Moon. Its risings and settings have been noted in many ancient records, including Babylonian clay tablets and Mayan codices. However, our ancestors did not understand these motions until the 15th century. At that time, the Copernican revolution in human thought acknowledged the Sun, not the Earth, to be the center of the Solar System, and that all the planets, including the Earth and Venus, revolve around it. The coming of the telescope in the 17th century revealed Venus as more than a star-like point of light. Now astronomers could measure the planet's apparent angular diameter and study its moon-like phases. These phases result from Venus' having an orbit that is inside that of the Earth. With a good pair of field glasses, you can see these phases yourself.

Venus is the one planet in our Solar System most similar to Earth in size and mass. Venus' mass, diameter, and density are all only slightly less than Earth's. There the resemblance ends. Its atmosphere is 100 times as dense as Earth's. Its surface is hot enough to melt lead. It rotates very slowly on its axis and has virtually no water. Its dense atmosphere consists mainly of carbon dioxide with clouds of sulfuric acid droplets. These differences intrigued planetary scientists, and they wondered why the two planets evolved along such different paths. Why is one capable of supporting life but not the other?

The image of Venus as seen through the best telescope is brilliant but uninteresting, and reveals little detail. During a relatively brief period in history, astronomers tried to measure the planet's rotation period and searched for some satellites, or moons, but they failed. Not too surprisingly, they shifted their interest to other, more revealing objects in the Solar System.

Eventually the development of new techniques spurred a revival of interest in Venus research. Beginning in the early 1900s, photographic and other instruments were developed along with powerful analytic methods. These could then be used to study Venus over a wide range of the electromagnetic spectrum.

Venus is visible from Earth, and humans have been observing the planet for thousands of years. The first sections in this chapter review our knowledge of Venus up to, but not including, Pioneer Venus. In these early pages, you learn about Venus' physical features and its place in our Solar System. The chapter concludes with descriptions of pre-Pioneer Venus space missions, their discoveries, and questions they left unanswered about the planet.

Figure 1-1. Because Venus orbits the Sun within Earth's orbit, it appears to stay close to the Sun as we observe it from Earth. At its greatest angular distance from the Sun, Venus is at eastern or western elongation. In this figure, the planet appears at eastern elongation when it sets after the Sun, and we see it as an evening "star." At western elongation, it is visible rising before the Sun as a morning "star."

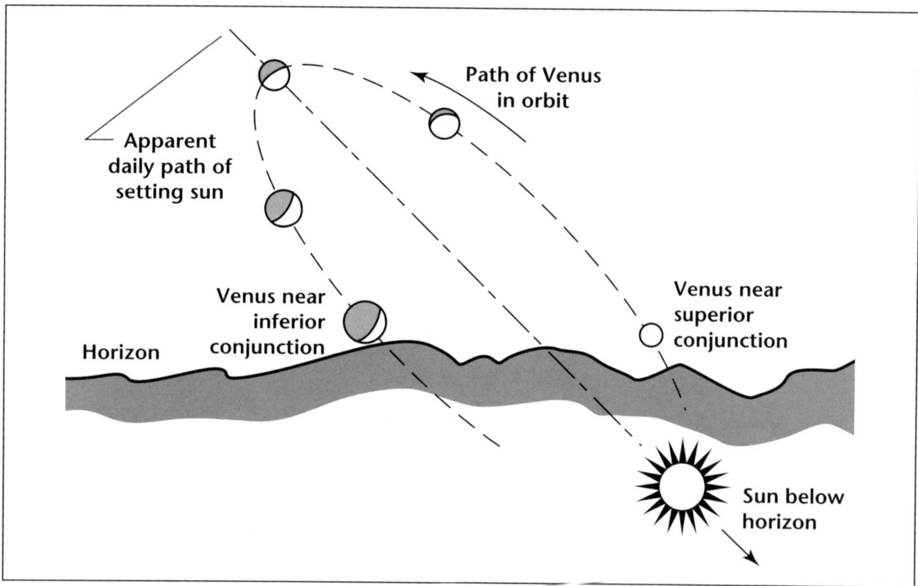

Scientists used infrared wavelengths to characterize the clouds and overlying atmospheric gases. Information about the surface and lower atmosphere came from microwave emissions. Analysis of radar signals that bounced off the planet determined its period of rotation. However, major discoveries about Venus had to wait until the 1960s when spacecraft became available to explore the planet. The first successful interplanetary probe, Mariner 2, flew by Venus in 1962. That flight began the space-age exploration of the second planet and our Solar System.

Venus as a Member of the Solar System

Astronomers call Venus an inferior planet because it revolves around the Sun inside Earth's orbit. (Its average distance from the Sun is 72.3% of Earth's average distance from the Sun). As a result, you see Venus as either a morning or an evening "star." Early peoples believed these two bright "wandering stars" were separate objects and gave them different names. The Greeks, for example, named them Phosphorus and Hesperus.

Venus appears to move through the constellations of the zodiac. It travels close to the ecliptic—the apparent yearly path of the Sun relative to the stars, which is the plane of Earth's orbit projected against the stars—and oscillates east and west of the Sun but never more than 48 degrees from it. We call the planet's positions at maximum angular distance east and west of the Sun the eastern and western elongations, respectively. At eastern elongation, Venus is an evening object. Each day, it follows the Sun across the sky (Figure 1-1). At western elongation, Venus rises before the Sun each day. The planet passes from greatest eastern elongation to greatest western elongation in about 144 days and from western to eastern in about 440 days.

Because it reflects 71% of the sunlight that bathes it, Venus is bright enough to see at midday if you know where to look. It is brightest about one month before and one month after inferior conjunction. This is when the planet passes closest to Earth between Earth and Sun. As noted earlier, Venus exhibits phases like the Moon (Figure 1-2). When it is brightest in Earth's skies, Venus appears as a fat crescent.

Venus takes 224.7 days to revolve around the Sun in its almost circular orbit (the orbit has a mean radius of 108.2 million km, or 67.2 million miles). Because Earth also moves around the Sun, the periods when Venus is visible at elongations or at conjunctions repeat every 583.92 days. Opportunities to send spacecraft to Venus with minimum energy also repeat with this period.

When behind the Sun at superior conjunction, Venus is 257.3 million km (159.9 million

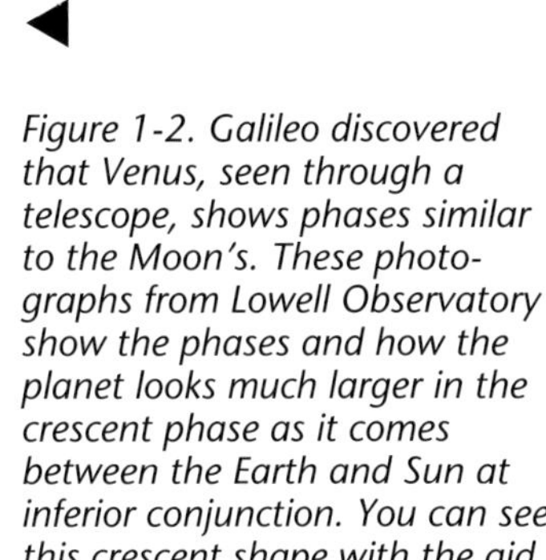

Figure 1-2. Galileo discovered that Venus, seen through a telescope, shows phases similar to the Moon's. These photographs from Lowell Observatory show the phases and how the planet looks much larger in the crescent phase as it comes between the Earth and Sun at inferior conjunction. You can see this crescent shape with the aid of a good pair of field glasses.

(Appendix A lists some major events in the exploration of Venus by Earth-based observations and from theoretical inferences.)

miles) from Earth. At inferior conjunction (Figure 1-3), Venus is 41.9 million km (26 million miles) from Earth. However, Earth's orbit is inclined 3.4 degrees to Venus' orbit, so Venus is nearly always slightly above or below the Sun at inferior conjunction. Only infrequently does the planet travel in front of the Sun (as we see it from Earth). Scientists refer to this movement as a transit. During transit, Venus is visible as a small black disk silhouetted on the bright face of the Sun. Transits of Venus occur in pairs 8 years apart with over a century intervening between successive pairs. The most recent transits occurred in 1874 and 1882, and the next pair will be on June 7, 2004, and June 5, 2012.

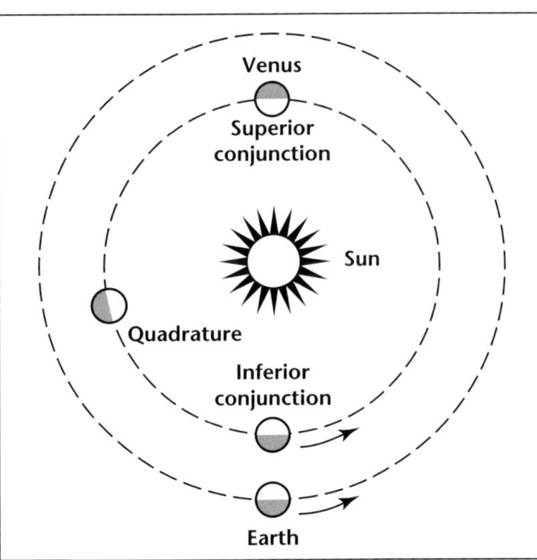

Figure 1-3. When Venus is closest to Earth at inferior conjunction, the planet is between the Earth and Sun. On the far side of the Sun, and most distant from Earth, Venus is at superior conjunction. Sometimes at inferior conjunction the positions of Venus and Earth on their orbits are such that Venus passes in front of the Sun's disc. Astronomers call this passage a transit. The next pair of transits occurs early in the 21st century.

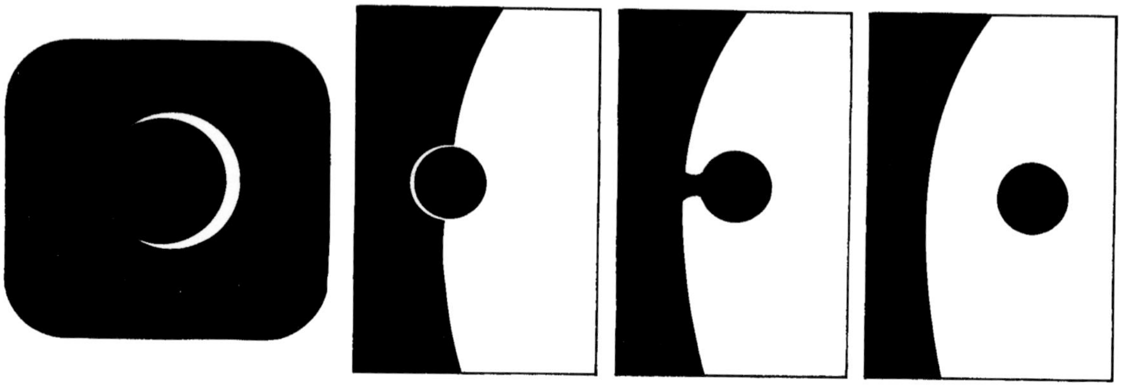

Figure 1-4. When Venus transits the Sun's disk, the planet's atmosphere distorts the black spot silhouetted against the bright solar photosphere. This optical effect reveals that Venus has an atmosphere. Also, when Venus is close to the Sun, as you observe it from Earth, its bright, thin crescent extends around the dark globe. This occurs because of the effects of the planet's atmosphere (see the leftmost diagram).

In the past, astronomers used Venus transit times to help determine the Earth's distance from the Sun. In 1874 and 1892, astronomers therefore journeyed to remote regions of the globe to observe Venus' transit with sensitive instruments. However, their efforts were foiled by a strange optical effect (Figure 1-4). As transit started, the planet's black disk would appear to remain connected to the dark sky beyond the limb of the Sun (i.e., the edge of the Sun's optical disk). The connection would thin to a mere thread, then snap. The Russian chemist M. V. Lomonosov transformed this annoying effect into an important discovery when he correctly attributed it to an atmosphere around Venus.

Table 1-1 summarizes the characteristics of Venus' orbit.

Table 1-1. Orbit of Venus

Mean distance from Sun	0.723 AU 108.2 million km 67.2 million miles
Inclination of orbit to plane of the ecliptic	3.39°
Sidereal period (period with respect to the stars)	224.7 Earth days
Mean synodic period (period with respect to Earth)	583.92 Earth days
Mean orbital velocity	35.05 km/sec 21.78 miles/sec
Closest approach to Earth	41.9 million km 26.0 million miles

Venus as a Planet

Why is Venus so different from Earth? The environment on Venus today differs significantly from our planet's. Its surface is much hotter, and its atmosphere is nearly 100 times as dense. Also, its rotation is much slower and is retrograde, meaning it is in the direction opposite to Earth's rotation and the general motions of the planets around the Sun. To make observation even more difficult, unbroken, planet-wide clouds hide Venus' surface. In ultraviolet light, these clouds show markings that appear to rotate about the planet in a period of 4-5 days. Astronomers had discovered that the mainly carbon-dioxide atmosphere contained only minute amounts of water vapor. Because Venus' magnetic field (if there is one) is small, the planet's interaction with the solar wind is different from Earth's.

Table 1-2. Physical Data on Venus

Diameter (solid surface)	12,100 km 7,519 miles 0.95 Earth's diameter
Diameter (top of clouds)	12,240 km 7,606 miles
Mass	48.8 × 10^{26} g 0.815 Earth masses
Density	5.269 gm/cm^3 0.96 Earth's density
Axial rotation period (retrograde)	243.1 Earth days
Rotation period, cloud tops (retrograde)	4.0 Earth days (approximately)
Period of solar day	116.8 Earth days
Inclination of rotation axis	177.0°
Surface gravity	888 cm/sec^2 0.907 g
Surface atmospheric pressure	9,616 kPa 1,396 psi 95 Earth atmospheres
Surface temperature	750 K (approximate) 480°C (approximate) 900°F (approximate)
Reflecting capability (albedo)	0.71 1.82 Earth's albedo
Stellar magnitude when brightest	-4.4

Venus also lacks a satellite. Physical data on the planet appear in Table 1-2.

Period of Rotation

Look through an optical telescope on Earth. Try as you might, you won't see clear details on Venus' brilliant, yellowish disk. Some early observers, though, claimed they saw faint, elusive markings. Did they really see them? We can't be sure. However, the markings they described were similar to those you would expect on extensive cloud systems.

As late as 1964, Earl C. Slipher, famous planetary photographer of Lowell Observatory, Flagstaff, Arizona, wrote, "All the early efforts to photograph Venus at Flagstaff (from 1904 on) . . . succeeded in registering only faint vague markings, too weak to add new information." The general absence of visible surface features prevented astronomers from measuring Venus' period of rotation. Wildly varying periods were claimed—from 24 Earth hours to a period equal to the Venus year (224.7 Earth days).

On May 10, 1961, a radar signal from a NASA Deep Space Network antenna at Goldstone, California, was bounced off Venus. Analysis of the returned echo indicated that the planet rotated extremely slowly. Later, radar astronomers determined that Venus rotates about its axis in 243.1 Earth days in the direction opposite to Earth's. Because its axial rotation and orbital revolution are of comparable periods, a solar day on Venus is 116.8 Earth days. Just imagine: 58 Earth days of daytime and an equally long nighttime. And the Sun rises in the west and sets in the east!

Strangely, Venus' period of rotation is almost locked to the periods of revolution of Earth and Venus around the Sun. The result: Venus turns very nearly the same hemisphere to Earth each time the planet passes between Earth and Sun at inferior conjunction.

Why Venus rotates so slowly is still an unsolved mystery—most other planets rotate in periods of hours rather than days. While scientists attribute Mercury's slow rotation to

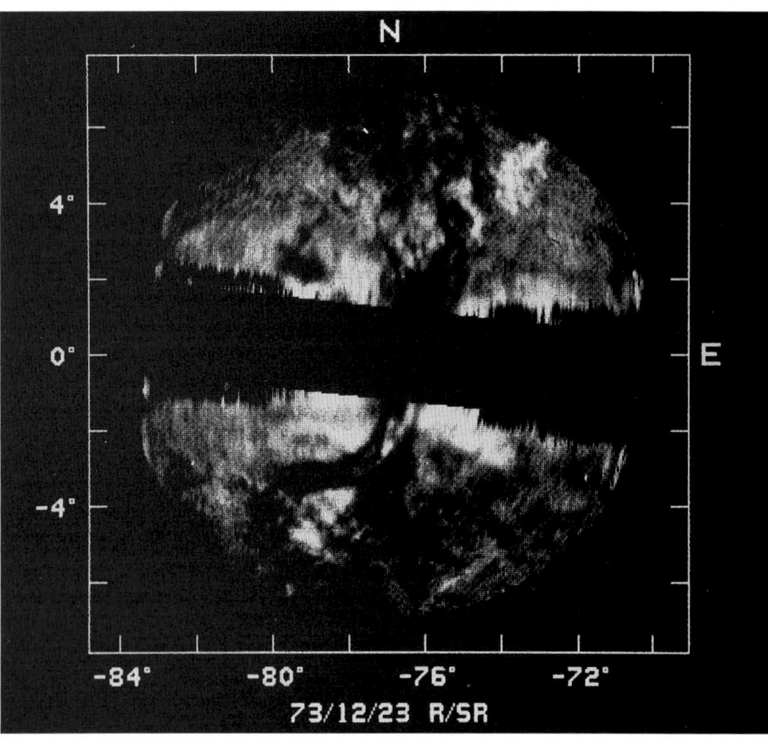

Figure 1-5. While it is impossible to see through the clouds of Venus at optical wavelengths, radar can penetrate to the surface. Radar maps of Venus show many surface features. An early radar picture of the planet's surface appears in this figure. It is one of a series that R. M. Goldstein of the Jet Propulsion Laboratory obtained (he used equipment from NASA Deep Space Network with a large antenna at Goldstone, California).

the Sun's tidal effects, Venus is too far from the Sun for such effects to be significant over the planet's 4.6 billion year lifetime. One speculation is that a grazing collision with an asteroid-sized body slowed Venus' rotation.

Shape of Venus

Scientists have used the Earth's moon and artificial Earth satellites to explore its gravitational field. This information, along with the Earth's deviation from perfect sphericity (i.e., its oblateness), can be used to help develop models of the Earth's interior. However, Venus is almost a perfect sphere. Its lack of oblateness and lack of a satellite prevented astronomers from developing good models of the planet's internal structure and composition. Most planetologists assumed that Venus' interior was similar to Earth's: a liquid core, a solid mantle, and a solid crust.

Surface Features

Venus' surface remained a mystery until radar probed through its dense atmosphere. Using radar, scientists discovered large but shallow circular features in its equatorial regions. Scientists believed these were most likely craters. Stretching 1000 km (621 miles) north and south across the equator (Figure 1-5) was a major chasm. Radar observations also showed a large-scale granular surface structure, which might be a rock-strewn desert. Planetologists interpreted some areas of high radar reflectivity as vast lava flows and mountainous areas.

Despite Venus' dense atmosphere and clouds, some sunlight does penetrate to the surface. At these locations, solar flux is about equal to an overcast day in midlatitudes at Earth's surface. Instruments measured the amount of solar radiation at the surface at an integrated flux of about 14,000 lux (when the Sun was at about a 30° angle from overhead). Photographs from one Soviet lander spacecraft (Figure 1-6) confirmed a dry, rocky surface that unknown processes have fractured and moved about. A second lander produced a picture of rocks with rounded edges and pitted surfaces. Measurements from the spacecraft indicated that surface rocks have a density between 2.7 and 2.9 g/cm^3, typical of basaltic rocks on Earth. This information supported earlier theories that Venus had separated into a core, mantle, and crust.

Other early spacecraft results showed that Venus had little water. Did Venus ever have oceans? If it did, what happened to them? Some researchers speculated that water rose as vapor into the high atmosphere, where solar radiation broke it down into hydrogen and oxygen. The hydrogen then escaped into space from the top of Venus' atmosphere while heavier oxygen remained and oxidized crustal rocks. Others hypothesized that Venus might have formed so close to the Sun that high temperatures within the solar nebula prevented water from condensing and becoming part of the planet. If so, Venus would never have had enough water within its rocks to form early,

deep oceans like Earth's. Our oceans played a role in clearing the atmosphere of most of Earth's carbon dioxide through the reaction of the carbon dioxide with water to form carbonate rock. By contrast, Venus' carbon dioxide has remained mainly in its atmosphere.

On Venus, because of high surface temperatures, scientists expected chemical reactions between the atmosphere and the minerals in rocks to occur much faster than on Earth. However, on our wet planet, running water continually exposes rocks to the atmosphere and speeds chemical reactions. But without water, it seemed unlikely that such processes would take place. Of course, unless fresh rocks were continually exposed, Venus' atmosphere would never achieve equilibrium with surface materials.

Atmospheric Composition

Although astronomers discovered Venus' atmosphere in the 16th century, its extent and composition remained a mystery until recently. The planet's atmosphere consists of three distinct regions: the part above the visible cloud tops, consisting of the ionosphere and exosphere; the clouds; and the region extending from the base of the clouds to the surface.

In the 1930s, infrared spectroscopy revealed carbon dioxide absorption bands in Venus' spectrum. Carbon dioxide appeared to be much more abundant in Venus' atmosphere than in Earth's. Later, high-resolution spectroscopy confirmed that carbon dioxide is the dominant gas. It also found traces of water vapor, carbon monoxide, hydrochloric acid, and hydrogen fluoride. Unfortunately, spectroscopy could not reveal the exact amount of carbon dioxide.

Soviet space probes that penetrated the Venusian atmosphere (see Chapter 7) confirmed Earth-based observations, and Veneras 4 and 5 suggested a concentration of 97% carbon dioxide. Radio-occultation data confirmed these probe measurements. However, temperature and pressure measurements from probes differed from radio-occultation measurements in a way that seemed best explained by supposing that Venus' atmosphere contained only

▲

Figure 1-6. The first picture from the surface of Venus, obtained by the Soviet spacecraft Venera 9 in 1975, shows a rocky surface and a clear view to the horizon. The rocks appear to have been fractured and broken in a geologically recent time.

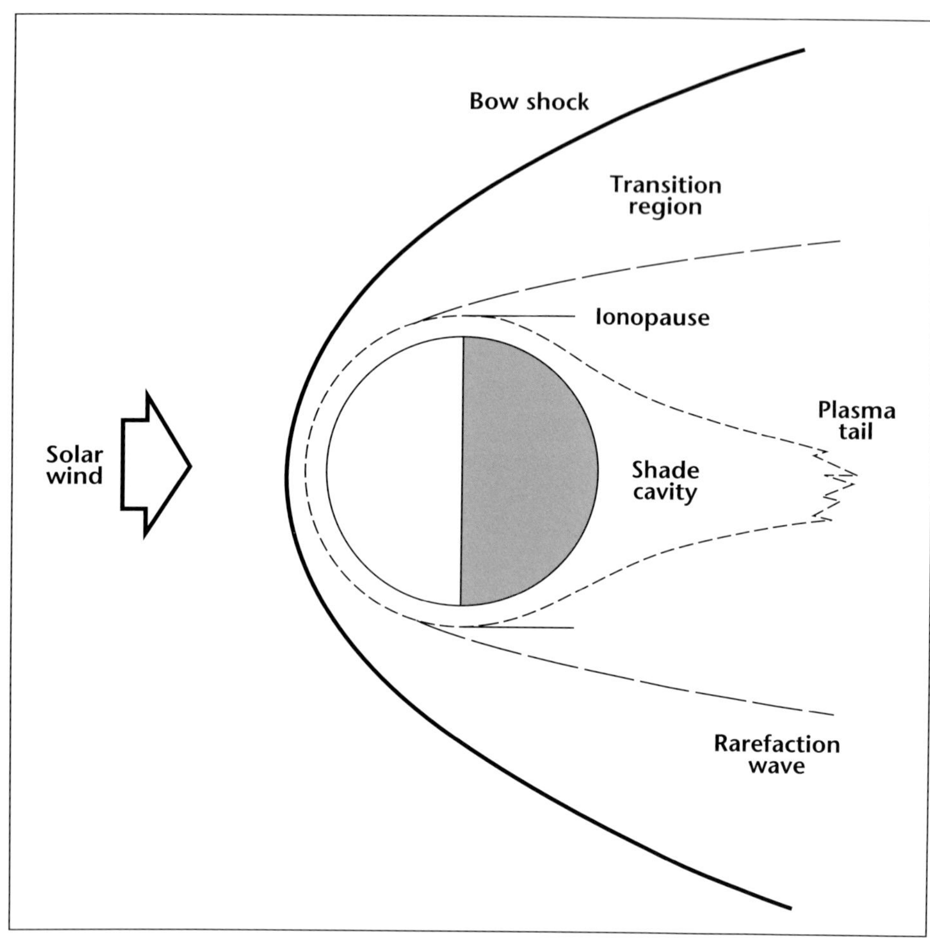

▲

Figure 1-7. Because Venus does not have a magnetic field, the planet interacts much differently with the solar wind than does Earth. This simplified diagram shows the expected configuration as scientists understood it before the Pioneer Venus mission.

70% carbon dioxide. Also, if there were large amounts of argon in the atmosphere, carbon dioxide could be as low as 25% and still satisfy all the measurements astronomers made from Earth.

The amount of carbon dioxide in a planetary atmosphere affects how scientists interpret the planet's microwave spectrum. With accepted percentages of carbon dioxide, microwave observations indicated as much as 0.5% water vapor below Venus' clouds. Instruments on Veneras 9 and 10 provided data that suggested 0.1% water vapor below the clouds. At the cloud tops, however, they indicated only 0.0001% water vapor. Of course, there was the chance that if the atmosphere contained another gas that was a poor absorber of microwaves, the planet's atmosphere could contain even more water. If that were true, scientists might account for the larger amounts of water that Veneras 4 and 5 measured at the surface.

On the other hand, the spacecrafts' measurements might have been flawed—passage through Venus' sulfuric acid clouds could have contaminated their instruments.

Carbon dioxide has also played an important role in the evolution of the planet's atmosphere. And it affects the radiative properties and dynamic traits of the present atmosphere. Despite carbon dioxide's preponderance, the total amount of the gas seems to be about the same as that locked up in carbonate rocks in Earth's crust.

Upper Atmosphere

Observations from Earth and from flyby and orbiting spacecraft provided data on the atmospheric region above the cloud tops. In contrast with the lower atmosphere, this region was colder and, above 150 km (93 miles), more rarefied than Earth's atmosphere.

Because Venus lacks a significant magnetic field, the solar wind interacts directly with the upper atmosphere and ionosphere (Figure 1-7). Venus' ionosphere is thinner and closer to the planet's surface than is Earth's. Like our ionosphere, Venus' has layers where the electron density peaks (Figure 1-8). Peak electron density in Earth's ionosphere is about 100,000 to 1,000,000 electrons/cm^3 at about 250 to 300 km (155 to 186 miles). The major ion is atomic oxygen. On Venus, by contrast, scientists measured a peak of about 600,000 electrons/cm^3 at about 142 km (88 miles). The major ion there appeared to be molecular oxygen.

NASA's Mariner 10 spacecraft, which in 1973 flew by Venus on its way to Mercury, found two clearly defined layers in the nighttime ionosphere (see Figure 1-8): a main layer at 142 km (88 miles) and a lesser layer at 124 km (77 miles). The lower layer had a peak density

about 75% of the higher layer. Spacecraft data revealed a sharp boundary (ionopause) in the dayside ionosphere at 350 km (217 miles). Measurements from the 1967 Mariner 5 spacecraft had placed the boundary at 500 km (311 miles). On the planet's nightside, the ionosphere was found to extend high into space, probably into a long plasma tail stretching away from the Sun.

Radio occultation data—measurements of a spacecraft's radio signal as it disappears behind the planet—allowed researchers to determine temperatures in the region just above the cloud tops. At higher altitudes, in the exosphere, temperatures were determined from measurements of radiated ultraviolet radiation (airglow). Temperatures at the top of the Venusian ionosphere required a gas much lighter than carbon dioxide. Scientists speculated that this gas might be helium, because (1) at 127°C (260°F) or so, the thermal escape of helium from the atmosphere would be small, and (2) if helium had outgassed from Venus' rocks early in its history, as had occurred on Earth, then some of the helium would likely have collected in Venus' upper atmosphere. Finally, from both infrared and ultraviolet emission measurements, researchers discovered a corona of hydrogen atoms beginning at about 800 km (497 miles) altitude, containing up to 10,000 atoms/cm^3.

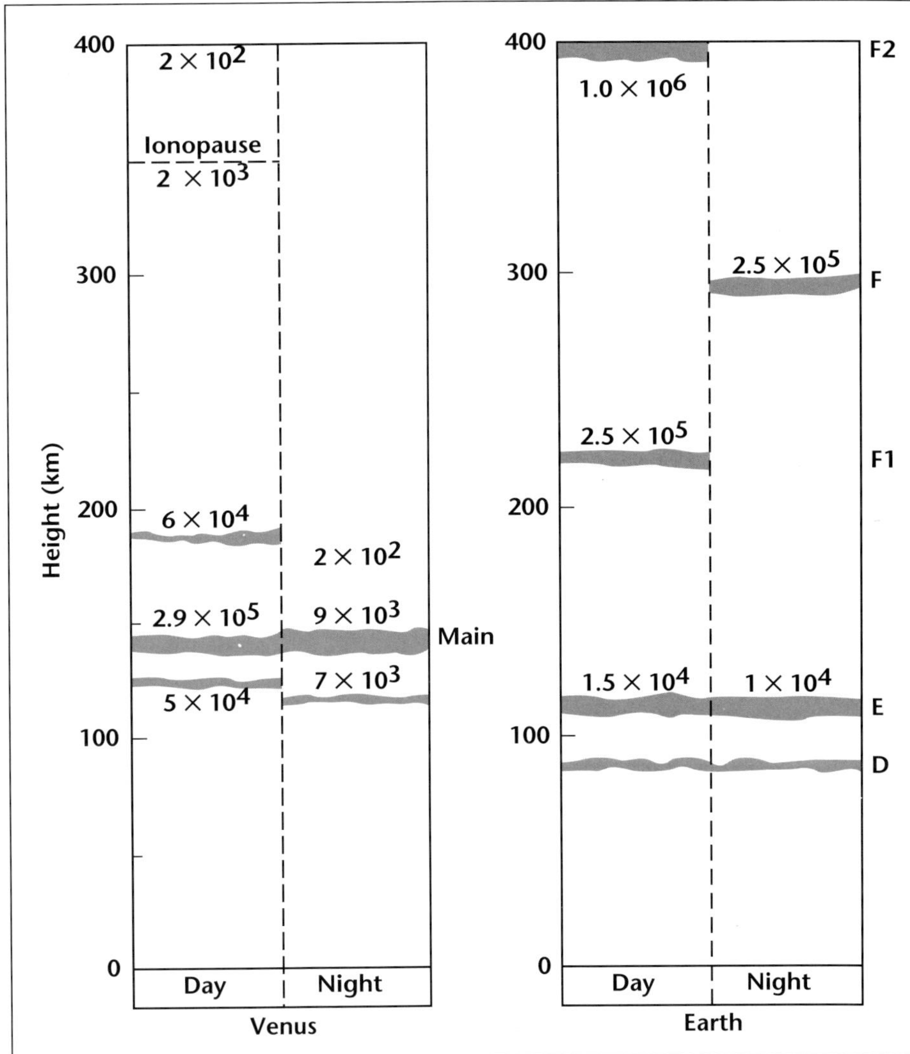

Figure 1-8. Made before the Pioneer Venus mission, these graphs compare the ionosphere layers of Earth and Venus.

Clouds

Mariner 10 photographed at least two layers of extremely wispy haze above the main cloud deck—probably layers of aerosols—80 to 90 km (50 to 56 miles) above the planet's surface. The layers extended from equatorial regions to higher latitudes.

Scientists did not understand the main cloud layers' composition. In fact, the clouds remained controversial until the early 1990s. At one time, astronomers speculated they were dust and extended down to the surface. Another speculation was that they were condensation clouds with a clear atmosphere beneath them. Suggested components included ammonium nitrate, carbon suboxide, formaldehyde, nitrogen dioxide, polymers of hydrocarbonamide, and hydrochloric acid.

From polarization studies, scientists had concluded by 1971 that the cloud particles had to be spherical, about 1 to 2 microns in diameter, and were not grains of dust. Neither

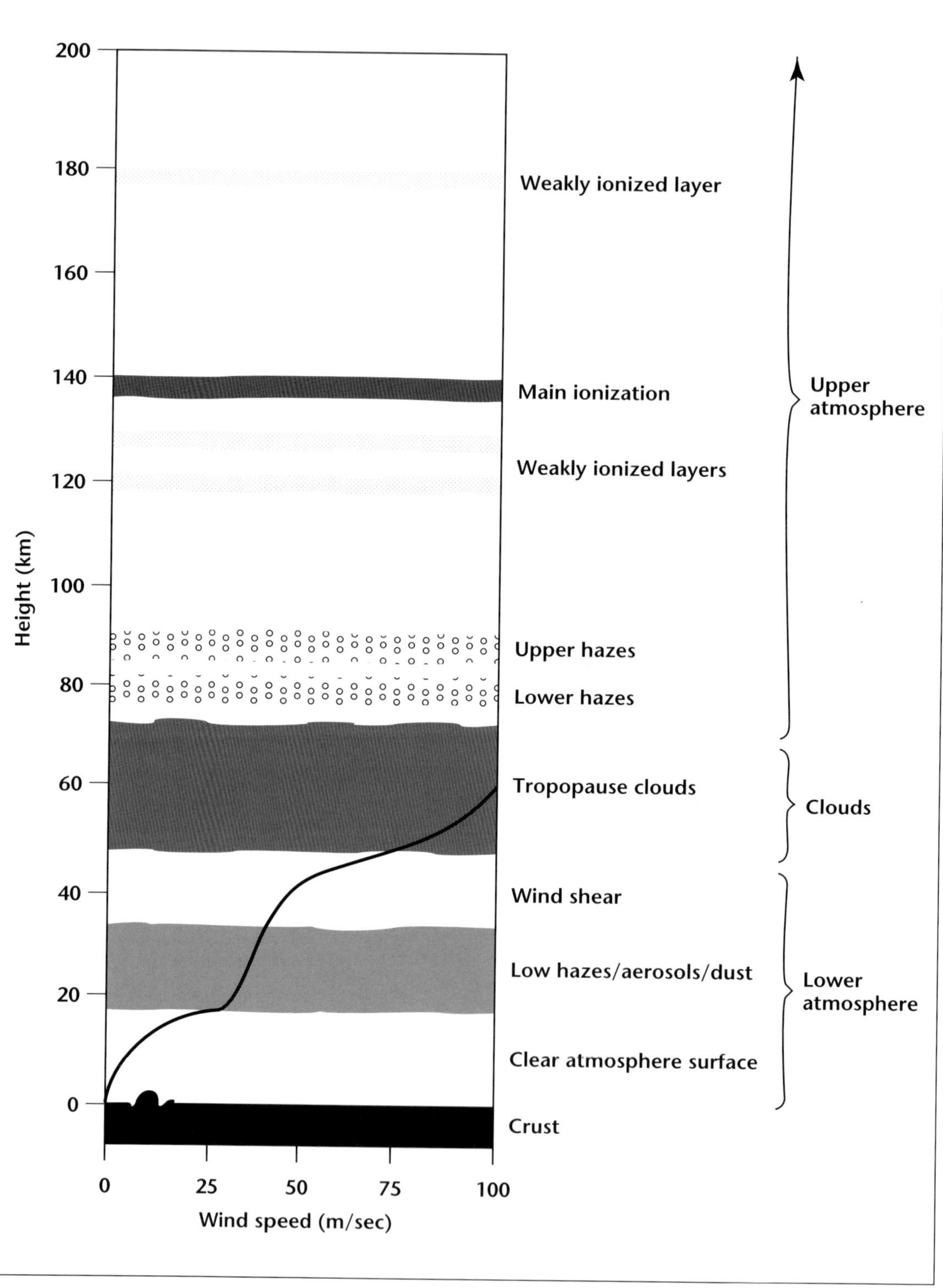

Figure 1-9. Scientists inferred three distinct regions of Venus' atmosphere from several sources. These included earlier spacecraft flyby missions, Soviet entry probes, and Earth-based observations. You can see these regions in the diagram. They are the high atmosphere above the clouds, the thick layer of clouds, and the clear atmosphere beneath the clouds. The diagram also shows a wind velocity profile to illustrate how the wind decreases abruptly at the base of clouds.

did they seem to be ice or water droplets, nor droplets of hydrochloric acid or carbon suboxide.

Scientists now accept that the cloud droplets are composed of sulfuric acid. They reached this conclusion in 1973 after studying measurements of Venus' infrared spectrum that had been made with instruments aboard a Learjet high in Earth's atmosphere. Two theorists had suggested this composition earlier, pointing out that concentrated sulfuric acid is a very effective drying agent and could account for the atmosphere's dryness above the cloud tops.

The droplets consist of about a 75% acid-water solution and are about 1 micron in diameter. Sulfuric-acid clouds can remain as clouds over a wider range of temperatures than water clouds. Below the bottom of main cloud layers, the temperature is high enough for sulfuric-acid droplets to evaporate into water and sulfuric-acid vapors.

While Venus' clouds seem opaque from Earth, they are, in fact, very tenuous but deep layers. Veneras 9 and 10 determined that visibility within clouds is between 1 and 3 km (0.6 to 1.9 miles). These clouds are more like thin hazes than typical clouds on Earth. They form a very deep region some 15 to 20 km (9 to 12 miles) thick (Figure 1-9). This is more than twice the thickness of cloud layers on Earth. Venera spacecraft passed through several layers and emerged from the cloud deck's lower boundary at about 49 km (30 miles). Scientists on Earth have studied distinctive, dark ultraviolet markings on the clouds. These are probably the same optical markings that early observers had noted. In Figure 1-10 you can see horizontal ψ-shaped features. They have an extension of the equatorial bar through arms that are sometimes angular and at other times circular. Features that look like a reversed letter C appear more often near the evening terminator than the morning terminator. Horizontal Y-shaped features sometimes have a tail stretching round the planet. Sometimes there are two parallel equatorial bands. Patterns are mostly symmetrical about the equator. Arms of the various features open in the direction of their retrograde motion, which varies between 50 and 130 m/sec (164 and 427 ft/sec). However, a major question about cloud motions remained unsolved: did they result from actual movement of atmospheric masses? Or were they merely a wave motion?

Winds

Even before Pioneer's in-depth exploration of Venus, astronomers had determined that Venus' stratosphere appears to have a

Figure 1-10. Characteristic cloud markings on Venus appear in three drawings at the right and a photograph at the left. Astronomers observed the C-, Y-, and ψ-shaped markings from Earth. Mariner 10 images, returned as the spacecraft flew past the planet, confirmed the markings.

continuous zonal motion averaging 100 m/sec (328 ft/sec). This speed indicates a rotation period of approximately 4 days, which is 60 times faster than the planet's own spin. This difference in speed, relative to the planet's surface, causes high-velocity winds to blow continually in the high atmosphere. Deeper in the atmosphere, wind velocities decrease greatly, dwindling to a relative calm near the surface. The Soviet probes showed an abrupt change between high- and low-wind velocities at about 56 km (35 miles) altitude. This change occurs near the base of clouds. Over the whole of the planet, meridional winds of much less velocity blow, with the atmosphere rising at low latitudes and sinking toward the poles.

Thermal emission from the upper atmosphere differed little between night and day and between low and high latitudes. This showed there is strong dynamic activity within the atmosphere, and heat in large amounts is transferred around the planet horizontally from day to night and from equator to poles. While diurnal, or daytime, heating is important above 56 km (35 miles), dynamic effects prevail below that altitude.

Magnetic Field

Venus' lack of a magnetic field is another important difference between it and Earth. Earth's field is strong, amounting to about 0.5 gauss at its surface. In 1962, Mariner 2, the first spacecraft to fly by Venus, discovered that Venus has no significant field. In fact, Venus' field strength is less than 1/10,000 of Earth's.

Scientists still do not completely understand how planets generate and maintain their magnetic fields. They believe a self-sustaining dynamo in a fluid core accounts for Earth's field. Convection currents in the core cause electric currents, and these produce the external magnetic field. This theory, which seems to apply to Jupiter, Saturn, Uranus, and Neptune, predicted that slow-spinning satellites and planets without molten cores do not have magnetic fields. However, this dynamo theory failed to predict slow-spinning Mercury's magnetic field, discovered by Mariner 10.

Lack of a Satellite

Several astronomers in the 1800s claimed discovery of a "moon" of Venus. However, their satellites turned out to be faint stars. Venus does not have a satellite.

Early Spacecraft Missions to Venus

Before the Pioneer Venus mission, Venus had been the target for 13 spacecraft. Three of these were American and 10 were Russian. Five were flybys and 8 were landers. Several Russian missions were flybys and landers that separated before reaching Venus.

Initial Soviet attempts to reach Venus with spacecraft failed. Then came the spectacular 190-day voyage of NASA's Mariner 2 in 1962. Mariner 2 was America's first interplanetary spacecraft, and it flew within 34,833 km (21,645 miles) of the planet.

During the rest of the 1960s, Russia and America used two different methods to explore Venus. The Russians flew probe and lander missions as well as flybys. The United States used flybys only. The two countries sometimes obtained conflicting information about Venus. For example, a Soviet Venera 4 lander recorded a surface temperature of 265°C (510°F) in 1967. In the same year, Mariner 5 flyby experiments indicated a surface temperature of 527°C (981°F). Atmospheric pressure calculations did not agree either. Later, scientists learned why. Atmospheric pressure had crushed Venera 4 at an altitude of about 34 km (21 miles)—the probe had never reached the surface.

The 1969 Soviet landers were structurally tougher, but even they failed to survive the atmosphere's intense pressure. The Soviets finally tasted success in 1970 when Venera 7 landed on Venus and returned data for 23 minutes. Later in the 1970s, other Soviet landers returned pictures of the rock-strewn surface. For more information on the pre-Pioneer Soviet program, turn to Chapter 7. Descriptions of major findings for three American flybys appear next.

Mariner 2

A flyby spacecraft, Mariner 2 blasted off on August 27, 1962. The spacecraft flew within 34,833 km (21,645 miles) of Venus on December 14, 1962. Among the mission's discoveries were that (1) Venus is blanketed by cold, dense clouds about 25 km (15.5 miles) thick with a top at or about 80 km (50 miles); (2) the surface temperature is at least 425°C (800°F) on both day and night hemispheres; and, (3) the planet has virtually no magnetic field or radiation belts.

Mariner 5

A flyby spacecraft, launched June 14, 1967, Mariner 5 passed Venus at 3391 km (2107 miles) on October 19, 1967. Occultation experiments provided readings that helped scientists calculate temperatures of 527°C (981°F) and pressures of 100 atmospheres on the surface. Researchers also determined detailed ionospheric structure at two locations on the planet. Using an ultraviolet photometer, they observed very low exospheric temperatures that were unexpected and difficult to explain.

Mariner 10

A spacecraft bound for Mercury, Mariner 10 passed Venus en route. NASA launched it on November 3, 1973, and it flew past Venus at 5793 km (3600 miles) on February 5, 1974. Mariner 10 was the first spacecraft to photograph Venus' clouds. Taken in ultraviolet light, the photographs revealed the clouds' structural details. Mariner 10 also confirmed the reality of the C-, Y-, and ψ-shaped markings and verified the 4-day rotation period of the ultraviolet markings. The spacecraft found significant amounts of helium and hydrogen in the upper atmosphere. Using optical limb scanning, scientists detected high altitude haze layers in the upper atmosphere above the cloud tops. Mariner 10 confirmed that Venus lacks a magnetic field of any consequence, determined the structure of the ionosphere, and established temperature and pressure profiles into the upper atmosphere.

Unanswered Questions

Many questions about Venus' atmosphere remained unresolved at the time. How does the Venus weather machine work? What makes Venus so hot compared to Earth—a greenhouse effect? Or is there a significant dynamic contribution? How did the atmosphere of Venus evolve? Did Venus once have a more moderate surface temperature? What caused the dark ultraviolet markings in Venus' clouds? What are the constituents of the atmosphere at different levels?

Scientists believed that the answers to such questions would help us learn more about our own Earth. While many factors complicate Earth's meteorology (mixing of oceanic and continental air masses, partial cloud cover, axial tilt, and rapid planetary rotation), Venus' meteorology appeared to be much simpler. The atmosphere has a basic composition of 97% carbon dioxide, with hardly any water. There are no oceans to complicate matters, and because the planet has a slow rotation, Coriolis forces are minor. Since its spin axis tilts only slightly, there are virtually no seasonal effects.

At the time of the Venera landings in 1975, Louis D. Friedman and John L. Lewis made several important observations. They pointed out that, despite all the missions to Venus, some of the most important and fundamental scientific questions remained unanswered. For example, very few early results helped explain why Venus differs so much from Earth.

Without answers to basic questions, how could we learn more about planetary processes and evolution? We needed to know more about Venus' global chemical composition, and its thermal and differentiation history. This required information about crustal composition, the planet's internal structure, and the ages of crustal rocks. We needed to know if there was evidence of tectonic activity, continental drift, and volcanism. Mapping of the gravitational field in local regions and other geodetic data also were important, as was the mapping of surface features to determine local geologic structure. We needed to know more about atmospheric composition, thermal structure, cloud structure, and atmospheric circulation. In short, early spacecraft observations had provided intriguing glimpses in some areas, but had not provided much reliable and quantitative information. By the early 1970s, the United States had two decades of developing reentry vehicles for intercontinental ballistic missiles (ICBMs). The space program was just beginning to use this technology. For example, ICBM research provided technology that would help spacecraft survive high temperatures and deceleration forces in Venus' atmosphere. This was a very important breakthrough. It allowed us to send highly sophisticated instruments, already demonstrated on other American space missions, through Venus' atmosphere to its surface. With this technology, scientists now could take a new approach to exploring the cloud-shrouded planet. The time was perfect for Pioneer Venus.

On March 15, 1973, Richard Goody of Harvard University appeared before the House Committee on Science and Astronautics to discuss the NASA budget authorization for fiscal year 1974. During his talk, he repeated a statement he had made before the Royal Society in London on the 500th anniversary of the birth of Copernicus: ". . . it is no longer possible to consider Earth entirely aside from the other planets—planetary science has grown to contain many aspects of the earth sciences and for some geophysicists the aim of inquiry has now become the nature of the entire inner Solar System." He stressed that observations of planets such as Mars and Venus could assist some current attempts to model and predict climatic changes on Earth.

Although no one expected Pioneer Venus to answer all the important quesions about Venus, it has taken us closer to understanding the planet and why it differs from Earth. Perhaps the most important aspect of planetary exploration is to learn about extreme cases of conditions that resemble those on Earth. Venus and Mars provide these needed comparisons with Earth. NASA's Pioneer Venus program and the Russian Venera program (before, during, and after Pioneer Venus), with data from NASA's Magellan program in the early 1990s, have provided much of the information needed to make these important comparisons.

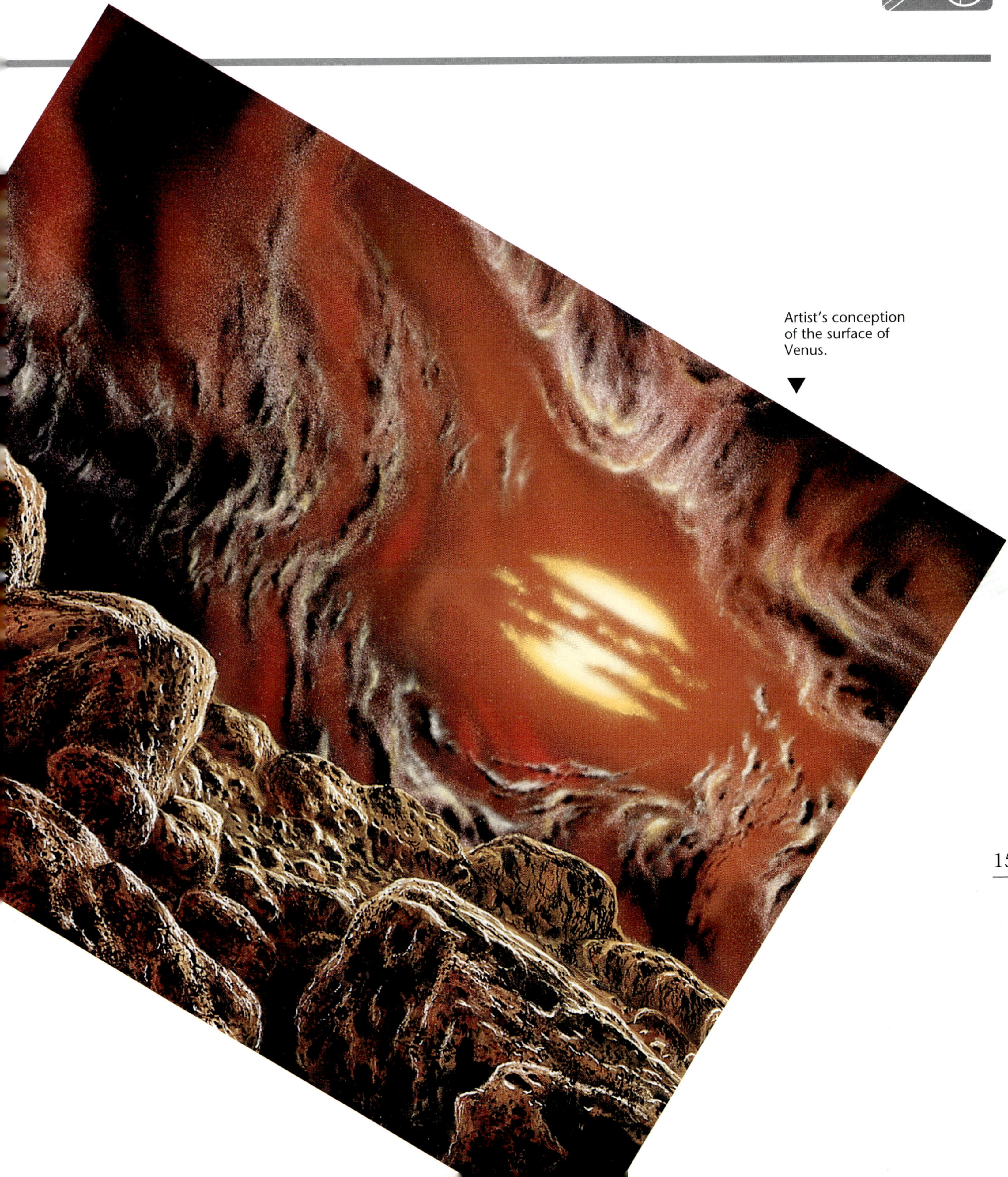

Artist's conception of the surface of Venus.

CHAPTER 2

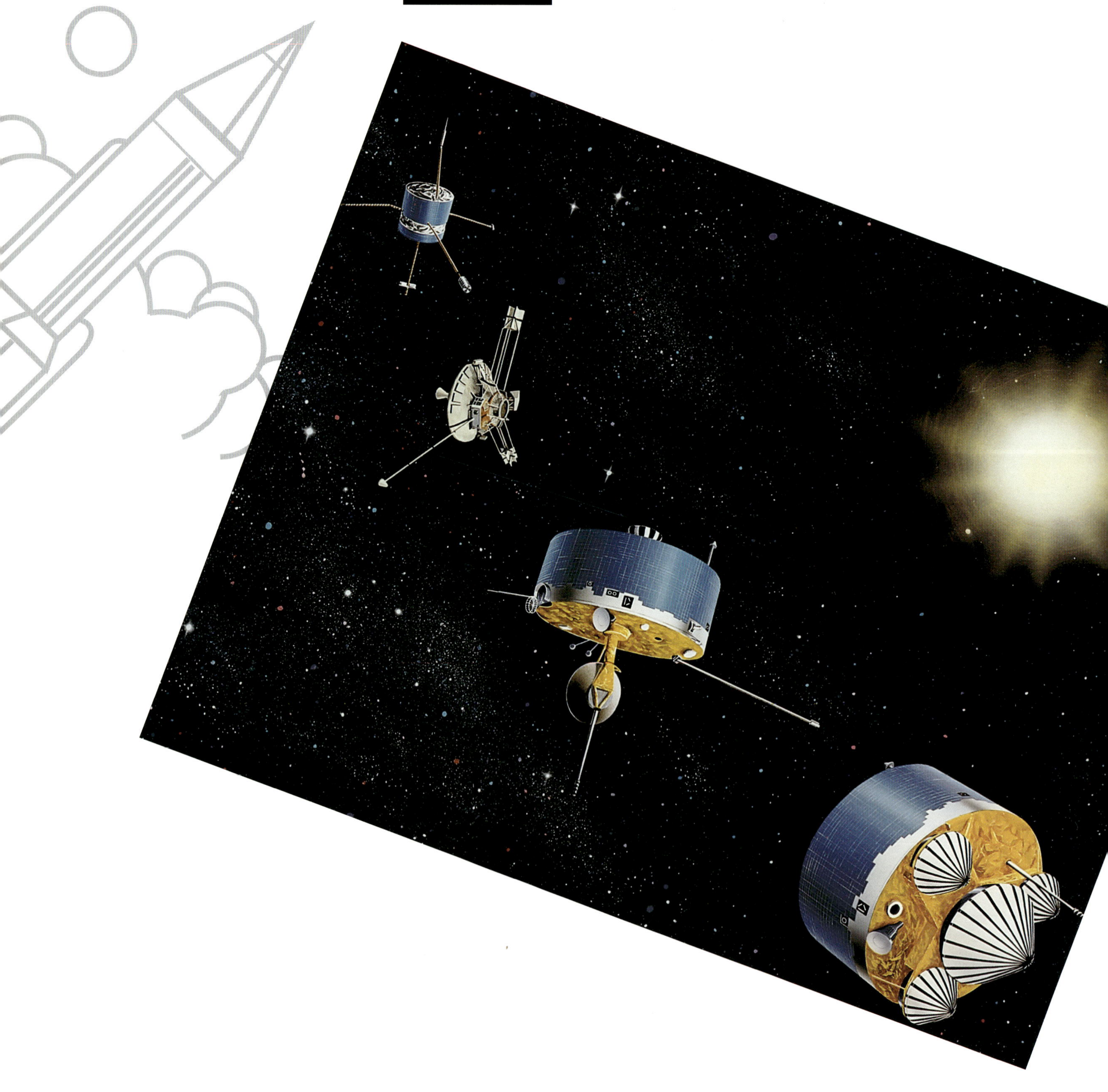

FROM CONCEPT TO LAUNCH

This chapter presents a behind-the-scenes look at Pioneer Venus' early days. It covers the years 1967 to the launch of Pioneers 12 and 13 in 1978. The text discusses the program's early studies and concerns. For example, what is the most complete scientific payload? Who will design the launch vehicle and the interplanetary spacecraft? How would the project be funded? When will be the best launch dates? As you read about these issues and watch them evolve into Pioneer Venus, you also meet the program's major players.

In March 1959, Warren H. Straly of the Army Ballistic Missile Agency presented a paper at the Hawthorne, California, meeting of the Lunar and Planetary Exploration Colloquium. The meeting took place at the Northrop Corporation. Earlier, considerable emphasis had been on the planet Mars as a target for interplanetary spacecraft. Straly compared Mars missions with missions to Venus. He concluded the latter were preferable in terms of overall energy requirements for the mission and for transmitting data back to Earth. He pointed out that astronomers had neglected Venus, basically because Mars was a more interesting planet to observe. With telescopes, astronomers could see the surface of Mars and observe interesting changes on its surface. A planet-wide cloud system, however, hid Venus' surface. A short while before this meeting, the December 8, 1958, issue of *Missiles and Rockets* magazine had a related article. It reported on a NASA plan to launch a spacecraft to Venus in June 1959. The article claimed the spacecraft would carry a spectrometer, a magnetometer, a microwave detector, and other instruments. The report said that the launch vehicle was to be a converted ICBM booster. Unfortunately, this mission never took place. However, a NASA spacecraft, Mariner 2, did fly by Venus in 1962. Another flyby, Mariner 5, followed it in 1967. The Soviets tried unsuccessfully to reach Venus with Sputnik 7 and Venera 1 in 1961 and with a number of different spacecraft in 1962 through 1964.

The Pioneer Venus project began shortly after NASA's Mariner 5 flew by Venus and Russia's first successful Venus mission, Venera 4, probed the planet's atmosphere. These events occurred in October 1967. Three scientists—R. M.Goody (Harvard University), D. M. Hunten (Kitt Peak National Observatory), and N. W. Spencer (NASA Goddard Space Flight Center)—formed a group to consider the possibility of a simple entry probe to investigate Venus' atmosphere. Goddard Space Flight Center awarded a study contract to AVCO Corporation. In 1968, the Center also began studying capabilities of small planetary orbiters using the Explorer Interplanetary Monitoring Platform (IMP) spacecraft. (Thor-Delta launch vehicles would carry these craft into space.) Scientists called the proposed mission the Planetary Explorer.

In the years 1967-1970, scientists had few scientific facts on which to base plans for a Venus mission. Ground-based observations had added very little to their knowledge of the planet. In addition, the few spacecraft that had flown near Venus had returned little new information.

Space officials admitted their methods for exploring Mars and our own Moon would be inadequate for Venus. Before Pioneer Venus, scientists designed spacecraft missions mainly within the limits of existing technology. Beginning with the Venus mission, they adopted a new view that looked beyond available technology for future missions. Researchers now asked key scientific questions about Venus and then defined missions and new technologies to give them answers. Using this new approach, Venus-mission scientists realized that spacecraft payloads should not consist of individual and often unrelated experiments (as they had in past missions). Instead, experiments would apply to a broad range of mission goals.

Early Studies

By June 1965, researchers had completed a significant study (*Planetary Exploration 1969-1975*) with backing from the National Academy of Sciences' Space Science Board. Their study concluded that planetary exploration should be wide-reaching. Rather than a space program to achieve single, isolated goals, the study's authors envisioned one that covered a broad range of interrelated scientific disciplines. Among recommended projects were explorations of Venus with low-cost spacecraft. Toward this goal, the Space Science Board recommended that NASA start a program of Pioneer/IMP-class spinning spacecraft to orbit Mars and Venus. The Board also suggested NASA should plan missions to other planets.

Also, during the summer of 1965, the Space Science Board mounted a summer study. They later issued a thick report entitled *Space Research: Directions for the Future.* R. M. Goody and J. Chamberlain were members of the Working Group on Planetary and Lunar Exploration, which G. MacDonald chaired. The panel on Venus consisted of R. M. Goody, V. Suomi, and G. Wasserburg. Their recommendations for space probes came under the headings geodetic measurements, surface profile (by radio altimetry), cloud structure, upper atmosphere, and dropsondes. The panel recommended specific dropsonde measurements. These were composition, especially water vapor (with a suggestion for a simple mass spectrometer), nature of clouds, and intensity of solar radiation and reradiated infrared radiation (to test the greenhouse theory). Their suggestions for dropsondes also included some sort of penetrometer, to distinguish between solid and liquid surfaces, and a seismometer. So, a quarter century ago, researchers had already earmarked nearly all the instruments for the Venus probe mission. Instruments for an orbiter also were clearly highlighted in these early studies.

Unfortunately, NASA did not enthusiastically receive these ideas. To move the project forward, R. M. Goody started a campaign. In 1966, he sent to D. M. Hunten a paper that was an exploratory proposal for a Venus dropsonde. By early 1967, Goody had enlisted D. M. Hunten and N. W. Spencer into an informal consortium to help define the mission and the instruments. Spencer was a pioneer in exploration of Earth's upper atmosphere. His specialties included sounding rockets and the Explorer series of satellites that sampled the top of the atmosphere. Hunten was a specialist in instruments and was well-versed in current knowledge about Venus. Under Goody's leadership, these three scientists recruited other experts into an energetic group that pushed strongly for an advanced mission to Venus. They envisioned a mission that would orbit the planet and send probes down to its surface, gathering data about the atmosphere as they descended.

Goddard Space Flight Center published its results in January 1969. The Center recommended the Venus project should begin during 1973. R. M. Goody, D. M. Hunten, V. Suomi, and N. W. Spencer wrote the plan, *A Venus Multiple-Entry Probe Direct-Impact Mission.* A consortium of Harvard University, Kitt Peak National Observatory, University of Wisconsin, and Goddard Space Flight Center proposed the study. Besides the authors, some 25 scientists added to the study. Goody pushed this report, sending copies to influential science writers. He appended a note that "despite its Goddard cover it is a piece of private enterprise done with the intention of pushing NASA into a rational planetary program based first and foremost on science

objectives. We wanted to demonstrate that the objectives on Venus could be rationally thought out, and that they point to a feasible mission, which I hope the U.S. may adopt."

Scientists considered several different approaches for a mission to Venus. These approaches included a buoyant Venus station (a balloon that would float in the planet's atmosphere), probes, and orbiters (Figure 2-1). Mission researchers evaluated pros and cons of three different scenarios: (1) a flyby mission with probe release, (2) a direct-impact bus with separate probes reaching Venus before the bus,

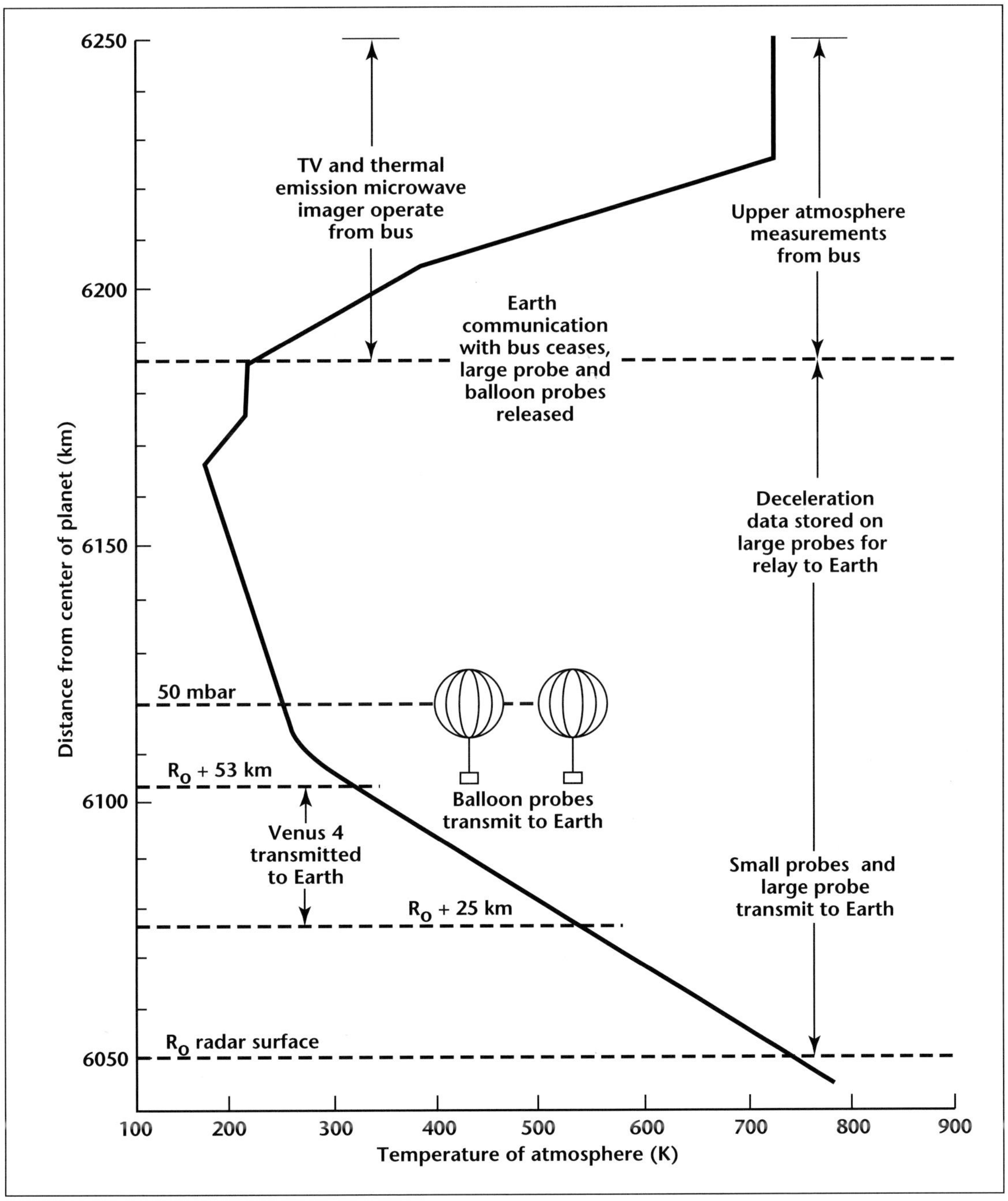

Figure 2-1. Regions of Venus' atmosphere that various probes could investigate (from the first plan for a comprehensive mission to the cloud-shrouded planet).

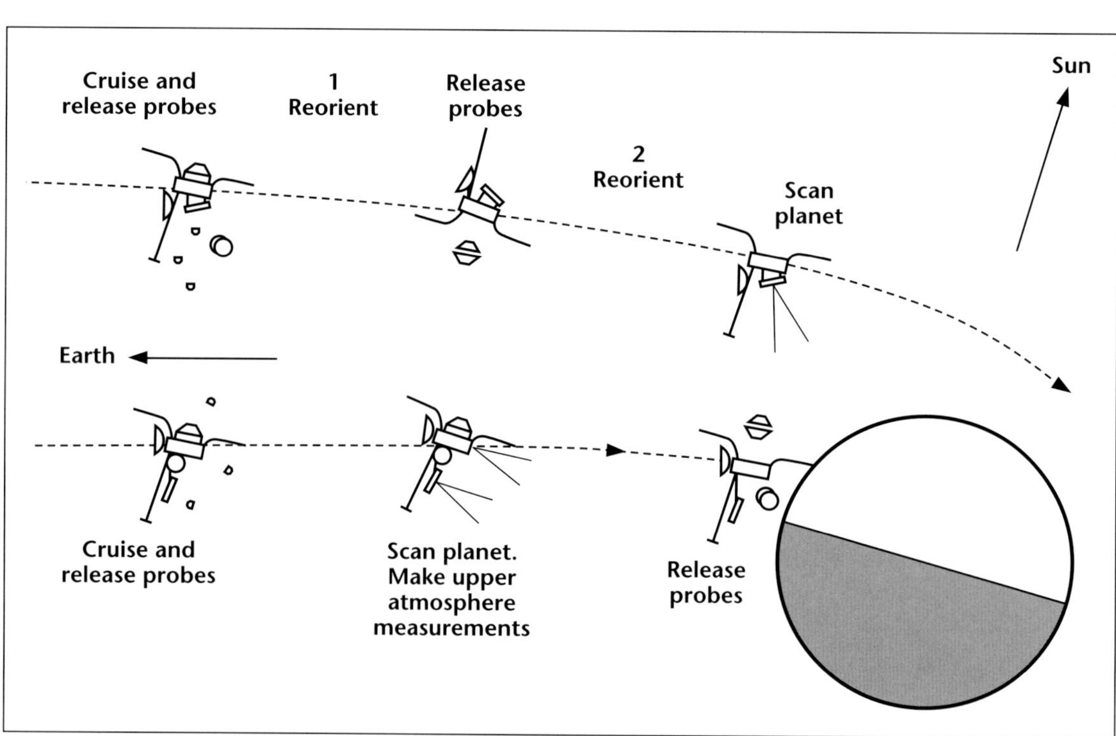

Figure 2-2. The early study compared several mission alternatives, such as a flyby and a direct impact mission for release of probes. The study concluded that the latter was more effective.

and (3) an orbiter that would release probes. After careful study, scientists concluded that the direct-impact bus mission had better chances of collecting scientific data than a flyby mission (Figure 2-2). A system relying on release of probes from a planetary orbiter had its advantages, too—it generated lower temperatures for probes entering the atmosphere. However, a planetary orbiter proved to be very expensive. To launch probes and an orbiter around Venus, such a system required too much propellant, which added to the spacecraft's weight. Complexity and cost also ruled out large, buoyant stations as an alternative (at least until more details of the Venusian atmosphere became available).

Venus' cloudy atmosphere was an effective barrier to its surface features. The Goddard report suggested that a system of three small and four large probes could solve crucial problems concerning the cloudy atmosphere. These included the nature of clouds and the structure, chemistry, and motions of the atmosphere. Ten days before encounter, three small probes could enter the planet's atmosphere near the subsolar point, the antisolar point, and the south pole. During a slow descent to the surface, the three probes could take specific measurements. These measurements would include atmospheric pressure, temperature, and a component of the horizontal wind. Ninety minutes before encounter, and at a distance of five Venus radii—about 30,000 km (18,642 miles)—from the surface, bus science measurements could begin. Probes could take television and microwave thermal emission pictures of the planet down to an altitude of 135 km (84 miles). They also could measure atmospheric density, electron density, temperature, day airglow, and ion and neutral particle composition.

The four large probes could leave the bus at an altitude of about 135 km (84 miles). This separation would happen just before its high-speed entry into the atmosphere destroyed the bus. Two large probes could be identical small balloons that would carry radar transponders. The balloons could float in the atmosphere where pressure is about 50 millibar, or about 70 km (43 miles) above Venus' surface. The radar transponders would make it possible to track the balloons from Earth. While scientists tracked them, the balloons would measure pressure, temperature, solar radiation flux, and upward thermal radiation flux. The other two large probes could penetrate toward the surface. They would measure pressure, temperature, gas composition, radiation fluxes, cloud particle composition, number density, and particle size. Perhaps they could even reveal physical features of the planet's surface.

The Goddard report stated probes were the only way to take measurements crucial for understanding Venus' atmosphere. For a given cost, the report concluded that the direct-impact probe could achieve a real advantage over orbiting and flyby spacecraft delivery systems. This advantage would translate into more complete atmospheric measurements and greater reliability in achieving science goals.

In 1969, Goddard awarded a follow-on contract to AVCO Corporation to study a probe mission to Venus using a Thor-Delta launch vehicle. By the end of that year, NASA had merged the concepts into a *universal bus* (a combination of the Venus probe spacecraft and the Planetary Explorer Orbiter spacecraft). Their idea was to develop a spacecraft that could either deliver multiple entry probes into the Venusian atmosphere or send a vehicle into orbit around the planet.

The "Purple Book"

In 1970, 21 scientists of the Space Science Board and the Lunar and Planetary Missions Board of NASA studied the scientific potential of missions to Venus based on the technology amassed from experience with Explorer spacecraft. They produced a final report, *Venus—Strategy for Exploration,* which became known as the "Purple Book" because of its purple cover.

The report recommended that exploration of Venus should be a NASA goal for the 1970s and 1980s. It also proposed the Delta-launched, spin-stabilized Planetary Explorer spacecraft as the main vehicle for initial missions. These missions would include orbiters, atmospheric probes, and landers. The report stated NASA could reduce the cost of these missions if the agency accepted some higher risks than in previous space missions.

The report outlined a strategy to explore Venus. No more than two missions would be tried at each launch opportunity when the relative positions of the planets made a mission possible on the basis of an available launch vehicle and the weight of the science payload. They also would avoid hybrid missions because of their complexity and cost. (A hybrid mission might be a spacecraft carrying both an orbiter and an atmospheric probe or a lander.) Missions would use identical payloads wherever possible.

The report recommended that project scientists carefully weigh the scientific value of results against mission costs. The strategy was to keep mission costs at a minimum (that is, under $200 million). This would allow NASA to plan a series of missions to Venus. The report suggested two multiprobe missions for the 1975 opportunity and two orbiters for the

1976/77 opportunity. Later opportunities were less clear; orbiters, landers, and balloons were all candidates. The report proposed that the 1978 opportunity should be a follow-on landing mission.

This 1970 study also pointed out the seeming paradox of differences and similarities in the evolution of the two planets. It claimed that exploring the second planet from the Sun promised to reveal new insights into planetary evolution.

Because of its opaque atmosphere and absence of satellites, scientists knew less about Venus in 1970 than they did about Mars. Ideally, they needed measurements of Venus to determine the chemical composition and mineralogy of the surface materials, the heat flux from the interior, the presence or absence of an iron-rich core, and the variation of elastic-wave velocity with depth and with wave intensity. Making such measurements on Venus would be extremely difficult because of the high temperature at the planet's surface—about 475°C (887°F). Nevertheless, a program of measurements on a scale proposed for Planetary Explorers would allow highly significant measurements to be made. Surface elevations could be measured with a radar altimeter on an orbiter, and some information about the distribution of mass in the planet could be obtained from the way in which the orbit of such an artificial satellite is perturbed.

The "Purple Book" made several other recommendations. It suggested that NASA continue to support and develop specific Earth-based studies of Venus to complement its spacecraft-based studies. Among these techniques were thermal mapping of the planet's surface by analysis of radio emissions from the surface, radar topographical mapping, and analysis of radiation from cloud tops. The report suggested that NASA set up and maintain a continuous group to (1) plan Venus explorations, (2) advise on strategy for these missions, and (3) recommend payloads for each mission. The study stressed the need for a wide range of novel scientific experiments for the missions, such as those for investigating Venus' clouds. In a summary statement, the authors wrote, "We believe that the combination of scientific goals and the feasibility of contributing to these goals makes the exploration of Venus one of the most important objectives for planetary exploration of the 1970s and 1980s."

Effect of the Soviet Venus Probe, Venera 7

In the fall of 1970, funding a new program for planetary exploration that could meet a 1975 launch date was unlikely. So, planners rescheduled the entire Venus exploration program. They revised the plan to launch two multiprobe spacecraft during the 1976/77 opportunity. In addition, they planned for a single orbiter spacecraft in 1978 and a single multiprobe (a floating balloon probe and a lander) in 1980.

Soviet scientists also were extremely interested in exploring Venus. During most launch opportunities, they sent spacecraft to the cloud-shrouded planet (see Chapter 7). They experienced many technical difficulties, and several early spacecraft failed. However, their efforts to study another planet's atmosphere were partially successful. The worldwide scientific interest they created more than offset their failures.

On December 15, 1971—soon after the Space Science Board published its 1970 report—a Soviet spacecraft, Venera 7, successfully entered Venus' atmosphere. For 23 minutes, it sent data from the surface. In view of these

new data, scientists asked whether the recommendations of the 1970 study still stood. A special panel of experts met to reassess the recommendations. The panel's conclusion:

> The Planetary Explorer program recommended in the Venus study would be a well-articulated, intensive study of the planet designed to attempt to answer a list of first-order questions. Among these are the number, thickness, and composition of cloud layers; the nature of the circulation; explanation of the high surface temperature; the reason for the lack of water and the remarkable stability of the carbon dioxide atmosphere; the nature of the interaction of the solar wind with the planet; the elemental composition of the surface; the distribution of mass and magnetic field strength; and the measurement of seismic activity. Venera 7 was a highly specialized probe designed to perform only two functions—to measure atmospheric temperature and pressure down to the surface of Venus. It succeeded in measuring the temperature and confirmed the most widely held expectation; that the surface temperature is high. It has in no way changed the conditions on which the Venus study was based or answered any of the questions that planetary explorers are designed to answer. We can find no reason, therefore, to recommend changes in the scientific objectives set forth in previous Board studies. . . .

Transfer of NASA's Venus Mission to Ames Research Center

Meanwhile, NASA had continued practical work on high-velocity entry of spacecraft into planetary atmospheres. By 1970, research scientists at NASA Ames Research Center had gathered much experimental data about effects on bodies moving at high speed in an atmosphere. Their technique was to photograph and analyze various entry shapes in hypervelocity free flight tunnels at speeds up to 50,000 km/hr (31,070 mph). These speeds were higher than the speed needed to enter Venus' atmosphere.

By 1971, Ames Research Center had designed, fabricated, and tested a spacecraft and most of its instrument systems. Engineers designed the equipment to demonstrate selected planetary experiments and instrumentation in Earth's atmosphere. The Planetary Atmosphere Experiments Test (PAET) was a vital step for future missions. It established a technical base for advanced planetary exploration of Mars, Venus, and eventually the outer planets. A Scout solid-propellant multistage rocket launched the test spacecraft. The launch vehicle's third and fourth stages carried the PAET spacecraft back into Earth's atmosphere at 24,000 km/hr (14,914 mph).

Launched at 3:31 p.m. EDT on June 20, 1971, the test spacecraft was highly successful. Just as experimenters planned, instruments scooped up atmospheric gases. Even more important, PAET demonstrated the capability of selected experiments to determine structure and composition of an unknown planetary atmosphere from a high-speed entry probe. This was the type of practical data researchers needed to design a probe that could enter Venus' atmosphere. The PAET program proved the capability of Ames Research Center personnel to participate in such a mission. Meanwhile, in July 1971, NASA issued an Announcement of Opportunity (AO) for scientists to participate in defining the Venus program. In November of that year, NASA discontinued the Planetary Explorer program at Goddard. By January 1972, the agency had transferred it to Ames Research Center, Moffett Field, California. At Ames, a study team quickly organized itself and the project was renamed Pioneer Venus.

This team defined the system and worked closely with a Pioneer Venus Science Steering Group made up of interested scientists to define the mission's scientific payloads.

Pioneer Venus Study Team

R. R. Nunamaker (chair), H. F. Matthews, M. Erickson, T. N. Canning, D. Chisel, R. A. Christiansen, L. Colin, J. Cowley, J. Givens, T. Grant, W. L. Jackson, T. Kato, J. Magan, J. Mulkern, L. Polaski, R. Ramos, S. Sommer, J. Sperans, T. Tenderland, N. Vojvodich, M. Wilkins, L. Yee, E. Zimmerman

Science Steering Group and the "Orange Book"

NASA established this Pioneer Venus Science Steering Group in January 1972. The group's purpose was to enlist widespread science community participation in the early selection of the mission's science requirements. The Science Steering Group met with Pioneer Venus project personnel from February through June 1972. They developed in detail the scientific rationale and objectives for the early Venus missions. The group also conceived and planned candidate payloads and spacecraft. Their efforts provided a useful guide for the NASA Payload Selection Committee and for the contractors who would later develop the payloads and spacecraft.

During the first five months of its operations, the Science Steering Group held several meetings. In 1972, the group published a comprehensive report that became the accepted guide to Venus exploration. Known as the "Orange Book" (again because of the cover's color), the report carefully reviewed and endorsed the scientific rationale for missions to Venus. It based its reviews and endorsements on developments since the earlier Space Science Board's 1970 report, *Venus—Strategy for Exploration*. These developments included delays in starting the program, scientific findings from the Soviet probe Venera 7, new Earth-based observations, new theoretical analyses, and continued analysis of data that earlier Soviet and American spacecraft gathered. The report recommended that missions continue with multiple probes in 1976/77. It also suggested a single orbiter in 1978 followed by a probe-type mission in 1980.

The Science Steering Group's report stated that most scientific questions about Venus required *in situ* atmospheric measurements. Measurements should start at the cloud tops and extend as far as possible toward the surface. The group defined 24 important questions about Venus (Table 2-1).

The required technology and scientific instruments needed for the mission were considered state-of-the-art at that time. Therefore, researchers believed a probe mission at the first opportunity was desirable. In case of a failure, they suggested a dual launch mission. If both spacecraft were successfully launched, they recommended retargeting the second probe based on what they learned from the first. To ensure the best chance for success, the group suggested a third probe for the final launch opportunity.

The study recommended that the first mission should consist of two identical spacecraft and payloads. These would be ready for launch from December 1976 through January 1977. Each spacecraft would consist of a bus, a large probe, and three small probes. The large probes would have parachutes; the small probes would be free-falling and identical. The spacecraft would be spin-stabilized and would use solar power. Cruise from Earth to Venus would take about 125 days. The probes would separate from the bus about 10 to 20 days before entry into the Venusian atmosphere. In addition to transporting the probes, the buses also would enter the Venusian atmosphere (at shallow angles) and send back data until they burned up. Their mission: to gather information about the upper atmosphere.

Table 2-1. Questions by Science Steering Group for Pioneer Venus Mission

1. Cloud layers: What is their number, and where are they located? Do they vary over the planet?
2. Cloud forms: Are they layered, turbulent, or merely hazes?
3. Cloud physics: Are the clouds opaque? What are the sizes of the cloud particles? How many particles are there per cubic centimeter?
4. Cloud composition: What is the chemical composition of the clouds? Is it different in the different layers?
5. Solar heating: Where is the solar radiation deposited within the atmosphere?
6. Deep circulation: What is the nature of the wind in the lower regions of the atmosphere? Is there any measurable wind close to the surface?
7. Deep driving forces: What are the horizontal differences in temperature in the deep atmosphere?
8. Driving force for the 4-day circulation: What are the horizontal temperature differences at the top layer of clouds that could cause the high winds there?
9. Loss of water: Has water been lost from Venus? If so, how?
10. Carbon dioxide stability: Why is molecular carbon dioxide stable in the upper atmosphere?
11. Surface composition: What is the composition of the crustal rocks of Venus?
12. Seismic activity: What is its level?
13. Earth tides: Do tidal effects from Earth exist at Venus, and if so, how strong are they?
14. Gravitational moments: What is the figure of the planet? What are the higher gravitational moments?
15. Extent of the 4-day circulation: How does this circulation vary with latitude on Venus and depth in the atmosphere?
16. Vertical temperature structure: Is there an isothermal region? Are there other departures from adiabaticity? What is the structure near the cloud tops?
17. Ionospheric motions: Are these motions sufficient to transport ionization from the day to the night hemisphere?
18. Turbulence: How much turbulence is there in the deep atmosphere of the planet?
19. Ion chemistry: What is the chemistry of the ionosphere?
20. Exospheric temperature: What is the temperature and does it vary over the planet?
21. Topography: What features exist on the surface of the planet? How do they relate to thermal maps?
22. Magnetic moment: Does the planet have any internal magnetism?
23. Bulk atmospheric composition: What are the major gases in the Venus atmosphere? How do they vary at different altitudes?
24. Anemopause: How does the solar wind interact with the planet?

The 1978 launch would be an orbiter mission. Generating electrical power from solar cells, the spacecraft would be spin-stabilized. It would be launched between May and August 1978. After its interplanetary cruise, the spacecraft would go into an elliptical orbit around Venus. Engineers would design the spacecraft to orbit the planet for a Venus sidereal day (243.1 Earth days). Major goals would be to (1) produce a global map of the Venusian atmosphere and ionosphere, (2) get measurements directly from the upper atmosphere and its ionosphere, (3) investigate interactions between solar wind and ionosphere, and (4) study the planet's surface by remote sensing.

The Steering Group still contemplated a third probe mission for 1980. They expected details of a 1980 mission to become clearer as two things happened: (1) as they more clearly defined the 1976/77 mission and (2) as they later reviewed that mission's results. The study made no recommendations for a launch in 1982.

Despite Russian entry probes and flybys, scientists knew very little about Venus' lower atmosphere in 1972. For example, they did not know how many cloud layers there were, how thick they were, or what was in them. There were at least three very different hypotheses to explain the planet's high surface temperature.

After an independent study of the Soviet Venus program, the Science Steering Group agreed with the Space Science Board's earlier assessment of the Venera program. The previous 11 years of Soviet exploration of Venus had produced direct measurements of the lower atmosphere. These measurements included pressure, temperature, density, and gross atmospheric composition. The National Academy of Sciences' Venus study, however, exposed a wide range of scientific problems that the Soviet programs had not tackled. Among them were questions about the magnetosphere, upper atmosphere, lower atmosphere, and the solid planet.

When it came to recommending instruments for the spacecraft, the Science Steering Group adopted a conservative approach to avoid increasing costs. The group decided that acceptable instruments should have already performed successfully in Earth's atmosphere. They also agreed experiments should not use novel concepts of measurement. Wherever possible, instruments should already qualify for spacecraft or aircraft use. If they did not already qualify, instruments had to be simple and rugged. Only if they performed satisfactorily in laboratory tests should they be considered for the Venus missions.

The Pioneer Venus Mission Crystallizes

The Pioneer Venus program began as a model, low-cost program. It developed around innovative approaches to management and an understanding that the total cost would remain below $200 million. The program crystallized as a single-opportunity mission. Consisting of a Multiprobe spacecraft and an Orbiter spacecraft, it reflected significant, major advances in the sophistication of spacecraft and instruments compared with earlier Venus spacecraft.

The Pioneer Venus Multiprobe would be an important step in answering questions about the planet's atmosphere. It would provide data about the cloud layers, their forms, physics, and composition. It would investigate the atmosphere's bulk composition, its solar heating, deep circulation and driving forces, its loss of water, the stability of carbon dioxide, and the vertical temperature structure. Also included would be data on ionospheric turbulence, ion chemistry, exospheric temperature, magnetic moment, and the anemopause where the solar wind reacts with the planet's atmosphere.

The Pioneer Venus Orbiter also would provide significant information about cloud forms, cause of the four-day circulation, loss of water, gravitational moments, extent of the four-day circulation, vertical temperature structure, ionospheric motions, ion chemistry exospheric temperature, topography, magnetic moment, bulk atmospheric composition, and the anemopause.

European Study

Early in 1972, members of the European Space Research Organization (ESRO) asked to participate in the 1975 Orbiter mission. A meeting took place in April 1972, which included NASA and ESRO members. Attendees decided to examine jointly how the two organizations could work together on the 1978 Venus Orbiter mission. NASA would produce and provide ESRO with the Orbiter version of the basic spacecraft, or Bus, together with common equipment. ESRO would then adapt the Bus, including a retromotor to slow the spacecraft as it approached Venus. (The retromotor would allow it to enter an orbit around the planet.) Also, ESRO would provide a high-gain antenna to allow communications at high data rates and would integrate scientific experiments. In addition, the European group would undertake qualification tests on the spacecraft and its payload. NASA would then accept the Orbiter for launch and flight operations.

To define the objectives for a Venus Orbiter launch in 1978, a Joint Working Group of European and U.S. scientists formed. The scientists met periodically and issued a report in January 1973, *Pioneer Venus Orbiter*. This report recalled that a series of missions had been proposed since the start of the NASA Venus exploration concept. A series combining orbiter and probe capabilities was the favored method for exploring Venus' environment. By mid-1972, the group had defined the present mission series. They called for a Multiprobe mission in the 1976/77 launch opportunity and for an Orbiter mission in 1978. The science experiments for the Orbiter mission required a highly inclined orbit plane—greater than 90° with respect to the ecliptic, the plane of Earth's orbit. According to the Working Group, a low periapsis (the point in its orbit where the Orbiter would be nearest Venus) of 200 km (125 miles) or less was desirable. The periapsis would be at about latitude 45°, initially in the sunlit hemisphere. Solar gravity would cause the periapsis altitude to increase. To keep the altitude in a desired range would require periodic orbital change maneuvers. Apoapsis (the point in its orbit where the Orbiter would be farthest from Venus) would be at 60,000 to 70,000 km (37,284 to 43,498 miles). Drag at periapsis would decrease the apoapsis altitude and reduce the period in orbit, which would initially be close to 24 hours. Maneuvers would be needed to maintain the period.

Researchers also defined experiments and specified required characteristics of scientific instruments. They described three science payloads, depending on how much scientific payload the spacecraft could carry.

The Working Group stated that, in general, a model payload should consist of instruments to measure four important areas of interest about Venus:

(1) Interaction of the solar wind with the ionosphere would be investigated by a magnetometer, a solar wind and photoelectron analyzer, an electric field detector, and an electron and ion temperature probe.

(2) Aeronomy and the airglow would be investigated by a neutral mass spectrometer, an ion mass spectrometer, and an ultraviolet spectrometer/photometer (aeronomy includes investigating atmospheric composition and photochemistry).

(3) The atmosphere's thermal structure and lower atmospheric density would be investigated by an infrared radiometer and a dual-frequency (S- and X-band) occultation experiment.

(4) Surface topography, reflectivity, and roughness would be investigated with a radar altimeter.

The Group considered several other instruments and experiments. These included a microwave radiometer to map thermal emission from the planet's surface, an electric field sensor to detect plasma waves generated by the interaction of the solar wind with the ionosphere, a solar ultraviolet occultation experiment, and a photopolarimeter.

Scientists were extremely interested in determining Venus' gravitational field and geometrical shape. Such information is important to our understanding of the origin and evolution of the Solar System's inner planets. It also helps us determine why Earth and Venus evolved so differently. Gravitational experiments require an orbiter with a periapsis high enough to avoid any atmospheric drag. They also require one capable of remaining in orbit long enough to gather many data points of tracking. Unfortunately, there was a conflict between in situ measurements, requiring a low periapsis, and gravitation measurements, requiring a high periapsis. To resolve this conflict, the Working Group recommended that the mission go beyond the nominal 243 days. The extra days would allow experimenters to make accurate gravity measurements.

Later, the Managing Executive Council for ESRO voted not to participate. But they made this vote only after the European Space Organization had made valuable contributions to the program's development. These contributions included important studies at Messerschmitt-Bölkow-Blohm and at the British Aerospace Company.

Pioneer Venus Science Payload

Meanwhile, during the ESRO study, NASA made a decision in August 1972 to restrict the program to two flights only. The flights would be a Multiprobe at the first opportunity (1977) and an Orbiter at the second opportunity (1978). In September 1972, NASA issued an AO for scientists to participate in the Multiprobe mission. In addition to investigators who would develop hardware for the scientific instruments, NASA, for the first time, invited interdisciplinary scientists and theoreticians to participate. After learning about the AO, the Science Steering Group disbanded. This decision freed the members from conflicts of interest so they could respond to the AO if they so chose.

Mission scientists selected the preliminary payload for the Multiprobe mission in April 1973. An AO for the Orbiter mission followed this selection in August 1973. During the following months, a NASA Instrument Review Committee reviewed instrument design studies for the Multiprobe mission. They also considered proposals for the Orbiter's scientific payloads. NASA headquarters received recommendations in May 1974 and finalized the payloads on June 4, 1974.

Scientists chose 12 instruments for the Orbiter, 7 for a Large Probe, 3 identical instruments for each of four Small Probes, and 2 for the Multiprobe Bus. In addition, they chose several radio-science experiments that were applicable to all spacecraft (Table 2-2).

During the early program, a total of 114 scientists were involved. Science management, however, was restricted to a smaller group. This group consisted of the principal investigators, a radio-science team leader, a radar team leader, interdisciplinary scientists, and program and project scientists. These individuals comprised a new Science Steering Group under the chairmanship of T. M. Donahue and co-chairmanship of D. M. Hunten, L. Colin, and R. F. Fellows. (On his retirement in 1978, the program scientist, R. F. Fellows, was succeeded by R. Murphy and then H. Brinton.)

Table 2-2. Science Instruments: Project Acronyms and Principal Investigators

Composition and Structure of the Atmosphere
- Large Probe Mass Spectrometer (LNMS), J. Hoffman
- Large Probe Gas Chromatograph (LGC), V. Oyama
- Bus Neutral Mass Spectrometer (BNMS), U. Von Zahn
- Orbiter Neutral Mass Spectrometer (ONMS), H. Niemann
- Orbiter Ultraviolet Spectrometer (OUVS), I. Stewart
- Large/Small Probe Atmosphere Structure (LAS/SAS), A. Seiff
- Atmospheric Propagation Experiments (OGPE), T. Croft
- Orbiter Atmospheric Drag Experiment (OAD), G. Keating

Clouds
- Large/Small Probe Nephelometer (LN/SN), B. Ragent
- Large Probe Cloud Particle Size Spectrometer (LCPS), R. Knollenberg
- Orbiter Cloud Photopolarimeter (OCPP), J. Hansen (later L. Travis)

Thermal Balance
- Large Probe Solar Flux Radiometer (LSFR), M. Tomasko
- Large Probe Infrared Radiometer (LIR), R. Boese
- Small Probe Net Flux Radiometer (SNFR), V. Suomi
- Orbiter Infrared Radiometer (OIR), F. Taylor

Dynamics
- Differential Long Baseline Interferometry (DLBI), C. Counselman
- Doppler Tracking of Probes (MWIN), A. Kliore
- Atmospheric Turbulence Experiments (MTUR/OTUR), R. Woo

Solar Wind and Ionosphere
- Bus Ion Mass Spectrometer (BIMS), H. Taylor
- Orbiter Ion Mass Spectrometer (OIMS), H. Taylor
- Orbiter Electron Temperature Probe (OETP), L. Brace
- Orbiter Retarding Potential Analyzer (ORPA), W. Knudsen
- Orbiter Magnetometer (OMAG), C. Russell
- Orbiter Plasma Analyzer (OPA), J. Wolfe (later A. Barnes)
- Orbiter Electric Field Detector (OEFD), F. Scarf
- Orbiter Dual-Frequency Occultation Experiments (ORO), A. Kliore

Surface and Interior
- Orbiter Radar Mapper (ORAD), G. Pettengill
- Orbiter Internal Density Distribution Experiments (OIDD), R. Phillips
- Orbiter Celestial Mechanics Experiments (OCM), I. Shapiro

High Energy Astronomy
- Orbiter Gamma Burst Detector (OGBD), W. Evans

To deal with specific subjects, various committees formed among the scientists. Several of these were long standing, including six Working Groups for each scientific area of investigation. Before launch, they developed key questions, and afterward they synthesized the results they received from the spacecraft. Another very active group was concerned with mission operations planning for the Orbiter. This group, called the OMOP Committee (for Orbiter Mission Operations Planning Committee) consisted of H. Masursky, L. Colin, T. M. Donahue, R. O. Fimmel, D. M. Hunten, G. H. Pettengill, C. J. Russell, N. W. Spencer, and A. I. Stewart. H. Masursky served as chairman of OMOP until his death, at which time D. Hunten succeeded him.

Six Working Groups developed key scientific questions. Chairmanship of these groups varied during the mission, but the major leaders were J. Hoffman, composition and structure of the Venus atmosphere; R. Knollenberg, clouds; M. Tomasko, thermal balance; G. Schubert, dynamics; S. Bauer, solar wind and ionosphere; and H. Masursky, surface and interior.

The mission procured instruments in several ways. Usually, the principal investigator was responsible for having a particular instrument built. He could either (1) build it in his own laboratory, (2) subcontract its construction, or (3) use a combination of these methods.

According to the second scenario, the Pioneer project office would contract some industrial firm and then monitor how the firm developed the instrument. During this process, the principal investigator still participated to assure that it met the requirements of his experiment.

As an example of the third scenario, the project office built the Orbiter's radar mapper for a radar team. Carl Keller, an Ames Research Center engineer, had overall decision-making responsibility. Hughes Aircraft built the radar as a result of an open bid procurement.

There was much talk at the beginning of the program, before the AO went out, that the mission would use only instruments that had flight-proven capability. The instruments had to have flown in other spacecraft or in Earth's atmosphere. This requirement was intended to save money and improve reliability. But in practice, very few items of "off-the-shelf" hardware were available. An instrument identical to one from an earlier mission usually had to have significant design changes to work on a new spacecraft. Most important, redesign is often necessary because manufacturers no longer make an "old" instrument's parts. Some instrument redesign was necessary for an even simpler reason: Pioneer Venus was NASA's first attempt to study another planet's atmosphere.

An example was the Orbiter electric-field detector. Because it had flown on earlier Pioneer spacecraft, mission planners thought it might fly without change on Pioneer Venus. But when engineers took a closer look, they realized they had to redesign the small ball-like antennas on the detector. As it turned out, the electric-field detector was not the only equipment to be redesigned. Engineers did a lot of redesigning, particularly for the Multiprobe, because no spacecraft like the Multiprobe had ever flown. Engineers were challenged to closely package instruments that would take many measurements never before taken. Consequently, many instruments were new designs that involved critical development tasks.

From the beginning, the neutral mass spectrometer for the Large Probe was the most difficult new design. Mission planners initially selected more instruments than they would use. The neutral mass spectrometer was a prime example. Two mass spectrometers were under development—one at Goddard Space Flight Center and the other at the University of Texas at Dallas. Both had funds for a year of continued in-house development. An instrument review committee reviewed all instruments, in particular the two mass spectrometer designs. Eventually, the NASA Headquarters Science Steering Committee chose the University of Texas instrument.

Planners chose two other instruments that did not fly. One was a radar altimeter for the Large Probe. After a year's work, it became clear that

the instrument was too heavy, too complex, and too costly. Since scientists could derive altitude as a function of time from the probe's atmospheric structure experiment, NASA decided to remove the radar altimeter experiment. The other instrument was a photometer system from the University of Wisconsin. After a year of in-house study at the University, NASA Headquarters realized that the experiment was neither required nor far enough along for the mission.

Early in the program, NASA made preliminary choices about experiments and then amended them as more information became available. There was nothing unusual about preliminary selection of experiments and then making a final selection some 12 or 18 months afterward. For example, mission scientists never intended both mass spectrometers to fly on the Multiprobe. Also, they eliminated the photometer and radar altimeter on the grounds of payload weight and development studies. The Orbiter, however, was a different story. Mission planners approved for flight all instruments they had initially selected for it.

Challenges of Instrument Development: Probes

Many mission experiments were, indeed, unique. The big problem for the probes was packing all the instruments into a small pressure shell. The shell protected the instruments as they traveled through Venus' hostile environment. For the Orbiter, the most difficult task was ensuring reliability of operation for at least 243 Earth days.

The challenge in space missions is always meeting the scheduled launch date. For Pioneer Venus, all the instruments were ready on time. At one point, however, the Jet Propulsion Laboratory (JPL) encountered significant development problems with the infrared radiometer. Within a year of the launch date, mission planners were still concerned that they might have to scratch the instrument from the payload. JPL responded by intensifying its development effort and was able to test the completed instrument on time.

The neutral mass spectrometer was a principal development challenge. One main difficulty was to develop an inlet system for the instrument. While most mass spectrometers in space applications operate under quasi-vacuum conditions, the Pioneer Venus instrument had to operate at pressures 100 times Earth's atmosphere.

The ion source of every mass spectrometer has to operate within a narrow range of pressures. Therefore, the pressure within the instrument has to remain constant. Engineers needed an inlet system that would reduce the ambient pressure from 10^4 torr to the 10^{-5} torr (10^6 Pa to 10^{-3} Pa) that the ion source required. This was a tremendous pressure reduction. To achieve this reduction, engineers had to build the inlet system to admit very small quantities of gas. Yet these small quantities had to be large enough for analysis before the instrument purged itself for the next sample. The University of Texas designed an innovative system. It consisted of a ceramic microleak (CML) inlet and a variable conductance valve. Their design challenge was to change the instrument's conductance automatically. Their novel solution: let the ambient pressure of the Venusian atmosphere control the valve.

When engineers attempted to adapt the CML for the Pioneer Venus mission, they ran into snags. Initially the inlet was stainless steel. When engineers tested it in sulfuric acid vapor, the acid never entered the instrument for sampling. Instead, it became trapped in the oxide coating. Since one mission task was to check for acids in Venus' atmosphere, mission engineers had to correct this problem. It took two years to solve the problem. They developed a suitable ceramic coating with a passivated surface, and the inlet was made of tantalum instead of stainless steel.

Scientists realized, too, that aerosol particles in the planet's atmosphere might block the mass spectrometer's small inlet opening. The University of Texas developed a narrow slit design to minimize blockages, and engineers installed a heater coil around the inlet to vaporize such particles. (Despite their efforts, sulfuric acid droplets covered the inlet for a time during the mission and blocked the inlet.)

The mass spectrometer caused even more difficulties. A single inlet would be fine in the dense lower atmosphere. But in the upper rarefied atmosphere, this instrument needed an additional inlet to provide sufficient gas input. Engineers designed the second inlet to stay open until roughly the time of parachute release. Then a pyrotechnic device crushed the line and stopped further gas entry. Even if the cutoff device failed, there would still be a valid set of data, although somewhat degraded.

In addition to its novel inlet design, this instrument had several other firsts. It was the first mass spectrometer of its size to survive the forces from an entry deceleration of 400 g. It also used the first microprocessor to fly in space: an Intel 4004. The microprocessor allowed the spacecraft to take a full spectrum of data once every minute over the whole mass range of 200 amu. The microprocessor selected the true data point from several data points and adjusted for calibration changes. A high confidence factor was associated with the single data point transmitted. Without the microprocessor, it would have been possible to transmit a spectrum only once for every 10-km change in altitude. With the microprocessor, sampling occurred at every 1-km change in altitude.

Other instruments also posed some problems. For example, mission scientists had not originally proposed to fly a gas chromatograph. However, the original study team developed strong arguments in favor of the gas chromatograph, and they finally included it in the payload package. (A gas chromatograph is a high-pressure instrument while the mass spectrometer is a low-pressure instrument.)

At Ames Research Center, Vance Oyama had developed a gas chromatograph for the Viking landings on Mars. Engineers used his experience to design an instrument for Pioneer Venus. In the program's early days, mission planners considered the chromatograph a backup instrument. They would use it to provide some spectra of atmospheric composition if the mass spectrometer failed. However, they soon saw that the two instruments complemented each other. For example, the gas chromatograph could measure water vapor that the mass spectrometer could not measure reliably.

Robert Knollenberg, a cloud physicist, had developed a small spectrometer that the U.S. Air Force used to measure the number of ice particles in clouds. Knollenberg and Ball Brothers Research (Boulder, Colorado) adapted this instrument for Pioneer Venus. Their Cloud Particle Size Spectrometer was essentially an optical bench with a laser at one end and a prism at the other. Part of the optical bench had to be outside the pressure hull of the spacecraft. This design had a drawback. It exposed the bench to twisting and other distortions that would occur as the pressure vessel heated in Venus' atmosphere. Lou Polaski, Ames Research Center, was responsible for developing probe instruments. After studying

the problem, he realized how to correct the problem. The pressure vessel needed heaters on the window in the vessel and on the prism outside the window.

Hughes Aircraft Company was the contractor for the spacecraft and probes. Hughes also built the faceplate through which the optical bench would penetrate the probe's wall. To this faceplate, Hughes attached the instrument parts that Ball Brothers supplied. Ball Brothers then aligned the complete unit. Said Polaski: "It was a tremendous challenge to get a very precise optical bench through a wall that was changing relative to the rest of the optical bench. The instrument really worked well but only as a result of a lot of good engineering work."

Another unique instrument the probe carried was a solar flux radiometer. It was unique because engineers developed the sensor portion separately from all the electronics. Martin Marietta, Denver, built the electronics. The University of Arizona's Optical Science Center designed and built the optical head with the sensors.

The infrared radiometer used warm infrared detectors that had to remain at a constant temperature. To maintain this temperature, the detectors were packaged in phase-change material (the "blue ice" in recreational refrigeration). Technically, this material is a eutectic salt. The gas chromatograph also controlled the temperature of its columns with "blue ice." To control the temperature of its optical head, the solar flux radiometer used it, too. Scientists picked salts that would keep the temperature at the required value (like ice floating in water will keep the water at a constant temperature until all the ice has melted). But it was not that simple. All salts had to be frozen before the probe entered the Venus atmosphere. Mission planners had to prove conclusively that from the probe's release from the Bus to its arrival at the Venus atmosphere—about three weeks—the phase-change material would remain frozen. That proof took considerable time and effort.

The net flux radiometer that flew on each Small Probe had a flux plate that flipped back and forth to measure the up and down flux. This radiometer required a diamond window that was smaller than the Large Probe's infrared radiometer window. Two diamond windows were on each side, and the radiometer hung out over the back of the probe. Its strange appearance earned it the nickname of "The Lollipop." The diamond windows came from the same stone as the big window, ensuring identical infrared transmission characteristics. They also made data correlation between the two instruments easier. Seven diamonds thus traveled to Venus—two diamond windows for each of three Small Probes and a single large window in the Large Probe.

Challenge of Instrument Development: Orbiter

For the Orbiter, the most significant instrument under development was the radar mapper. Hughes Aircraft, Culver City, California, built the mapper with a team led by Gordon Pettengill of Massachusetts Institute of Technology. The complete instrument used more than 1,000 microcircuits, weighed only about 10.9 kg (24 lb), and consumed a mere 30 W. This was the first time engineers had assembled a complex instrument for radar mapping in such a compact package. The responsible project engineer at Ames Research Center was Carl Keller, who played a key role in the instrument's development.

The imaging system aboard the Orbiter was a second generation imaging photopolarimeter that flew on the Pioneer spacecraft to Jupiter and Saturn. For the Pioneer Venus mission, scientists fitted it with an improved telescope and a new interface. The plasma analyzer also was an outgrowth of past programs.

The overall program cost for instrument development was within estimates. Some instruments ran over budget because problems were met in development. However, others came in below cost because problems the mission budgeted for did not materialize. Only one instrument was very late in delivery, and all were ready in time for the mission. Despite the instruments' complexity, the mission's financial management was remarkable in controlling costs to meet budgets.

Designing the Mission and Developing the Spacecraft

Paralleling the development of the science payload, the project had been busily developing the spacecraft. It awarded two concurrent study contracts of $500,000 each on October 2, 1972. One contract went to Hughes Aircraft Company Space and Communications Group, teamed with General Electric Company. The other went to TRW Systems Group, teamed with Martin Marietta. The contracts called for system definition by June 30, 1973. After the contractors defined the system, NASA would select a single contractor to design, develop, and fabricate the spacecraft.

The two contractors took different approaches. TRW considered the use of different basic spacecraft types for the Multiprobe Bus and Orbiter. Hughes preferred a single spacecraft design that would serve the dual purpose. The probe designs of the two contractor teams were similar in essentials, although the Orbiter configurations differed significantly. The TRW design aligned the Orbiter's spin axis parallel to the plane of the ecliptic and pointed toward Earth. The fixed high-gain antenna also pointed to Earth like the TRW-built Pioneer Jupiter/Saturn spacecraft's antenna. In this design, several instruments were mounted on a movable platform so they could scan the surface of Venus. The Hughes design had the spacecraft's spin axis perpendicular to the ecliptic plane, with the spin of the spacecraft sweeping the field of view across Venus. It also was to despin a high-gain antenna and point it toward Earth. The Hughes design won the contract.

Amid the challenge of solving technical problems came a major political disappointment. Congress did not authorize a mission start in the 1974 fiscal year. As a result, it was not possible to meet launch dates for the 1976/77 Multiprobe mission. At this point, August 1972, mission officials changed the launch series. They planned two launches, and both would fall back to the next launch opportunity. Both the Multiprobe and the Orbiter would use launch opportunities in 1978 and arrive at Venus about the same time, near the end of 1978.

Overview of the Mission

The two Pioneer flights to Venus were to explore the atmosphere of the planet, to study its surface using radar, and to determine its global shape and internal density distribution. The Orbiter would operate for eight months or more, making direct and remote sensing measurements. NASA designed the Multiprobe spacecraft to separate into five atmospheric entry craft some 12.9 million km (8 million miles) before reaching Venus. Each probe craft would measure characteristics of the atmosphere from its highest regions to the surface of the planet. These measurements would occur in periods of a little more than two

hours at points spread over the planet's Earth-facing hemisphere.

In celestial mechanics, there are two classifications of transfer ellipse trajectories for traveling between planets. A trajectory that carries a spacecraft less than 180° around the Sun on a voyage from one planetary orbit to another is a Type I trajectory. One that travels more than 180° is a Type II trajectory.

For Pioneer Venus, navigators wanted the Orbiter to fly a Type II trajectory to reduce its velocity upon arrival at Venus. As a result, the spacecraft would need much less propellant to slow it into an orbit around Venus—about 180 kg (400 lb) of propellant out of the spacecraft's total weight of 545 kg (1200 lb). A Type I trajectory to Venus would have required 50% of the total spacecraft weight to be propellant. The plan had the Orbiter launch during the period May 20 through June 10, 1975. It would follow a seven-month flightpath to Venus along a trajectory of about 480 million km (300 million miles) (Figure 2-3). The long trajectory would reduce both the propellant's weight and the orbital insertion motor's weight and size. This path also permitted the periapsis to be about latitude 20° N on the planet.

For the first 82 days after launch, the Orbiter spacecraft would fly outside Earth's orbit. It would then cross Earth's orbit and plunge inward on a long curving path toward the Sun. It would arrive at Venus on December 4, 1975, five days before the arrival of the probes, which would follow a shorter flightpath. The Multiprobe spacecraft would be launched a few days after the Orbiter crossed Earth's orbit, during August 7 through September 3. This spacecraft would follow a shorter, Type I trajectory.

Observing Venus from Orbit

On the Orbiter's arrival at Venus, the mission plan called for the spacecraft's motor to thrust for 28 seconds. This was the first time mission controllers would use a solid-propellant motor stored in the vacuum of space so long (125 days) for an orbit insertion maneuver. Their aim was to reduce the spacecraft's velocity so it would enter an elliptical orbit with a 24-hour period. The orbit was oriented 75 degrees to the equator of Venus—somewhat more inclined than the January 1973 study report suggested. Navigators initially desired a periapsis of 300 km (186 miles). They also wanted an apoapsis of 66,000 km (41,012 miles). Later they would command the spacecraft into an orbit having a periapsis of 150 km (93 miles). The orbit's eccentricity and inclination would accomplish a variety of scientific goals and meet a number of engineering requirements.

The periapsis would allow a remote sounding radar mapper to study the planet's surface and several instruments to take measurements within the upper atmosphere and ionosphere. It also provided excellent viewing geometry for remote sensing atmospheric experiments.

The apoapsis would produce an orbit with a period of about 24 hours, which was beneficial to tracking and ground operations. It also produced a nearly one-to-one correspondence of orbit numbers with days into the mission. For science, it provided good viewing geometry for obtaining cloud images and a wide sampling region for solar wind interaction experiments.

The orbit's high inclination permitted the spacecraft to make measurements and direct observations over a wide range of latitudes.

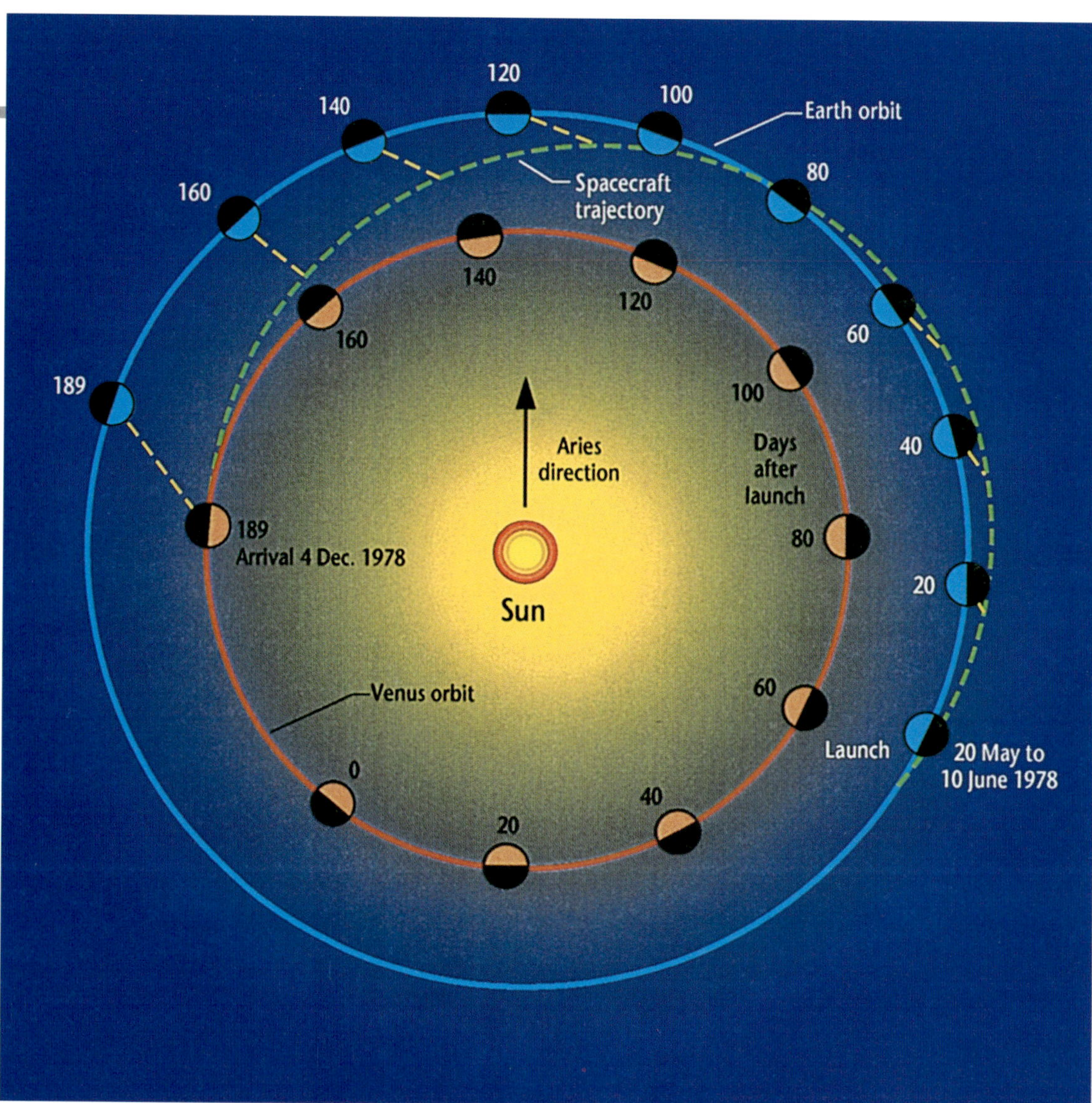

Figure 2-3. The Orbiter's trajectory carried it first outside the Earth's orbit for nearly half of its journey to Venus. This trajectory minimized the amount of propellant needed to enter orbit around Venus.

Perturbation of the spacecraft's orbit by solar gravity would change significantly the altitude and latitude of the periapsis. For the first 20 months of the mission (Phase I), flight controllers would use the spacecraft's thrusters to counteract altitude drift. Later, during Phase II, they would allow periapsis to rise to a maximum of 2310 km (1435 miles) by July 1986. At that time, solar perturbation would cause the periapsis to descend again. In the same period, the latitude of periapsis would move from 17° N to the equator.

Phase III would begin in late 1991 as periapsis again reached the lower thermosphere and ionosphere. Mission planners arbitrarily defined the transition from Phase II to Phase III to take place when the periapsis altitude reached 1000 km (621 miles). In 1992, flight controllers would use the remaining hydrazine propellant to keep the periapsis within 140 to 160 km (87 to 100 miles). During Phase III, the Orbiter would penetrate and sample deeper in the atmosphere than was acceptable during Phase I. Phase III would end in October 1992 when the hydrazine propellant was exhausted and the spacecraft's periapsis had descended into the atmosphere. By that time, drag would pull the spacecraft from orbit into a meteoric ending of a 14-year mission.

Probing Venus' Atmosphere

The Multiprobe's four-month trip to Venus resulted in the spacecraft approaching the planet at or about 1900 km/hr (1180 mph). The comparative trajectories for the Orbiter and the Multiprobe appear in Figure 2-4.

as simple as possible, they decided on a simultaneous launch from the Bus. In a one-firing episode, they could release all three Small Probes; separate launches would have been less reliable. However, this single launch episode demanded detailed computer analysis.

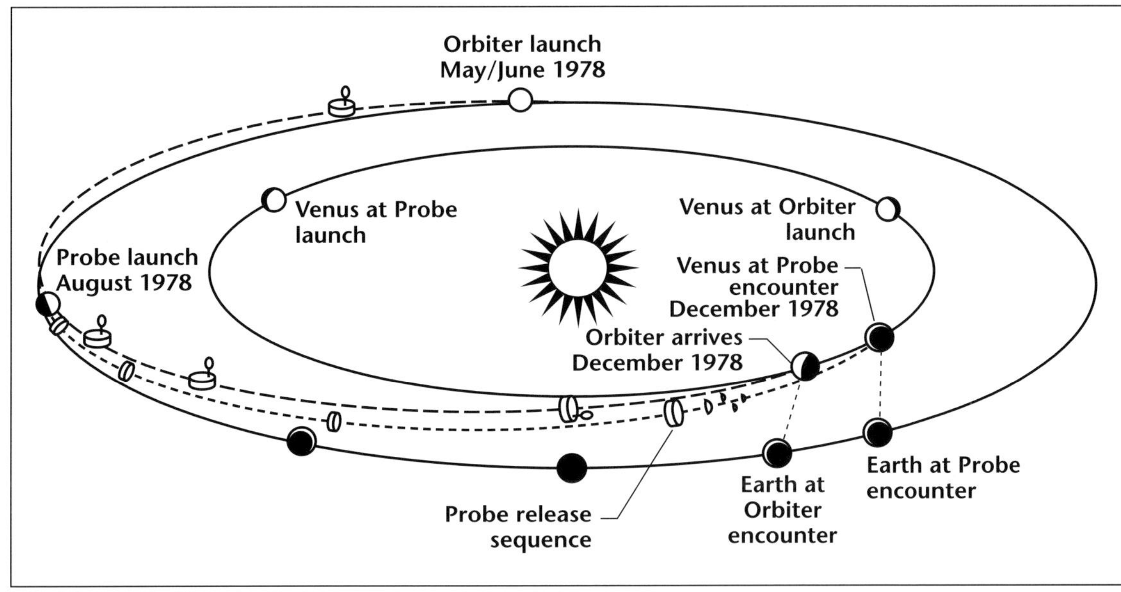

Figure 2-4. The Multiprobe followed a shorter trajectory and arrived at Venus a few days after the Orbiter. This drawing compares the two trajectories.

Twenty-four days before the probes entered Venus' atmosphere, the Multiprobe's axis would lie along the trajectory that the Large Probe would follow to Venus. The probe was then launched to follow its own path to the planet. Next, flight controllers changed the Bus' path to point toward the center of Venus. This change allowed the Small Probes to leave the spinning Bus when it was 20 days from the planet. The spin insured that the Small Probes separated along paths that would take them to their individual targets on the planet (Figure 2-5).

Originally, mission planners had discussed an alternative concept. In this scenario, they would have individually targeted the three Small Probes and each would have separated individually from the Bus. To keep the system

Where should the spin axis be pointed? At what spin rate should the spacecraft operate for the release? Mission planners needed these answers so they could direct the probes to enter Venus' atmosphere near the planet's limb regions (as viewed from Earth). But the entry could not be too close to the limb to limit slant-range communications through the planet's atmosphere. A computer program displayed different targeting options. Specifically, the program determined the angle of attack of each probe's entry into the Venusian atmosphere. All the probes were stabilized by their rotation, but what would happen if one entered the atmosphere sideways? Its heat shield would not have protected it from the heat of entry. The probe would have been destroyed.

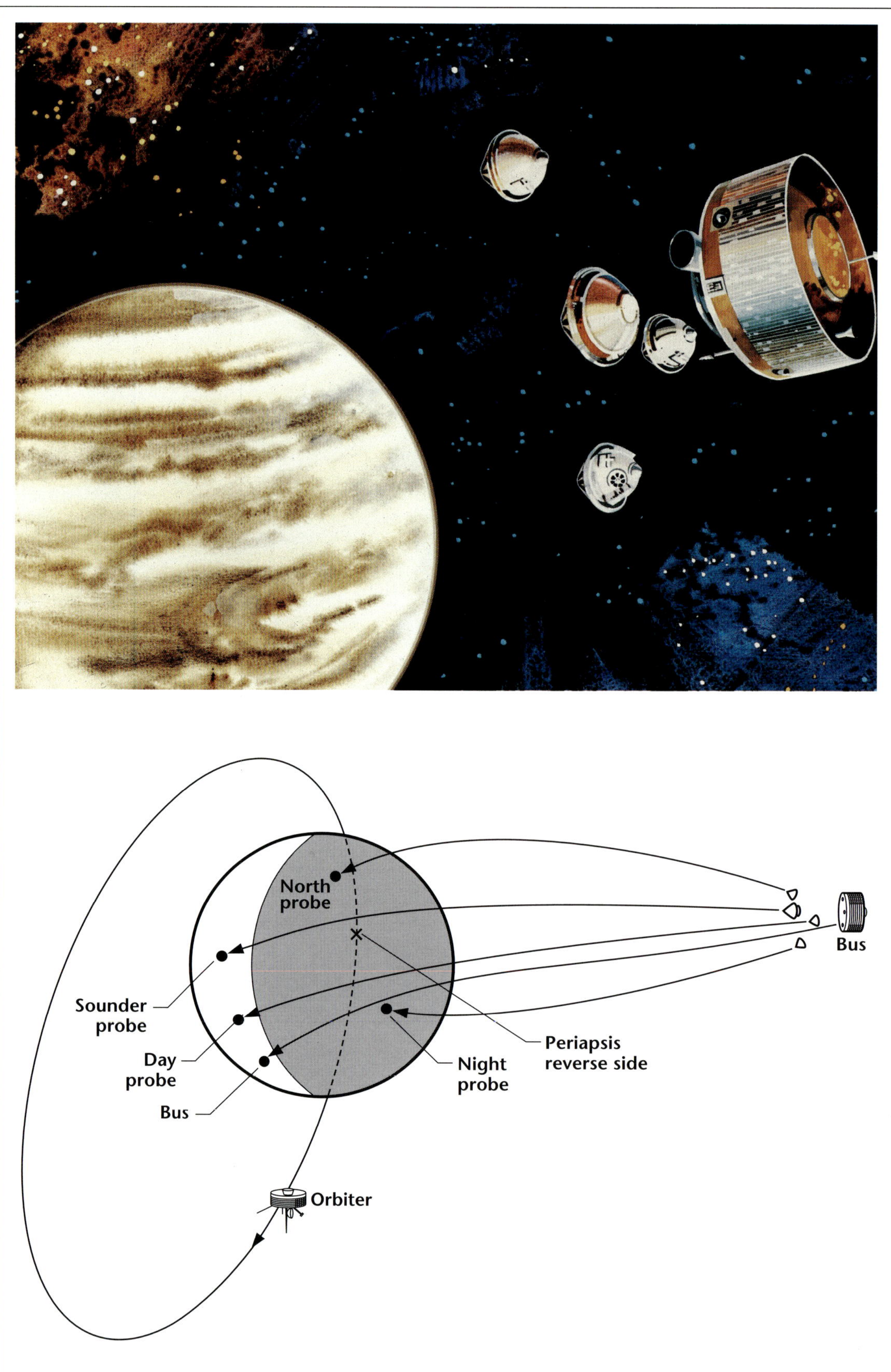

Figure 2-5. Approaching Venus, the Multiprobe released its four probes toward different target areas on the planet. (Top) Artist's concept of the probes and the Bus shortly after their release. (Bottom) Diagram of the paths of the probes and their entry points on the planet in relation to the orbit of the Orbiter spacecraft.

On arrival at Venus, the four probes entered the atmosphere. The Large Probe took about 55 minutes to descend to the surface, the three Small Probes, about 57 minutes. None of the probes were designed to withstand impact with the surface, each hitting it at about 36 km/hr (22 mph). The Bus itself hurtled into the upper atmosphere about 80 minutes after the probes. Unlike the probes, the Bus carried no heat shield; its task was to provide data only on the atmosphere's highest part.

All probes sent their data directly to Earth as they penetrated Venus' atmosphere on the hemisphere that faced Earth.

Launch Vehicle

Originally, the project planned to use the Thor-Delta launch vehicle for the Pioneer Venus flight. The system definition studies began with this launch capability as a design criterion for the two spacecraft. However, very early in the study it became clear that costs were rapidly rising as subsystem designs were severely restricted in weight and size. To reverse this trend, the project team asked competing contractors to study an alternative design that removed the weight and size restrictions. The contractors did this by comparing design and cost estimate results of an Atlas-Centaur launch with the launch capabilities of the Thor-Delta.

Based on these analyses, mission planners determined that the additional $10 million for the Atlas-Centaur launch vehicle would be acceptable. (The costs would at least equal the increased costs required to cover the miniaturization of the Multiprobe and Orbiter spacecraft designs for the Thor-Delta requirements.) NASA, therefore, approved use of the Atlas-Centaur—NASA's standard launch vehicle for payloads of intermediate weight (Figure 2-6). The Atlas-Centaur launch vehicle stands about 40 m (131 ft) high. It consists of an Atlas SLV-3D booster with a Centaur D-1A second stage. The nation's first high-energy launch vehicle, Atlas-Centaur used liquid hydrogen and liquid oxygen propellants for its upper Centaur stage. Engineers enclosed each spacecraft in a fiberglass nose fairing to protect it as the launch vehicle sped through Earth's atmosphere.

"New Start" Approved for Fiscal Year 1975

By July 1973, the system definition studies were completed and each team received a holding contract. The next step involved competitive bidding following issue of a Request for Proposal in June 1973. Based on the bidding results, the project team selected Hughes Aircraft Company in February 1974 to negotiate a cost-plus-award-fee (CPAF)

Figure 2-6. An Atlas-Centaur, NASA's standard launcher for payloads of intermediate weight, carried each of the Pioneer Venus spacecraft.

contract. This contract covered the initial conceptual design phase of the system. The proposed cost of design work for this phase was $3 million. There also was an option for final design, development, fabrication, testing of two flight spacecraft, and launch support at $55 million. NASA awarded a contract in May 1974, but not for the hardware. The mission still waited for Congressional approval of a "new start"—a new authorized space mission for fiscal year 1975.

In August 1974, Congress finally approved a new start for Pioneer Venus for fiscal year 1975. Further negotiations took place with the contractor. In November 1974, NASA made a final award, including hardware, to Hughes Aircraft Company.

System specifications were completed by February 1975. By the beginning of calendar year 1975, work was well under way. But the program still had to face major hurdles before launch. Said Charles Hall, project manager: "It always seems you don't have enough time and you are trying to find ways to do things faster. You are always having trouble with funding. You may have a total amount of funds that is enough for the program but you never seem to have enough for any particular year. So you are always making small perturbations to your plans to work around funding difficulties."

New Funding Problems

In June of 1975, during the budget hearings for fiscal year 1976, Pioneer Venus suffered a serious setback. The House of Representatives voted to cut $48 million from the NASA appropriations for the Venus mission. NASA had already spent $50 million on the program. The House vote was based on misinformation and a lack of understanding about the technical problems associated with a delay. Suppose NASA had delayed the launch to the 1980 opportunity (the most likely scenario if Congress had withheld funds). What would have happened? Engineers would have had to redesign the spacecraft because the 1980 launch opportunity was not as favorable as 1978. More launch energy or a lesser payload would have been the result. That might have been the end of the program. NASA would have needed as much as $50 million extra (over what they originally requested) for a mission to Venus at the less favorable launch opportunity.

However, scientists, the national press, and many organizations rallied to Pioneer Venus. Important scientific groups lent their support. The Nation's most famous climatologists and meteorologists stressed the importance of more and better information about the weather and climate of Venus and Mars. Increases in the world's population make it increasingly important that we understand Earth's climate better. We all would benefit by an ability to predict accurately long-term changes that might lead to droughts and poor harvests. Scientists from many fields pointed out the mission's importance to Earth sciences and to finding ways to lessen the effects of climate changes on food production. (For example, one scientist pointed out that a change in Earth's average temperature of only 1.5°C could wipe out Canada's entire wheat production. If such a change should occur unexpectedly, the effects on world food supplies could be disastrous.) Scientists stressed that understanding weather systems on Venus and Mars was essential to a better understanding of Earth's weather systems.

A Senate subcommittee restored funds for Pioneer Venus in July 1975. This action reversed the House move to slash all but $9.2 million from the project. But NASA's worries weren't over, yet. The project still

faced high hurdles. The Senate Appropriations Committee and then the full Senate had to approve the funds. If they did, a joint committee would still have to work out a compromise with the House. The Senate committee acted on the bill later in July and gave support to the mission. Early the next month, the Senate also recognized the program's importance. The Senators gave their approval to NASA's requested funding of $57 million for Pioneer Venus during fiscal year 1976.

During September 1975, the go-ahead finally came. The Senate-House conference committee restored all but $1 million of the funds to send the two Pioneer spacecraft to Venus in 1978. The Earth-based part of the mission was back on course. Scientists and engineers could again concentrate on their main task: having the spacecraft and their scientific instruments ready for launch opportunities.

Parachute Development

By June 1975, final contracts for scientific instruments were ready for signing. By July, engineers had studied and resolved most problems of integrating instruments into the spacecraft. The first tests of the parachute system, needed for the descent of the Large Probe into the Venusian atmosphere, had started. This aspect of the Pioneer Venus program made use of the largest structure of its type in the world: the Vertical Assembly Building at NASA's Kennedy Space Center, Florida. (NASA originally built the structure for final assembly of the huge Saturn V boosters that launched Apollo spacecraft to the Moon.) NASA used the building to test the Large Probe's parachute. In the test series, engineers dropped full-size parachutes with pressure vessels of various weights 135 m (443 ft) in the wind-free environment of the building. The series helped them determine the parachute's aerodynamic trim characteristics (Figure 2-7).

The Large Probe parachute was an important development item. It was essential because it would delay the descent of the Large Probe long enough to make many measurements as it settled through the clouds.

Figure 2-7. The parachute for the Large Probe was tested initially in drop tests within the large Vertical Assembly Building at NASA's Kennedy Space Center, Florida.

"For a time it almost looked as though we were never going to get a parachute," commented Charles Hall after the mission. He related how they had taken a newly designed parachute to the desert near El Centro, California, for a drop test from an F-4 airplane. Personnel attached the parachute to a pointed cylinder that carried high-speed (200 frames/sec) cameras and test instruments. When the airplane was traveling at high speed and proper altitude, the cylinder would drop and the parachute deploy, a drogue chute pulling out the main chute.

The day of the test arrived. As everyone had expected, the cylinder dropped and the drogue chute deployed. Observers were appalled, however, to see no trace of the main chute opening.

It literally disappeared. Hall described how, when investigators examined the film records, they found the parachute starting to open and then disintegrating into shreds. The camera speed was 200 frames/sec, and they could view the film one frame at a time. Said Hall: "You wouldn't believe it, but on one frame the parachute would be intact and on the next frame there would be nothing there. It was not that it was breaking away from the shrouds, the material itself was just ripped to shreds."

Engineers thought that the test environment was too severe. So they planned another test in which a lower dynamic pressure was exerted on the parachute. The results were equally bad.

A third try also failed. But Hughes engineers inspected the pictures more closely. They noticed that when the parachute was still intact, in the frame just before complete failure, many of the parachute gores (the angular sections of the parachute) were missing, although the chute fully deployed. They suspected this was the cause of the trouble since the part that opened would experience greater stresses than its design allowed.

Engineers next deployed one of the parachutes in Ames Research Center's 40- by 80-Foot Wind Tunnel. Even there all the gores did not open. The low wind speed in the tunnel was then reduced to a relative breeze so an engineer could walk inside and watch the opening. When the parachute opened and the gores still stayed folded, he tried to pull them apart but could not. The chute's design allowed the wind load to effectively hold the gores together. As a result, NASA abandoned this parachute design in favor of an earlier conical ribbon design.

Time was running out, and NASA had to take some chances. When the new parachute was ready, engineers put it through a final system drop test. There was not even time to try it with airplane drops first. In the earlier tests, the falling body had not been a sphere with a heat shield. Because of time constraints, NASA had to test the parachute, the heat shield release mechanism, and other hardware on one drop from a high-altitude balloon.

In December 1976, mission engineers tested the system in a balloon drop at the Army's White Sands Missile Range in New Mexico. The parachute deployed at an altitude of 16 km (10 miles). At that altitude, atmospheric temperature and density and the probe's speed would be close to the conditions on Venus just before its descent into that planet's dense, hot, lower atmosphere. Personnel designed the tests to confirm several features: deployment of the probe parachute, separation of the atmospheric entry heat shield, and, after 17 minutes of parachute descent, separation of the pressure vessel for its free-fall plunge. The fast descent after the parachute's release would let the probe penetrate deeply into the Venusian atmosphere. This plunge would be completed before high temperatures could destroy the probe's instruments.

The sky was clear at 4:00 a.m. when the balloon gently lifted its load from White Sands Proving Grounds, New Mexico. Ponderously, the great plastic bag carried the test vehicle to an altitude of 31 km (19 miles). At Ames Research Center, project leaders waited for the test results. Says Hall: "We got the phone call . . . 'It has been a complete failure.'" When staff gave the radio command to release the vehicle, it dropped swiftly from the gondola beneath the balloon, just as mission scientists had planned. However, as the probe released from the balloon, it hit the gondola, which caused the probe to turn upside down. Thus, when the parachute released, it pulled against

the parachute clevises in the wrong direction and broke them. The test vehicle plunged to the desert floor. "We were in trouble," Hall said. "We did not have a parachute."

When investigators studied the photographs at Ames Research Center and later carefully inspected recovered parts of the test vehicle, they discovered the reason for the failure. There were structural breakages all over the test vehicle. And these breaks had all occurred before impact. Obviously, the way the parachute released had caused the structural damage. At first it seemed that not only had the parachute failed, but also the whole system had not been properly stressed.

Engineers studied the photographs in detail. They saw that the test vehicle had been tumbling before the parachute deployed at 18,000 m (60,000 ft). In fact, after tumbling part of the way down, the test vehicle became stable, but fell tail first instead of nose first. When the parachute deployed, it came off at an angle that it was not designed for. The pictures showed the chute being deployed, and, in the next split second, the chute breaking away from the body of the Large Probe because of its wrong attitude.

Why had the test vehicle tumbled during its fall from the balloon? For the journey upward, it had been in a container about 3 m square (10 ft square). At the last minute, a test engineer became worried. He feared that, in the gondola's ascent to 30,500 m (100,000 ft), the temperature would drop too low, and equipment in the Large Probe would fail to operate correctly. As a result, he taped a protective blanket, made of 1.3-cm (0.5-in.) fibrous padding, beneath the box. When the probe fell through the blanket, one edge caught on the blanket, and the probe tumbled in its fall.

Engineers built another test vehicle, made another drop, and finally achieved success. Pioneer Venus finally had a working parachute for its Large Probe.

Spacecraft Development Challenges

Test engineers suffered anxious moments during a thermal vacuum test of the probe only seven months before the launch date. During the test, the batteries within the spacecraft failed completely. With the launch date so close, this looked like a major disaster for the program.

"In retrospect," said Charles Hall, "all these things look simple, but at the time we had no idea whether it was the test environment or the battery at fault. We made many side tests and had experts give their opinions, and as is generally the case with these problems you can get about as many people on one side as the other."

Investigation showed, however, that the batteries themselves were not at fault. It was the conditions of the test that had caused the failure. During the test, the spacecraft spun on an axis aligned horizontally. The g force thus varied in direction during each revolution. As a result, the electrolyte sloshed within the batteries, a condition that would not occur during an actual mission. This sloshing caused massive failures within the battery's cells.

The cable connections within the probes' confined space also led to difficulties. Within a spacecraft, the cable harness nearly always presents problems. This was especially true for the Venus probes. The harnesses for these probes were difficult to design because the probes had to be taken apart several times during testing. Assembling and disassembling the spacecraft often caused testing problems. Engineers installed equipment on two shelves and interconnected them by the harnesses. The

Figure 2-8. Many tests were necessary to ensure that the probes would be able to withstand the enormous pressures and high temperatures of Venus' atmosphere. (Top) The pressure vessel of the Small Probe appears partially assembled prior to running pressure descent tests on it. Engineers machined the two-piece structure from titanium forgings. It weighed approximately 18 kg (40 lb). Its diameter was 46 cm (18 in.), and the wall thickness averaged approximately 0.3 cm (1/8 in.). (Middle) Engineers constructed this full-scale mockup of the Large Probe's pressure vessel module during Phase B. They used it to study the packaging problems inherent in spherical geometry. (Bottom) A buckling indicator mandrel, used during buckling tests of probe vessel scale models, appears before assembly with a test model. In each test, the mandrel supported the inside surface of the pressure vessel model and recorded the imprint of the buckle pattern on its graphite coated surface. This procedure helped determine where failures occurred.

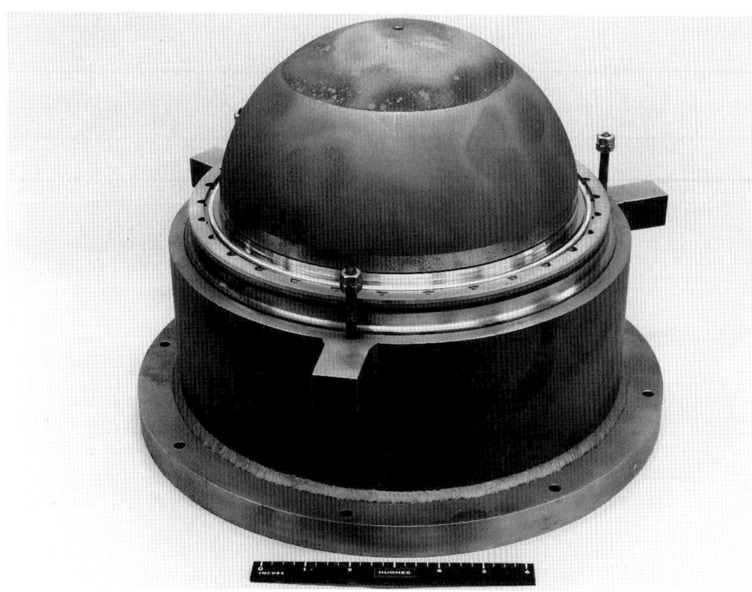

standard procedure with spacecraft was to assemble the whole package and test it. When they were done, they took it apart again so the principal investigators could make final calibrations on the instruments in their laboratories. They had to replace instruments in the spacecraft before shipping it to the launch area for mating with the launch vehicle.

From a systems integration and test standpoint, the Multiprobe Bus and Orbiter's common features helped simplify testing problems and software development for the test programs.

Another major problem was sealing the probes. On the probes' way to Venus, internal pressure had to be maintained against leakages into the vacuum of space. When a probe entered Venus' atmosphere, it had to resist the tremendous pressures and prevent inward leaks. During development of the spacecraft, engineers made many pressure tests (Figure 2-8). They wanted to make sure that the titanium shell could withstand the pressure and that the seals did not leak. Two types of seals were necessary for these opposing conditions. They both required a unique design and many more tests (Figure 2-9). For the vacuum of space, an O-ring type of seal was best. To resist the high pressure of the Venusian atmosphere, engineers used flat Graphoil seals (made of graphite fibers) between flat surfaces on flanges of the spacecraft parts.

The system worked well. One probe actually sent data after it had landed on Venus' surface, and these data showed no evidence of any damage.

Sealing the various spacecraft windows presented another series of problems. Engineers made many tests to ensure the seals would withstand both high pressure and temperatures (Figure 2-10). Significant development problems occurred, however, in making a suitable seal for the diamond window (Figure 2-11). Engineers decided early not to braze the window to the shell of the pressure vessel. Later, they reversed this decision and decided to coat the edge and then braze it to the diamond. As the program continued, window sealing remained a very difficult fabrication problem. In fact, it became a pacing problem that prevented testing the flight diamond window, with its full assembly, with the instruments in the spacecraft.

There were many more disappointments with the windows. Engineers would think they had a solution, but when they tried it, it would fail. They would try again, but just when they thought they had completed a successful test, the window seal would spring another leak. Eventually, they had to use a mechanical flat seal.

As time for shipment of the Large Probe approached, engineers decided that the internal pressure might be too low when the probe entered Venus' atmosphere. To increase pressure by 6 psia, they decided to add a nitrogen pressure bottle to the payload. This nitrogen bottle had a volume of 110 cm^3 (6.7 in.^{3}) which, with the nitrogen stored at 4,000 psia, would increase the internal pressure by 6 psia. With attachments, the bottle added 3.5 kg (7.8 lb) to the Large Probe's weight. An electrically fired squib valve punctured a sealing diaphragm and opened the bottle before the Large Probe entered the atmosphere. The rate of release was 5 psia/min. This addition required wiring changes. At the eleventh hour, the squib valve did not puncture

Figure 2-9. A metal pressure seal for a probe pressure vessel appears with its disassembled testing fixture. The seal has undergone a sealing test in a simulated environment that approximates a descent through Venus' atmosphere.

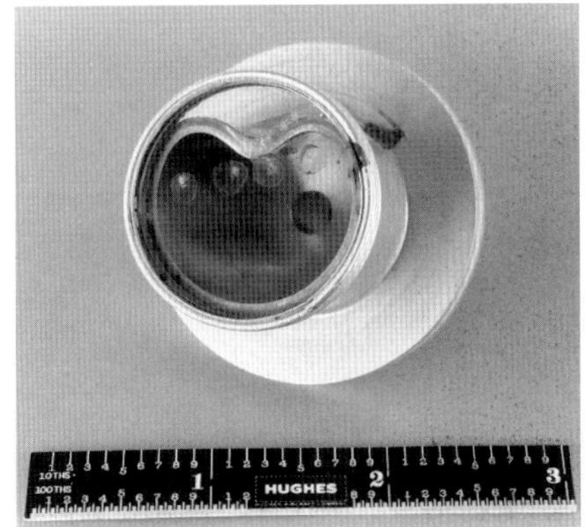

Figure 2-10. (Top) A side view of the 3.175-cm (1.25-in.) diameter sapphire window assembly after a test in a simulated Venus atmosphere pressure and temperature. The assembly consisted of an Inconel housing with a Kovar-sheathed heater wrapped around a brazed sapphire window. (Bottom) Result of pressure testing a sapphire window and its mount that were representative of the probe's windows. Note that the mount failed before the window itself failed. The failure pressure was approximately three times the maximum amount scientists expected at Venus' surface.

Figure 2-11. This photograph depicts an early configuration of the diamond window and its heater assembly. The test article incorporated a 10-mm window and demonstrated a technology for brazing the diamond and heater assembly to a mallory metal mount. Pressure tests to 2500 psi showed no leak or structural problems at that stage of window seal development.

the diaphragm, and modifications had to be made to the ram and valve body.

Mission Operations

Pioneer Venus mission controllers had to operate two different spacecraft at the same time. Keep in mind that all Pioneers are relatively unautomated spacecraft, designed to minimize costs. This required mission operations to maintain 24-hours-a-day control with careful analysis and planning at short notice. Although ground-controlled spacecraft have flexibility to change plans and objectives during a mission, they require constant monitoring and control. Pioneer Venus control and spacecraft operations were at the Pioneer Mission Operations Center (PMOC) (Figure 2-12) at Ames Research Center. Bendix Field Engineering Corporation provided support services.

Continued operation of previously launched Pioneer spacecraft made activities at the Mission Operations Center somewhat more complicated. Pioneers 6, 7, 8, and 9 continued to circle the Sun and to return interplanetary data. Pioneer 10, which flew past Jupiter in 1973, was heading out of the Solar System. In its travels, it continued to transmit important information from previously unexplored regions of space. Pioneer 11, which flew by Jupiter in 1974, was on its way to the first rendezvous of a spacecraft with Saturn.

All command information originated from the PMOC. The Center received telemetry data required for control of the mission and displayed the information as needed. Computers allowed personnel to enter commands and rapidly interpret the spacecraft's data stream for flight controllers. The integrated team working at the Center was made up of dedicated individuals from NASA and its support contractor, Bendix.

Because two spacecraft with separate missions were involved, two flight operations groups were on hand: an Orbiter group and a Multiprobe group. Both groups had a science-analysis team that determined each instrument's status and formulated command sequences for that mission. They also each had a spacecraft performance analysis team. These teams analyzed and evaluated spacecraft performance and predicted how it would respond to commands. A third group served both spacecraft. This was the navigation and maneuvers group that took care of spacecraft navigation, orbital injection and trim, and probe targeting.

To determine spacecraft trajectories, JPL provided computer analysis of the tracking information from the Deep Space Network (DSN). Support groups at Ames Research

Figure 2-12. The Pioneer Mission Operations Center (PMOC) at NASA Ames Research Center, California, commanded and controlled all spacecraft.

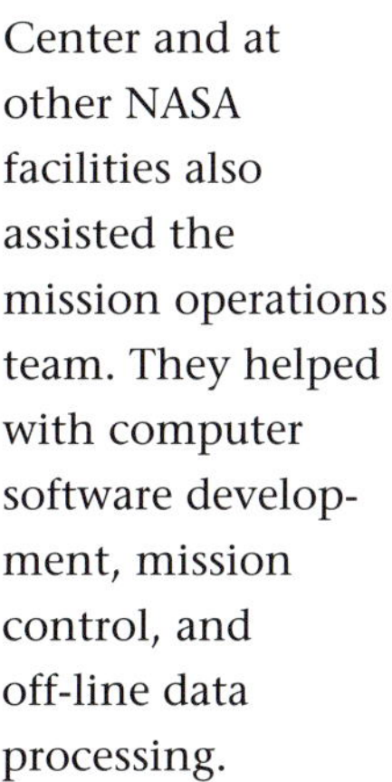

Center and at other NASA facilities also assisted the mission operations team. They helped with computer software development, mission control, and off-line data processing.

Data Return, Command and Tracking

To track all six spacecraft—four probes, Bus, and Orbiter—NASA used the DSN's global system of large parabolic dish antennas. The large antennas at each site were essential for critical phases of the mission. These events included reorientation of the spacecraft, velocity corrections, orbit insertion, and entry of the four probes into Venus' atmosphere. The large antennas also were involved in special science events such as radio-occultation experiments.

The DSN, which JPL managed, had facilities located at approximately 120° intervals around Earth (Figure 2-13). As the Orbiter and the Multiprobe appeared to set at one station due

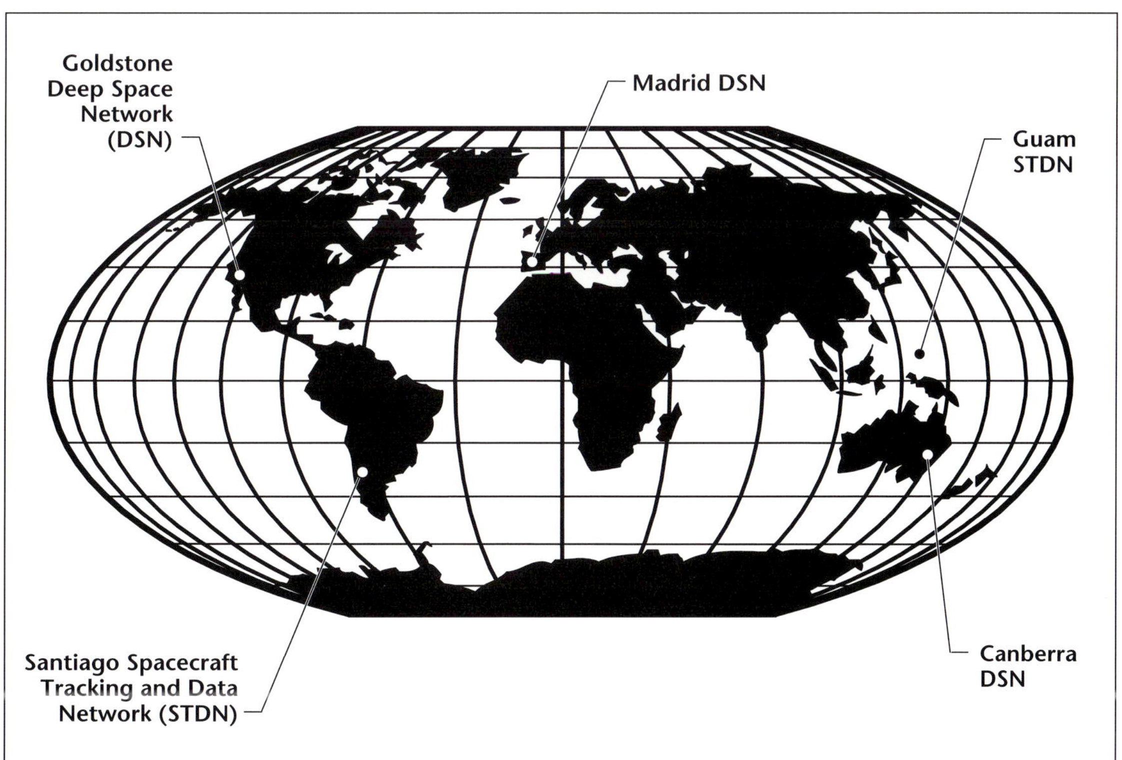

Figure 2-13. NASA used the worldwide system of the Deep Space Network (DSN) to communicate with the spacecraft during the mission, to issue commands to some of them, and to receive scientific data telemetered from all the spacecraft.

Figure 2-14. (Top) Two of these big antennas at Goldstone, California, and Canberra, Australia, maintained contact with the probes during their penetration of Venus' atmosphere. NASA used the two antennas at the same time to ensure that none of the data was missed during this one-hour descent. (Bottom) During probe and bus penetration of Venus' atmosphere, the Deep Space Network handled six spacecraft at once. All the spacecraft transmitted their information directly to Earth, as this diagram shows.

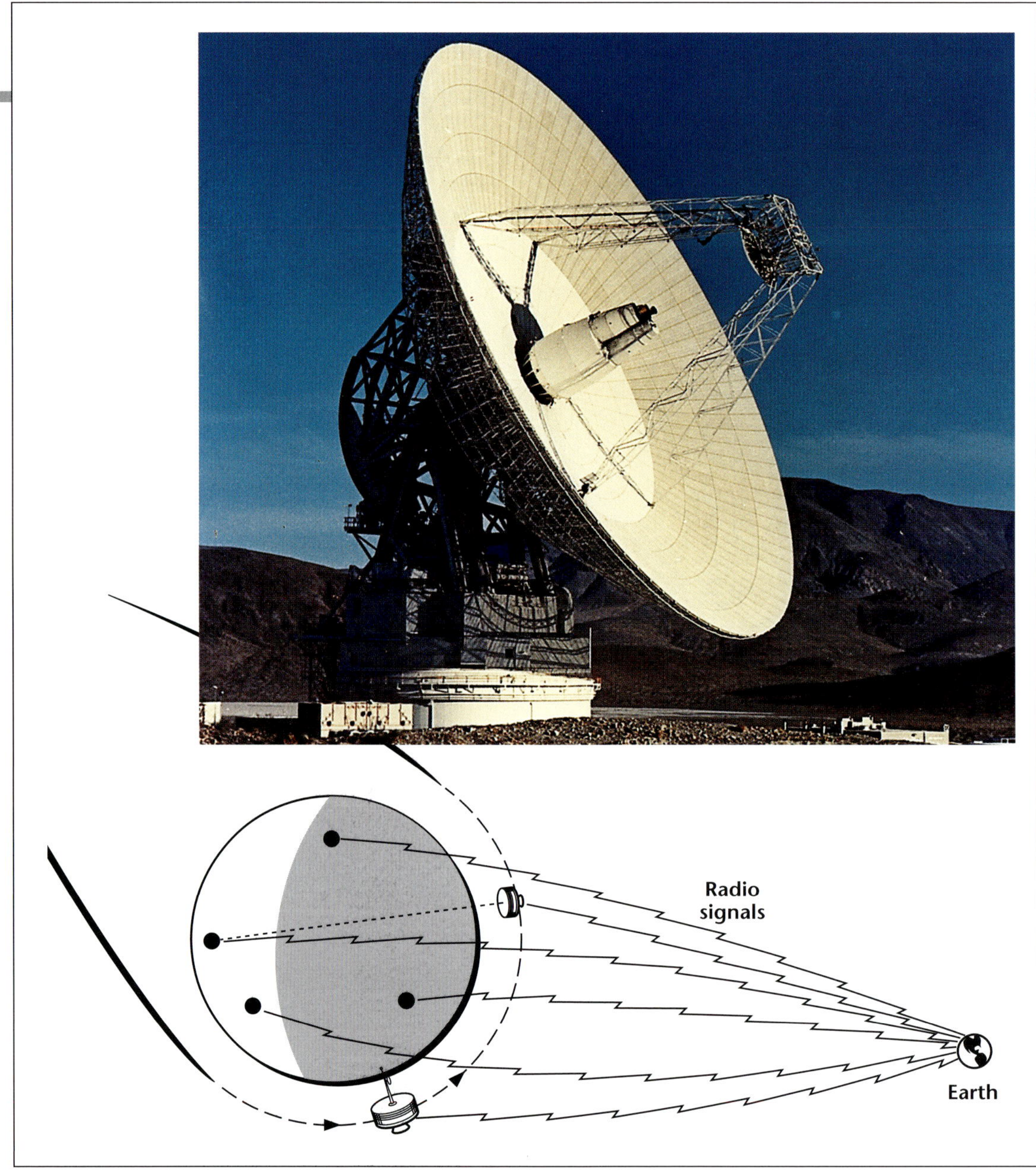

to the rotation of the Earth, they were rising at the next station. The DSN had six 26-m (85-ft) antennas. Two were at Goldstone, in California's Mojave Desert, two at Madrid, Spain, and two at Canberra, Australia. (Officials later upgraded one at each location to 34 m, or 112 ft, and shut down the remaining 26-m antennas during 1981 budget cuts.) There also were three 64-m (210-ft) antennas (Figure 2-14 top), one each at the three locations. During the Pioneer Venus extended mission, officials upgraded these to 70 m (230 ft). In the critical 2-hour period when the Bus entered the atmosphere and the four probes descended to the surface, scientists relied on the 64-m (210-ft) antennas at Goldstone and Canberra. They used these antennas to receive and record data from all five spacecraft at the same time (Figure 2-14). Two additional tracking stations provided special data gathering for the probes' Differential Long Baseline Interferometry (DLBI) experiment. These were the 9-m antenna stations that were part of the Spaceflight Tracking and Data Network (STDN) at Santiago, Chile, and on Guam.

During launch, the DSN, with the help of other facilities, tracked each spacecraft. These other facilities were the Air Force Eastern Test Range tracking antennas and elements of NASA's Spacecraft Tracking Data Network. Four instrumented aircraft from Wright Patterson Air Force Base also provided tracking support.

The Deep Space Network Stations formatted incoming telemetry. From there, it traveled over the high-speed circuits of the NASA Communications System (NASCOM) to the Pioneer Mission Computing Center (PMCC). There computers processed it to supply various types of real-time display information for all spacecraft and their experiments. The computers checked for unexpected or critical changes in data. They also provided information to specialists experienced in all details of the spacecraft, experiments, and the ground system. Their analyses ensured that spacecraft were always controlled correctly for the best science results. Computers at Ames Research Center verified outgoing commands. They then sent these commands to the Deep Space Network Stations where computers again verified the commands before relaying them to the spacecraft. JPL furnished navigation data and trajectory computations for the Pioneer spacecraft.

Mission specialists made several changes to the DSN for its use in the Pioneer Venus mission. They added receivers to handle five different data streams at the same time. To cope with large frequency drifts, they installed special wide band recorders. Two events caused the drifts: (1) shifts in probe velocity as the probes entered Venus' atmosphere and (2) atmospheric effects on signal propagation as probes descended through the dense, hot atmosphere. To make sure that no data were lost as the probes plunged through the atmosphere, the DSN took special precautions. They provided special equipment to tune receivers to each probe's signals and to record data in unsynchronized form for special off-line processing.

The PMCC did more than provide telemetry for mission operations and quick looks at scientific data. The Center also processed all telemetry to supply experiment data records to principal investigators for distribution to their team members.

Countdown to Launches

During February 1978, pre-shipment reviews took place at the Hughes Aircraft Company plant in El Segundo, California. Following these reviews, NASA shipped the spacecraft to the launch site at Kennedy Space Center, Florida. The main body of the Orbiter and the high-gain antenna were shipped separately. When they arrived in Florida, the first task was to mate the antenna and the spacecraft. Later, in the checkout area, engineers tested the complete Orbiter extensively to make certain that all subsystems and scientific instruments were operating correctly.

After these tests were complete, engineers installed class B ordnance (ordnance that would not harm the spacecraft or test personnel if it were fired by mistake). The spacecraft was then moved to Building SAFE-2 where staff loaded the rest of the ordnance and 32 kg (70 lb) of hydrazine propellant. Hydrazine was the fuel used for trajectory corrections and orientation maneuvers. Mission personnel then mated the spacecraft with an adapter that attached it to the launch vehicle. Following this procedure, they transferred the spacecraft to launch pad 36 where they mated it to the waiting Atlas-Centaur.

Figure 2-15. NASA successfully launched the first Pioneer Venus spacecraft, the Orbiter, on 20 May 1978. The launch took place from NASA Kennedy Space Center, Florida.

spacecraft were correct. After 10 days of various tests, mission officials gave the "go" for launch, and the countdown began.

There were no holds. The big Atlas-Centaur lifted the Orbiter into the Florida skies on its way to Venus (Figure 2-15). The launch was precisely on schedule—May 20, 1978, at 1313 UT.

Meanwhile, during April, the Multiprobe completed its pre-shipment review at Hughes. The Large Probe was shipped separately from the Bus and the three Small Probes. As soon as the spacecraft arrived at the Kennedy Space Center checkout area, the Small Probes were removed from the Bus. Each was thoroughly checked, as was the Large Probe. During checkout, the flight batteries remained under strict thermal control. They could not be on board the probe for testing because engineers could not put them through charge/discharge cycles. Other batteries were used for the tests, and the flight batteries were the last items installed on the probe at the end of checkout.

Tests in the thermal vacuum chamber at Martin Orlando verified pressure vessel seals. The gas within each probe contained a trace of helium. For 24 hours, test engineers sampled the vacuum chamber contents. Helium in the vacuum chamber samples would have revealed

Once the spacecraft was on the launch vehicle, another series of tests verified that the spacecraft and its systems had not degraded in any way. Then followed a series of radio frequency interference (RFI) tests. These tests verified that the radar for tracking the vehicle during launch would not create problems. Specifically, they ensured the radar neither interfered with the spacecraft nor affected the data coming from it.

After several practice countdowns, a final test with the DSN determined if signals from the

a seal failure. None appeared for any of the separately tested probes. All passed the leakage test satisfactorily.

Back at the Kennedy Space Center, technicians placed the probes on the Bus. Next, they installed pyrotechnics—explosive bolts for the Large Probe and bolt cutters for each Small Probe. These devices would release the probes from the Bus. Also, hydrazine was loaded into the Bus.

Figure 2-16. On 8 August 1978, slightly less than three months after the Orbiter left Earth, NASA launched the second spacecraft, the Multiprobe, from the Kennedy Space Center, Florida. It followed the Orbiter to a rendezvous with Venus in December of that year.

The Multiprobe was then moved to the launch pad, mated with the Atlas-Centaur, and underwent a final checkout. Engineers could only make a very brief check of the radio frequency link to each probe. The probes were all warm, near the ambient Florida temperature. However, when they reached the atmosphere of Venus, they would be very cold. To avoid exceeding temperature limits for equipment within the heavily insulated probes, tests had to be extremely brief. Personnel turned on each probe's radio frequency transmitters for a very short time to verify their signals. For the Large Probe, because the antenna was in a support cone, engineers had to connect a pickup antenna to an outside antenna. Without this, the probe could not relay its signal to the DSN for the test.

Finally, the Multiprobe countdown began. All went well until close to the scheduled launch date of August 6, 1978. Then, as technicians loaded the liquid helium into the Centaur, they discovered that the helium truck carried less liquefied gas than it should have. As a result, the countdown went on hold. It resumed on August 7 at 1830 EDT. NASA finally sent the Multiprobe on its way to Venus at 0733 UT on August 8, 1978 (Figure 2-16).

The first U.S. mission into the cloud-shrouded atmosphere of Earth's mysterious sister planet was successfully on its way.

CHAPTER 3

PIONEER VENUS SPACECRAFT

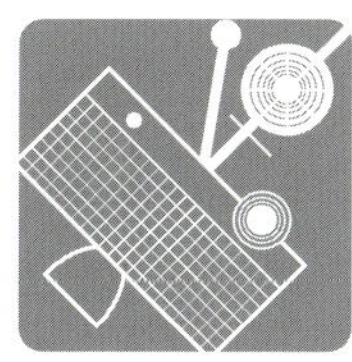

A new era dawned with an announcement from the Commander, Air Force Missile Test Center, Cape Canaveral, Florida, October 11, 1958: "The United States launched a three-stage experimental space vehicle at the Atlantic Missile Range at Cape Canaveral, Florida, at 0342 EST this morning. The launching was accomplished by the Air Force under the direction of the National Aeronautics and Space Administration (NASA). It was the second flight test of a number of small unmanned space vehicles designed to gather scientific data as a part of the U.S. International Geophysical Year program which is sponsored by the National Academy of Sciences with the support of the National Science Foundation. The vehicle is composed of the Thor intermediate range ballistic missile as the first stage (or booster), a modified Vanguard second stage, and an advanced version of the Vanguard third stage. Topping this vehicle is a highly instrumented scientific payload."

A short while later, another announcement followed: "The Department of Defense gave the name 'Pioneer' today to the payload of the successfully launched U.S. lunar probe rocket, the first man-made object known to escape the Earth's gravitational field."

This then was the genesis of an interplanetary spacecraft series bearing the name Pioneer. Their missions were many: explore beyond Earth and be first to visit Jupiter and Saturn, probe into the Solar System's outermost reaches, and penetrate the atmosphere of mysterious cloud-shrouded Venus and observe that planet from orbit for 14 years.

Several studies in the years after the first Air Force lunar probes showed how unmanned spacecraft might explore the Solar System. In early 1960, NASA transferred the solar probe study program to Ames Research Center. There it continued under the leadership of Charles F. Hall and a team appointed September 14 by Smith J. DeFrance, Director of the Center. Other members of the team were J. Dimeff, C. F. Hansen, W. A. Mersman, R. T. Jones, H. F. Matthews, H. Hornby, W. J. Kerwin, and C. A. Hermach. In these startup years, scientists envisioned a spacecraft approaching within 44,850,000 km (27,870,000 miles) of the Sun.

In succeeding years, Hall sought support from NASA Headquarters for this idea. He won approval from Edgar M. Cortright, then Deputy Director of the Office of Space Science, to develop an interplanetary Pioneer as a step toward a solar probe. Ames management concurred and, in April 1962, Space Technology Laboratories of Redondo Beach, California, completed a feasibility study. This study developed a concept for a spin-stabilized spacecraft with special features. Specifically, it would meet design constraints of low weight, low cost, and quick design and fabrication for various missions to explore interplanetary space and its environment.

Contracts were awarded following competitive bidding, and project personnel planned the

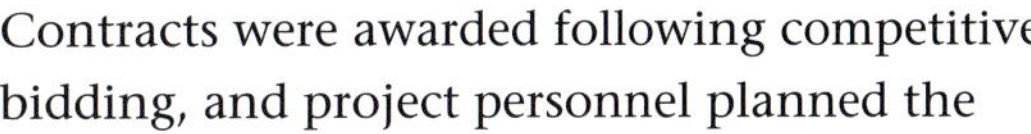

With the launch of Pioneer 1 on October 11, 1958, the Pioneer series of spacecraft began. The first part of this chapter gives a history of the series, leading up to Pioneer Venus. A detailed review of the interplanetary mission's two main components—the Orbiter and Multiprobe—appears in the remaining chapter sections. In addition to the spacecraft's structural details, the text also covers these features: data handling, commands, antennas, power sources, and communications from Earth.

Figure 3-1. The Pioneer spacecraft series started with the Interplanetary Pioneer 6, launched in 1965. (The Air Force had originally used the name for an earlier lunar probe series.) The Pioneer missions culminated in the Pioneer Venus spacecraft, launched in 1978.

first launch for 1965. The Pioneer program originally consisted of five spacecraft and their experiments. Ames Research Center managed the project, TRW Systems built the spacecraft, and experimenters provided the scientific instruments. Engineers designed all these spacecraft to orbit the Sun in approximately the plane of Earth's orbit (the ecliptic plane), some initially directed inside Earth's orbit, some outside.

The first of the Pioneer series spacecraft launched by NASA was Pioneer 6 (Figure 3-1) on December 15, 1965. On August 17 of the following year, Pioneer 7 was launched successfully. Pioneer 8 followed on December 13, 1967, and Pioneer 9 on November 8, 1968. The final launch in the series was on August 27, 1969. After 214 seconds, however, the Delta booster on this flight developed problems. The Range Safety Officer ordered the booster destroyed 484 seconds into the flight. The spacecraft was lost.

These Pioneer spacecraft showed the practicality of spin-stabilizing the spacecraft to steady it and simplify control of its orientation. The spacecraft also proved reliable and operable far beyond their initial missions. The scientific results were impressive. Pioneer missions confirmed that there was a spiral solar magnetic field imbedded in the plasma that streams outward from the Sun. They also confirmed the structure of Earth's bow shock and of the magnetopause. They mapped a geomagnetic tail and provided insights into what happens in interplanetary space when a solar flare erupts. The missions recorded energy spectra of solar electrons and positive ions and showed the solar wind's average electron temperature to be about 100,000 K during times of low solar activity. Cosmic ray telescopes showed that, during solar minimum, most high energy cosmic ray particles originated outside the Solar System. However, even at solar minimum, the telescopes showed low energy cosmic rays were mainly of solar origin. The spacecraft also measured shapes of plasma clouds and the electric fields in interplanetary space.

An important Pioneer discovery was that cosmic dust is not a serious hazard to man or spacecraft operating outside Earth's atmosphere. Also, astronomers improved Solar System constants and ephemerides (tables giving a celestial body's coordinates at a number of specific times during a given period). They accomplished this by accurately tracking Pioneer spacecraft in their heliocentric orbits. The precision of other astronomical measurements also improved. These included gravitational constants for Earth and the Moon, the mass ratio of Earth and the Moon, and the distance of Earth from the Sun (the astronomical unit).

In 1969, engineers designed a new class of Pioneer spacecraft: a low-cost, lightweight, spin-stabilized spacecraft for flybys of other planets. The first two Pioneers of this class were Pioneers 10 and 11, originally designed to fly by Jupiter. These spacecraft were highly successful in withstanding the intense radiation as they passed through the radiation belts of Jupiter in 1973 and 1974. They also succeeded in maintaining contact with Earth from the enormous distances of the outer Solar System. As a result, planners added a new task to Pioneer 11's mission. The craft now would

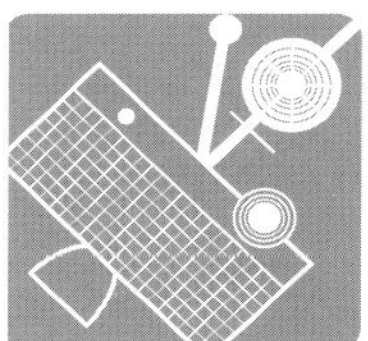

Table 3-1. The Pioneer Explorers

Name	Launch date	Mission	Status
Pioneer 1	11 Oct 1958	Moon	Reached 72,765 miles from Earth
Pioneer 2	8 Nov 1958	Moon	Reached 963 miles only
Pioneer 3	6 Dec 1958	Moon	Reached 63,580 miles only
Pioneer 4	3 Mar 1959	Moon	Passed 37,300 miles from Moon and entered solar orbit
Pioneer 5	11 Mar 1960	Solar orbit	Entered solar orbit
Pioneer 6	16 Dec 1965	Solar orbit	Still operating
Pioneer 7	17 Aug 1966	Solar orbit	Still operating
Pioneer 8	13 Dec 1967	Solar orbit	Still operating
Pioneer 9	8 Nov 1968	Solar orbit	Signal lost in 1983
Pioneer E	7 Aug 1969	Solar orbit	Launch failure
Pioneer 10	2 Mar 1972	Jupiter	Flew by Jupiter in 1973; still operating in outer solar system
Pioneer 11	5 Apr 1973	Jupiter	Flew by Jupiter in 1974; flew by Saturn in 1979; still operating in outer solar system
Pioneer 12	20 May 1978	Venus	Entered orbit around Venus on Dec 4, 1978. Orbited successfully until Oct 8, 1992
Pioneer 13	8 Aug 1978	Venus	Reached Venus Dec 9, 1979 and all probes and bus entered the atmosphere successfully

speed across the Solar System high above the ecliptic plane. Then, in 1979, it would fly by Saturn before following Pioneer 10 into interplanetary space in the outer Solar System, but in the opposite direction. Pioneer 11 proceeded ahead of the Sun (called the Solar Apex direction), and Pioneer 10 traveled in the Anti-Solar Apex direction.

Pioneers 10 and 11 showed that spacecraft could safely pass through the asteroid belt and through the Jovian radiation belts. They also made significant discoveries about the Solar System's two largest planets. They found that Jupiter must be a liquid planet and that its atmosphere is heated uniformly from equator to pole and in day and night hemispheres. They revealed that Jupiter's magnetosphere is a pulsating volume of particles and fields that are stirred up by its inner satellites. The spacecraft found three distinct regions and showed that the planet is the source of energetic particles hurtled across the Solar System. After they had confirmed the intensity and orientation of Jupiter's magnetic field, Pioneer 11 detected Saturn's magnetic field.

The spacecraft imaged Jupiter's polar regions for the first time, and Pioneer 11 observed Saturn's rings from the shadowed side. Pioneer 11 uncovered several new features in the Saturn ring system, including a thin A-ring beyond the F-ring. It also discovered additional satellites of Saturn. The spacecraft measured magnetic field strengths, and Pioneer 10 was the first to obtain images of the Galilean satellites and Pioneer 11 the first of Titan.

Pioneers 10 and 11 were, without question, highly successful precursors to the Voyager 1 and 2 outer-planet spacecraft that the Jet Propulsion Laboratory developed and launched in 1977.

The Pioneer Venus spacecraft—the Orbiter and the Multiprobe—were the next steps in the

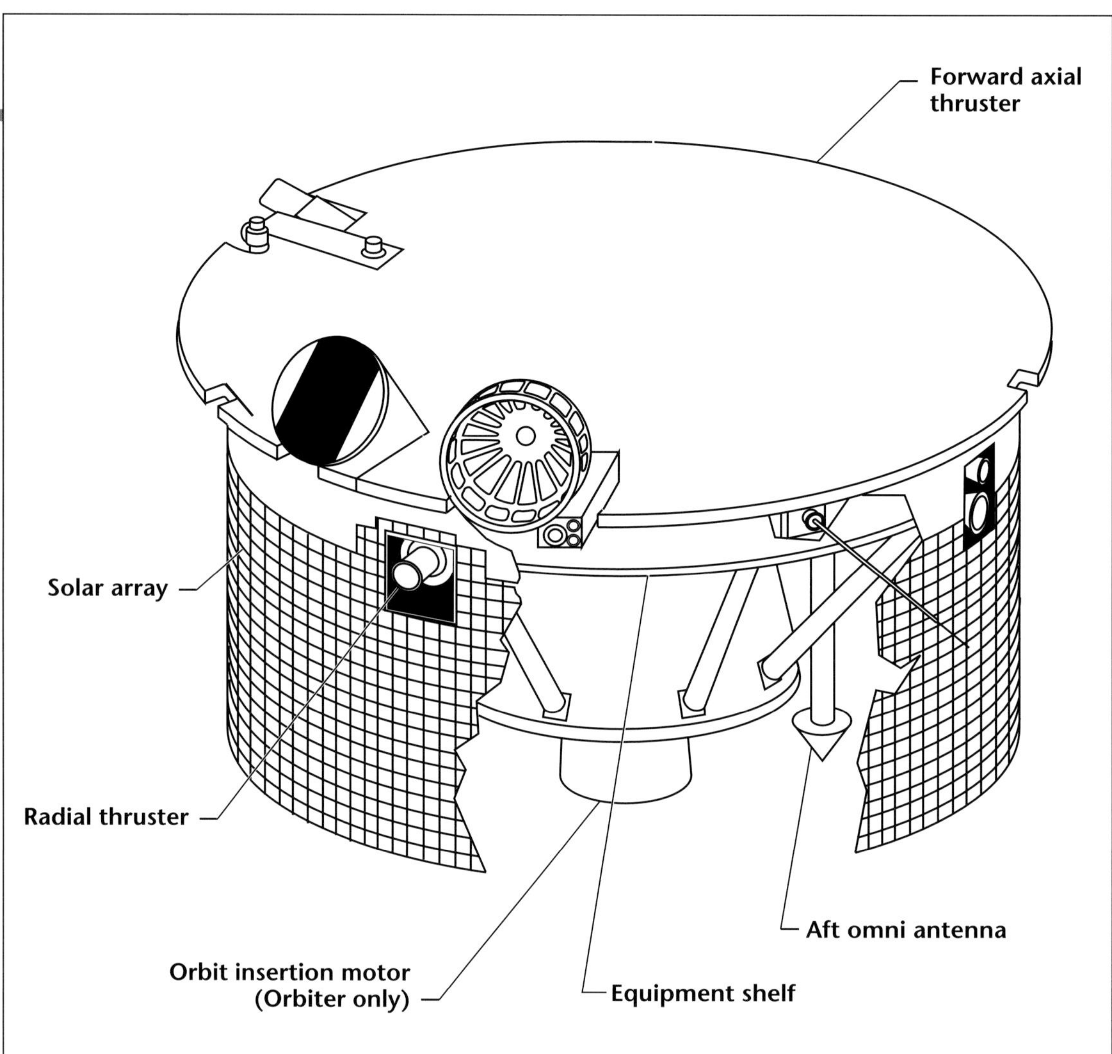

Figure 3-2. The main body of the Pioneer Venus spacecraft, a simple cylinder, was the common design for both Orbiter and Multiprobe. It had shelves for equipment, thrusters for maneuvering, an omni antenna, and, for the Orbiter only, a solid-propellant, orbit-insertion rocket motor.

evolution of this highly successful line of trailblazing interplanetary Pioneer probes. One of the Venus spacecraft became a true planetary probe when it carried several spacecraft into the Venusian atmosphere, rather than just flying by or orbiting the planet. Whereas TRW Systems built the previous Pioneers, Hughes Aircraft Company built the Pioneer Venus spacecraft. Ames Research Center continued in the project management role.

The Pioneer exploring spacecraft appear in Table 3-1. Information includes spacecraft names, launch dates, intended missions, and a summary of their status.

The Orbiter

The Orbiter was the spin-stabilized platform for the orbital mission's 12 scientific instruments. To reduce the mission's cost, it used the basic Pioneer Bus, common to both the Orbiter and the Multiprobe.

The main body of the spacecraft (Figure 3-2) was a flat cylinder 2.5 m (8.2 ft) in diameter and 1.2 m (4 ft) high. In the upper or forward end of the cylinder was a circular equipment shelf with an area of 4.37 m^2 (47 ft^2). All the spacecraft's scientific instruments and electronic subsystems were on this shelf (see Chapter 4 for descriptions of these instruments). Engineers fastened the shelf on the forward end of a thrust tube that connected the spacecraft to the launch vehicle. Twelve equally spaced struts supported the periphery of the shelf from the lower part of the thrust tube. Below the shelf, 15 thermal louvers (Figure 3-3) controlled heat radiation from an

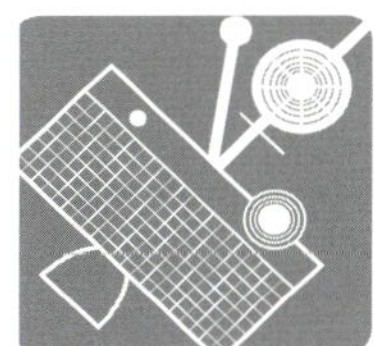

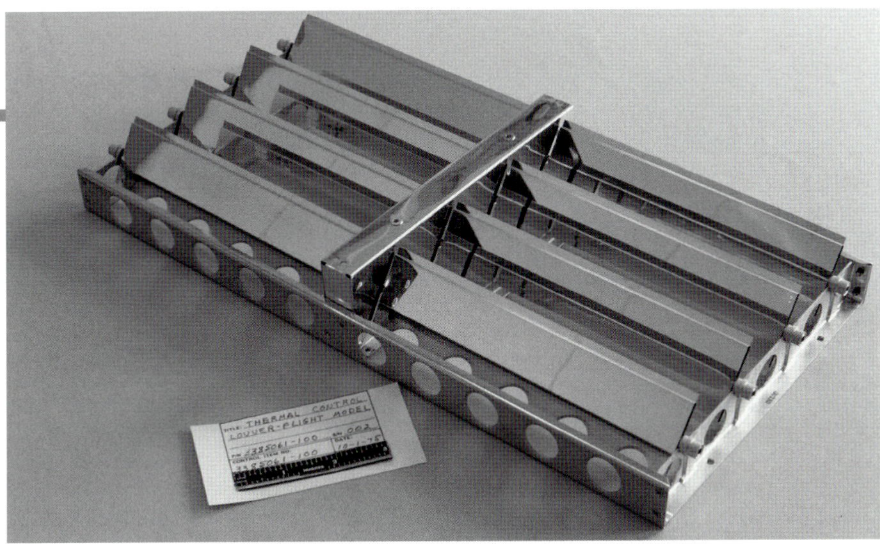

Figure 3-3. Thermal louvers controlled the internal temperature of the spacecraft. These flight model louvers appear partially open. Gold strips of Kapton film were added to the inboard side of the blades to radiate energy to the spacecraft. This reduced blade temperature as required.

equipment compartment that was between the shelf and the top of the spacecraft. A cylindrical solar array (Figure 3-4), attached to the shelf by 24 brackets, formed the circumference of the flat cylinder of the spacecraft.

On top of the spacecraft was a 1.09-m (3.6-ft) diameter, despun, high-gain, parabolic dish antenna (Figure 3-5). The antenna was on a mast so that its line of sight cleared equipment mounted outside the spacecraft. The despun design allowed the antenna to be mechanically directed to continuously face Earth from the spinning spacecraft. The antenna operated at S- and X-bands.

The spacecraft also carried a solid-propellant rocket motor (Figure 3-6) with 18,000 N (4046 lb) of thrust. This thrust would decelerate the spacecraft by 3816 km/hr (2371 mph) and place it into an orbit around Venus. Including the antenna mast, the Orbiter was almost 4.5 m (15 ft) high, and it weighed 553 kg (1219 lb) on Earth. Its launch weight included 45 kg (100 lb) of scientific instruments and 179 kg (395 lb) of rocket propellant.

A maneuvering system for the Orbiter's basic Bus controlled its rate of spin and made course and orbit corrections. The system also maintained the spin axis orientation, which was usually perpendicular to the ecliptic plane. Beneath the equipment compartment and attached to the thrust tube were two conical hemispheric propellant tanks (Figure 3-7). Each tank was 32.5 cm (12.8 in.) in diameter. Initially, these tanks stored 32 kg (70 lb) of hydrazine. This hydrazine was the propellant for three axial and four radial thrusters (Figure 3-8). These thrusters changed the attitude, velocity, or orbital period and spin rate of the spacecraft during the mission. Two axial thrusters aligned with the axis of spin and were at the top and bottom of the Bus cylinder. They were diagonally opposite each other and pointed in opposite directions. When ground controllers had to turn the spin

Figure 3-4. A cylindrical solar array formed the circumference of the Bus cylinder and provided electrical energy for the spacecraft and its payload of scientific instruments.

Figure 3-5. *A high-gain parabolic dish antenna, mounted on a mast, could be despun relative to the spacecraft so that it could point toward Earth.*

A third thruster unit was at the bottom of the thrust cylinder and permitted continuous firing of two bottom thrusters. This firing allowed moves in an axial direction so mission controllers could change the spacecraft's orbit.

The four radial thrusters were in two pairs, pointing in opposite directions. They were in a plane approximately perpendicular to the spin axis. (This plane passed through the center of gravity of the spacecraft.) The radial thrusters were used to change the spacecraft's velocity in a direction perpendicular to the spin axis. The thrusters also controlled the spin rate. They were equally positioned on the Bus cylinder's periphery. Firing two of them 180°

Figure 3-6. A solid-propellant rocket motor provided thrust to place the Orbiter into an elliptical orbit around Venus. Thiokol Corporation supplied this orbit insertion motor. The picture shows the motor case and the nozzle closure of titanium alloy forgings. The case consisted of two 0.028-inch thick hemispheres joined by a single weld. The nozzle closure, which fastened to the case, was the structural element of the nozzle. Integral bosses allowed attachment

axis, they fired the thrusters in pulses in opposite directions. To speed up or slow down the spacecraft along the direction of its spin axis, they fired only one thruster in pulses. They did this at two points 180° apart in the spacecraft's rotation, and chose the top or bottom thruster, depending on the direction required for the velocity change.

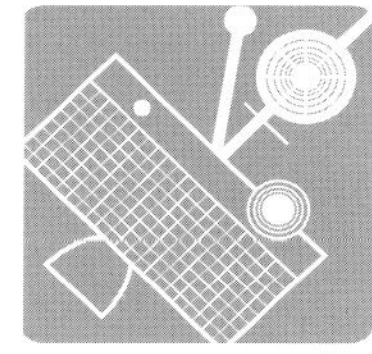

apart slowed the spin rate. Firing the other two increased the spin rate.

Sun sensors and a shelf-mounted star sensor provided attitude references to control the spacecraft. Each instrument had a slit aperture for its field of view.

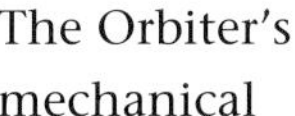

Figure 3-7. Hydrazine propellant for maneuvering thrusters was stored in two tanks within the spacecraft. Each tank, made from titanium alloy, held 16 kg (35 lb) of hydrazine under pressure.

The Orbiter's mechanical features consisted of six basic assemblies. These were the despun antenna assembly, the bearing and power transfer assembly and its support structure, the equipment shelf, the solar array, the orbit insertion motor and its case, and the thrust tube (Figure 3-9).

On Venus, the intensity of the Sun's radiation is nearly twice that on Earth. Pioneer Venus' thermal design isolated equipment from extremes of solar heat during the mission. To keep the spacecraft's critical elements at the right temperature, there were electric heaters that ground control could turn on. The solid propellant rocket, which inserted the spacecraft into orbit, and the safeing and arming devices required temperature control early in the mission. Also, throughout the mission, the hydrazine propellant had to stay unfrozen. But what would happen if equipment that developed heat during its operation was off for too long? Mission designers planned for this

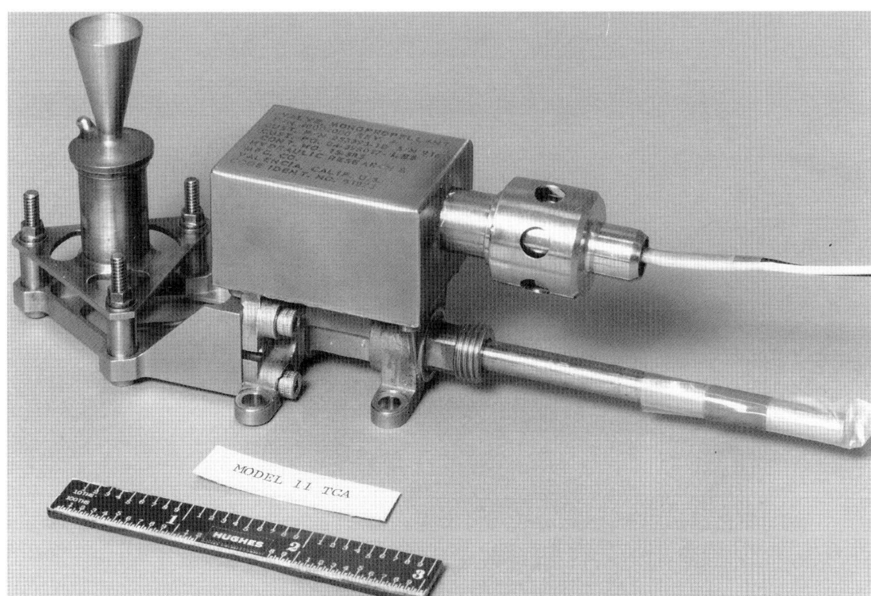

Figure 3-8. Small thrusters controlled the spacecraft's orientation and spin rate. They were in two redundant groups, positioned on the spacecraft so that ground control could change the spacecraft's velocity, spin rate, and attitude. This photograph shows a thruster assembly.

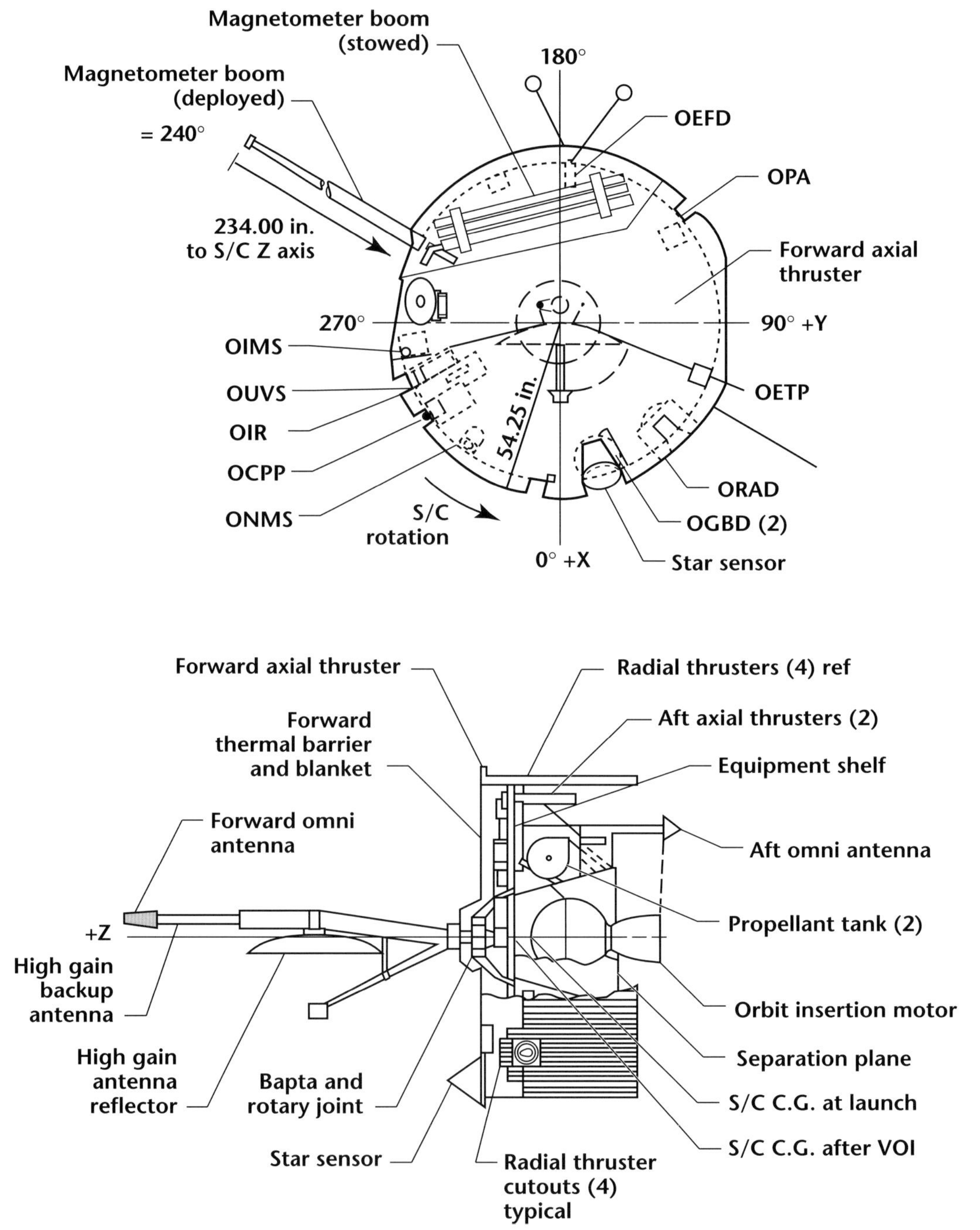

Note: Star sensor, tanks, axial thrusters, and aft omni antenna shown rotated for clarity

Figure 3-9. Principal elements of the Orbiter spacecraft appear on these cross-section and side views. (See Table 2-2, page 29, for instrument acronyms.)

situation by including other heaters that could raise the spacecraft's internal temperature.

Data-Handling Subsystem

A data-handling subsystem (Figure 3-10) within the Orbiter conditioned and integrated all analog and digital telemetry data into formats that ground control selected by radio command. Resulting information went to the communications subsystem for modulation of the downlink (spacecraft-to-Earth) S-band carrier. Twelve telemetry storage, playback, and real-

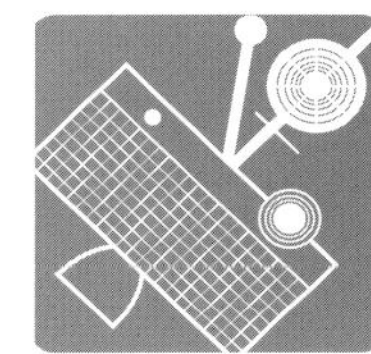

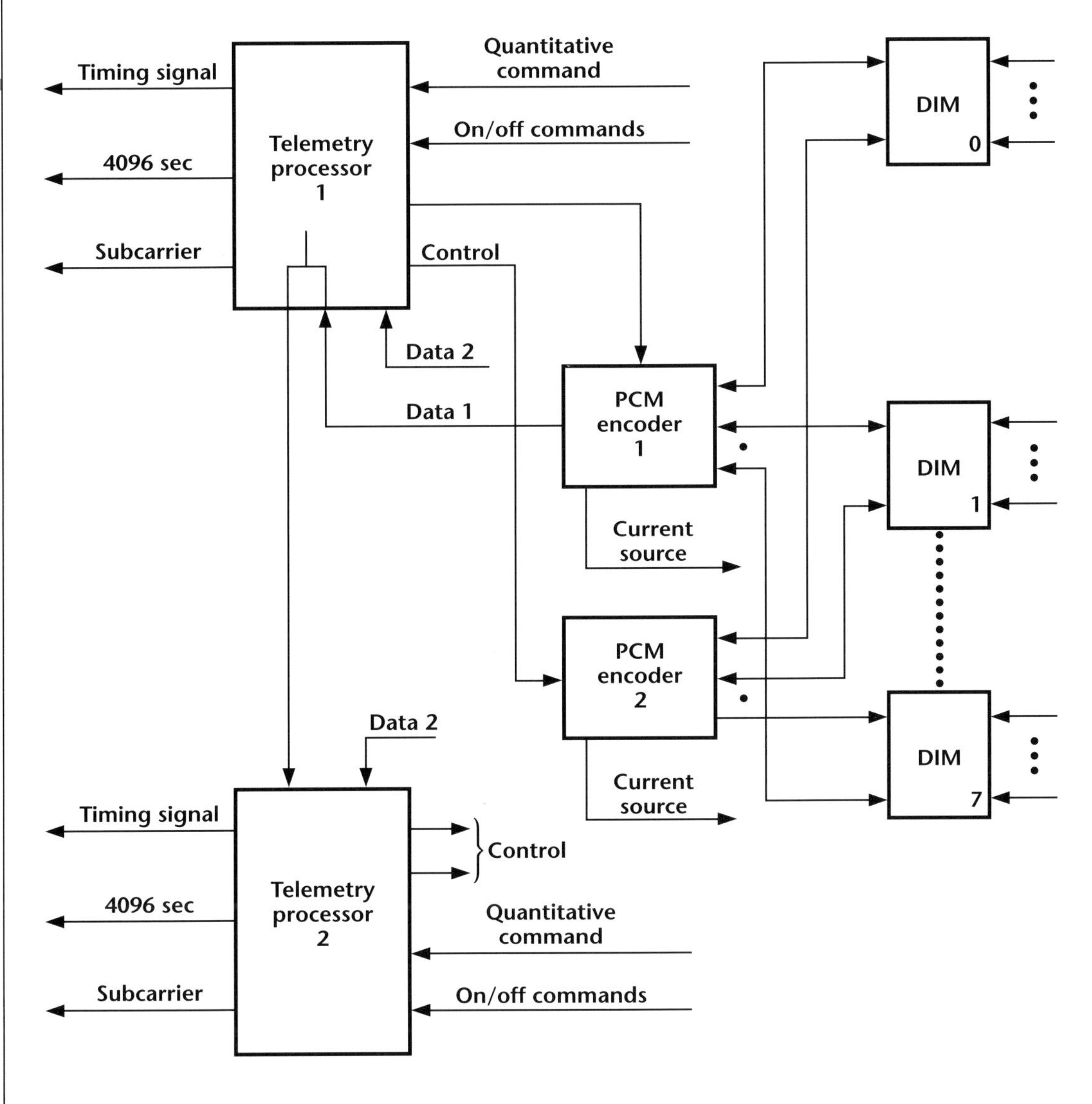

Figure 3-10. This block diagram shows the basic elements of the data handling subsystem common to the Orbiter and the Multiprobe.

time data rates between 8 and 2048 bits/sec were available. During interplanetary cruise, the Orbiter used a rate of 1024 bits/sec.

The data-handling subsystem included a data memory with two data storage units. Each unit had a capacity of 524,288 bits (equivalent to 1024 minor frames of telemetry). The subsystem was primarily for use during an Earth occultation when the spacecraft was behind Venus and not able to communicate with Earth. During this period, which could last up to 26 minutes, the data memory could store just over 1 million bits of data. That translated into an average maximum rate of 672 bits/sec.

For shorter occultation periods, the bit rates could be higher. Data were stored or read at the commanded bit rate. If, for any reason, the Deep Space Network could not receive data from the spacecraft, the data could remain in these data storage units.

The Orbiter data-handling system accepted information from spacecraft subsystems and the scientific experiments in several forms. These were serial digital, analog, and one-bit bilevel (on/off). The system converted analog and one-bit data to serial digital form and arranged all information in formats for transmission to Earth. This transmission was a

Figure 3-11. Assignment of data formats for the Orbiter appear in this figure. PER refers to the periapsis portion of the orbit, APO to apoapsis. PBK is for playback. The various scientific instruments appear by their project acronyms (see Table 2-3).

Pioneer Venus Orbiter format assignments

Project acronyms	Telemetry formats/Words per telemetry frame									Bits/frame
	PERA	PERB	PERC	PERD	PERE	LACR	PBK	APOA	APOB	Sub D
ORPA	5	3	3	—	—	1	—	—	2	42
OIMS	18	9	9	—	—	1	—	—	2	65
OETP	10	8	5	—	—	—	—	—	2	18
OUVS	11	7	7	—	18	7	1	—	1	41
ONMS	—	14	6	—	—	—	—	—	2	17
OCPP	—	8	—	8	—	—	—	43	—	33
OIR	—	—	4	47	4	—	—	—	—	41
OMAG	4	4	4	—	4	4	4	4	12	41
OPA	3	2	3	—	—	3	5	3	12	17
ORAD	—	—	10	—	28	—	—	—	1	54
OEFD	4	—	4	—	1	4	4	4	4	1
OGBD	—	—	—	—	—	1	1	1	17	82
Science subtotal	55	55	55	55	55	21	15	55	55	452
MRO	—	—	—	—	—	—	40	—	—	—
Engineering	—	—	—	—	—	34	—	—	—	—
Overhead	6	6	6	6	6	6	6	6	6	60*
Subcoms	3	3	3	3	3	3	3	3	3	—
Total	64	64	64	64	64	64	64	64	64	512

*Spare bits

continuous sequence of major telemetry frames, each with 64 minor frames. Each minor frame, in turn, contained 64 eight-bit words, or a total of 512 bits per minor frame. The words of a minor frame came in 1 of 13 formats that ground control could select by radio command. Each minor frame contained science and engineering data at the commanded bit rate, subcommutated data, spacecraft ID data, and frame synchronization data. Three subcommutated data formats belonged to each minor frame. One was for slowly changing science and science housekeeping data, and the other two were for slowly changing spacecraft engineering data.

The Orbiter had a total of 14 telemetry formats. Five formats, PERA, PERB, PERC, PERD, and PERE, were periapsis formats. These formats allowed mission controllers to change the emphasis for part or all of a periapsis period. Two formats, APOA and APOB, were for measurements during apoapsis when the spacecraft was relatively distant from Venus. The Launch-Cruise, or LACR, format furnished a higher rate of engineering data. At the same time, LACR permitted measurements by instruments capable of interplanetary observations. The Playback, or PBK, and the Data Memory Read Out, or DMRO, formats permitted reading out data from the Data Storage Unit, or DSU. However, PBK read with realtime scientific data, while DMRO excluded them. The Command Memory Read Out, or CMRO, format permitted a check of the command memory load. There also was an engineering format to furnish high rate engineering data for diagnostic purposes. To furnish high rate data from the attitude control system, there was an attitude control system format, or ACS. A 14th format could furnish high-rate data of a few selectable parameters for diagnostic purposes. The nine formats that include real-time scientific data appear in Figure 3-11.

Commands
The basic command system accepted a pulse-code-modulated, frequency shift-keyed, phase-modulated (PCM/FSK/PM) data stream. The stream was at a fixed rate of 4 bits/sec—the incoming commands from Earth via the radio receivers. Each command word consisted of 48 bits, including 13 bits for synchronization. This structure resulted in a one-in-a-million probability of the spacecraft accepting a false command. The system had a total of 192 pulse commands and 12 magnitude commands. Command demodulators activated the system, converted the signal to a usable binary bit stream, and passed it to cross-connected command processors. The spacecraft routed each command it received to the addressed destination for immediate action or stored it for later execution. Both command memories could store up to 128 commands or time delays. The command subsystem could completely decode each assigned command and generate an execution command. Or it could partially decode the command, which was completely decoded at its destination. Spacecraft units received commands from redundant command output modules.

Antenna Systems
The Orbiter carried a despun, high-gain, parabolic antenna. At S-band, this antenna directed a 7.6-degree beam toward the Earth throughout the mission. The antenna dish was 109 cm (43 in.) in diameter, and it concentrated the Orbiter's signal 316 times by directing it into the narrow beam. During the mission, the distance between Earth and Venus changed by 203 million km (126 million miles). Engineers designed the high-gain antenna to return data at the required rates over the greatest mission distance.

The high-gain antenna dish, a sleeve dipole antenna, and a forward omnidirectional antenna were all on a mast that projected 2.9 m (9.8 ft) along the spin axis from the top of the basic cylinder of the spacecraft (Figure 3-12). The sleeve dipole radiated in a flat pattern in a plane perpendicular to the spin axis. It provided backup if the despin mechanism failed and ground control could not point the dish antenna toward Earth.

Both omnidirectional antennas—one on the antenna mast and the other aft of the spacecraft—radiated in a hemispherical pattern. This design provided low-gain radiation in all directions around the spacecraft. At any orientation, the spacecraft could receive commands from Earth and communicate at low bit-rates.

One of two electric motors despun the three antennas on the mast relative to the spinning spacecraft. The mast was attached to a bearing assembly flange that was on the Bus thrust tube's upper end. A series of transfer switches electrically connected the three antennas to the spinning spacecraft's transmitters. These connections ran through a dual frequency rotary joint (Figure 3-13). Pulse commands, in turn, controlled the switches. The commands traveled through slip rings and brushes on the bearing and power-transfer assembly that supported and rotated the mast relative to the spacecraft.

A control system provided redundant electronics to control the despin mechanism. The system also drove either one of the two electric motors. Depending on signals from the Sun and star sensors, despin control electronics generated motor torque commands. The parabolic antenna could be pointed in elevation by a motor-driven jackscrew.

The Orbiter carried a 750-mW, X-band transmitter for radio experiments during occultation. The signal frequency of this transmitter was 1-1/3 times that of the main S-band

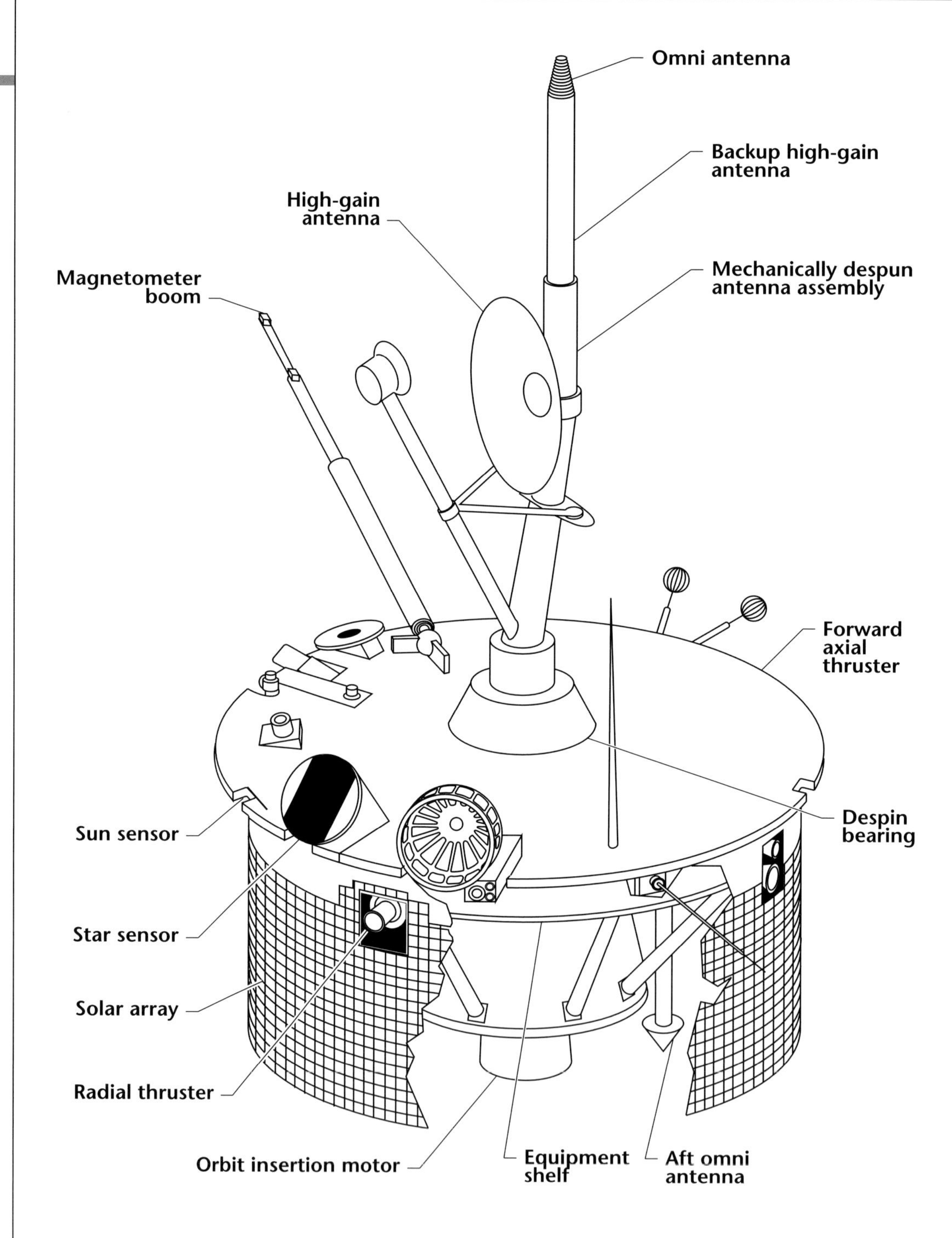

Figure 3-12. The antenna mast carried several antennas and the parabolic dish.

transmitter. The dish antenna transmitted both X- and S-band signals. Mission scientists could direct this antenna to point 15° from Earth-line as the Orbiter passed behind Venus. As the radio waves passed through Venus' atmosphere, they were refracted, or bent, toward Earth. Without repositioning the antenna again, the radio signal would have been refracted away from Earth. Repointing the antenna allowed the radio beam to dip deeply into Venus' atmosphere and still reach Earth despite refraction by the Venusian atmosphere. Radio occultation data were thus obtained at atmosphere levels closer to the planet's surface.

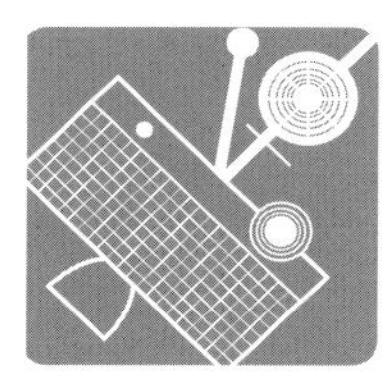

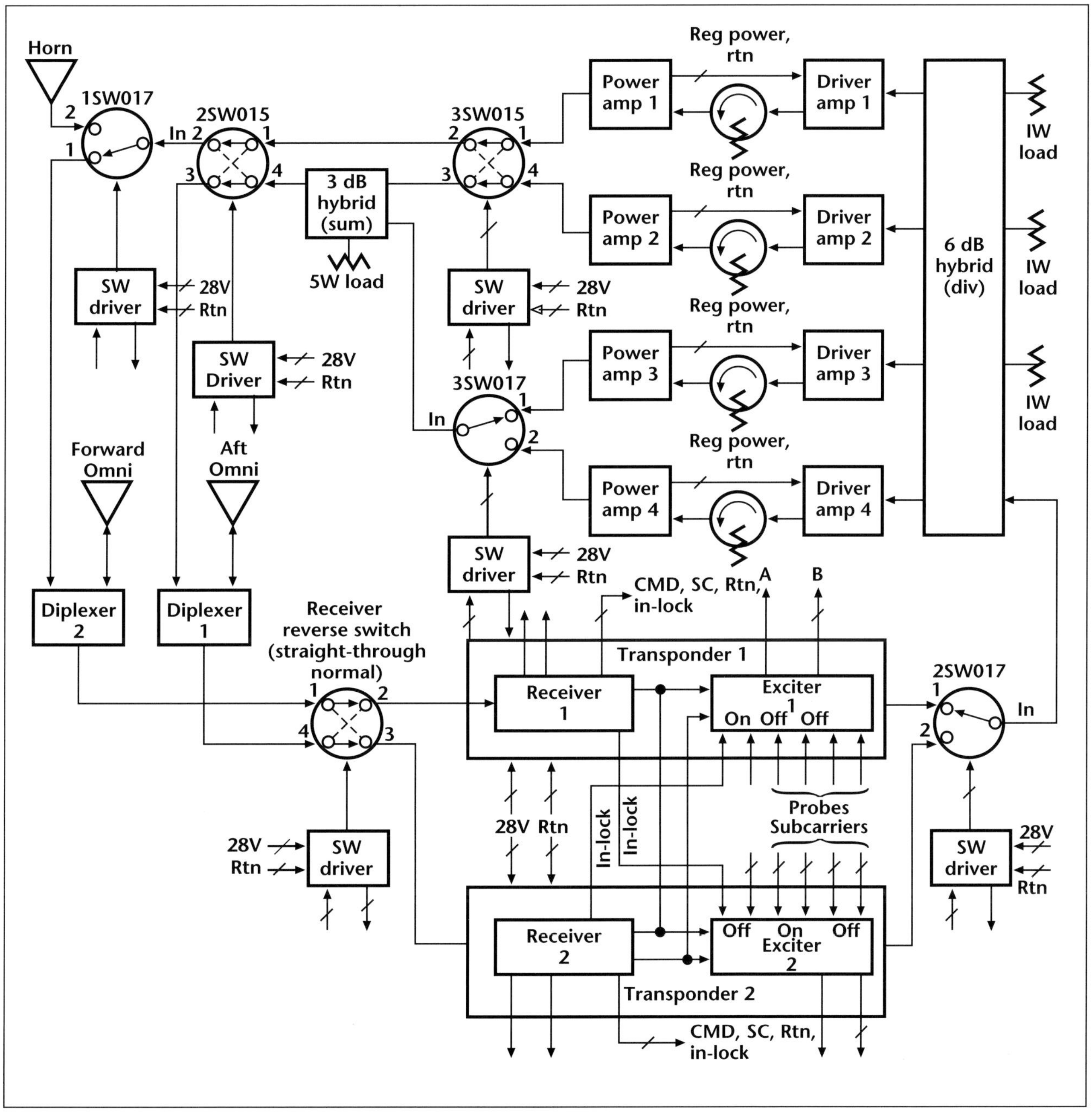

Figure 3-13. This diagram of the communication system for the common Bus on the Orbiter and Multiprobe shows how the antennas could connect to the redundant receivers and the transmitting power amplifiers.

The X-band signal could not be modulated, so it was used solely to study atmospheric effects on radio signals at two different frequencies. These studies provided many details about the planet's atmosphere.

Communications from Earth

Regardless of its orientation, the Orbiter could receive commands from Earth. It was able to do this through two redundant S-band transponders connected to its omnidirectional antennas. Each transponder received the radio signal from Earth. It then tuned the transmitter so that the outgoing radio signals' frequency from the spacecraft had a constant ratio to the incoming signals' frequency. This created a coherent mode of transponder operation. The coherence, in turn, allowed the system to measure precisely the Doppler shift in the radio frequency arising from the spacecraft's motion relative to Earth. This measurement was possible both on the outgoing and incoming radio signals. Thus, the spacecraft's velocity could be measured to 3 m/hr.

The receiver portion of each transponder responded only to certain frequencies. If they did not receive a command from Earth within 36 hours, the receivers automatically reversed. Thus, if one receiver failed, the other automatically took over within 36 hours.

The uplink (Earth-to-spacecraft) command capability was maintained by modulating the S-band carrier of approximately 2.115 GHz. The down-link telemetry modulated an S-band carrier of approximately 2.295 GHz.

Power

The Orbiter's power subsystem provided a semiregulated, 28 V direct current to all its electrical loads, including its science instruments. The primary source of power was the solar array, which had 7.4 m^2 (80 ft^2) of solar cells. Each cell was 2 cm^2 (0.79 in.^{2}). At Earth's orbit, the solar array provided 226 W, and at Venus it provided 312 W. When the array's output was insufficient, two nickel-cadmium batteries began operating automatically. This occurred when the bus voltage dropped below 27.8 V. (Passing through Venus' shadow or an inadequate angle of sunlight on the solar array could cause these voltage drops.) Each battery was rated at 7.5 A hr, and small solar arrays recharged the units. Seven shunt limiters dissipated excess solar power. This precaution kept the bus voltage at 30 V or below.

A power interface unit switched power to the Orbiter's propulsion unit heaters and other heaters as they needed it. This interface unit contained protective fuses. Power was distributed through the spacecraft on four separate power buses. If more current started to flow than was safe, the system removed loads to prevent a catastrophic failure. First, it disconnected the scientific instruments. Then it disconnected the switched loads, such as control and data-handling units. The transmitter was the final unit it took off line. The system left in a continuous power-on mode only those loads that were absolutely essential for the spacecraft. These loads included the command units, heaters, receivers, and power conditioning units.

Multiprobe Spacecraft

The Multiprobe (Figure 3-14) consisted of a basic Bus similar to the Orbiter's, a Large Probe, and three identical Small Probes. It did not carry a despun, high-gain antenna. The weight of the Multiprobe was 875 kg (1930 lb), including 32 kg (70 lb) of hydrazine. The Multiprobe used this propellant to correct its trajectory and orient its spin axis. The total weight of the four probes it carried was 585 kg (1289 lb). The Bus itself weighed 290 kg (639 lb).

The Multiprobe's basic Bus design was similar to the Orbiter's design. It also used a number of common subsystem designs. Mechanically, the Bus consisted of five subassemblies: (1) a support structure for the Large Probe, (2) a support structure for the Small Probes, (3) an equipment shelf, (4) a solar array around the periphery of the cylindrical basic Bus, and (5) a central thrust tube. The spacecraft

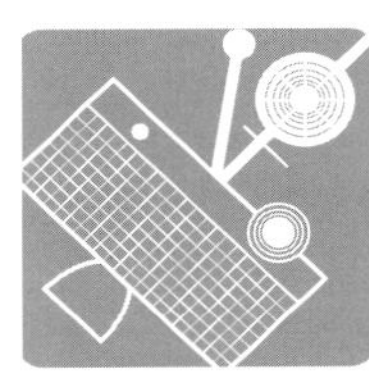

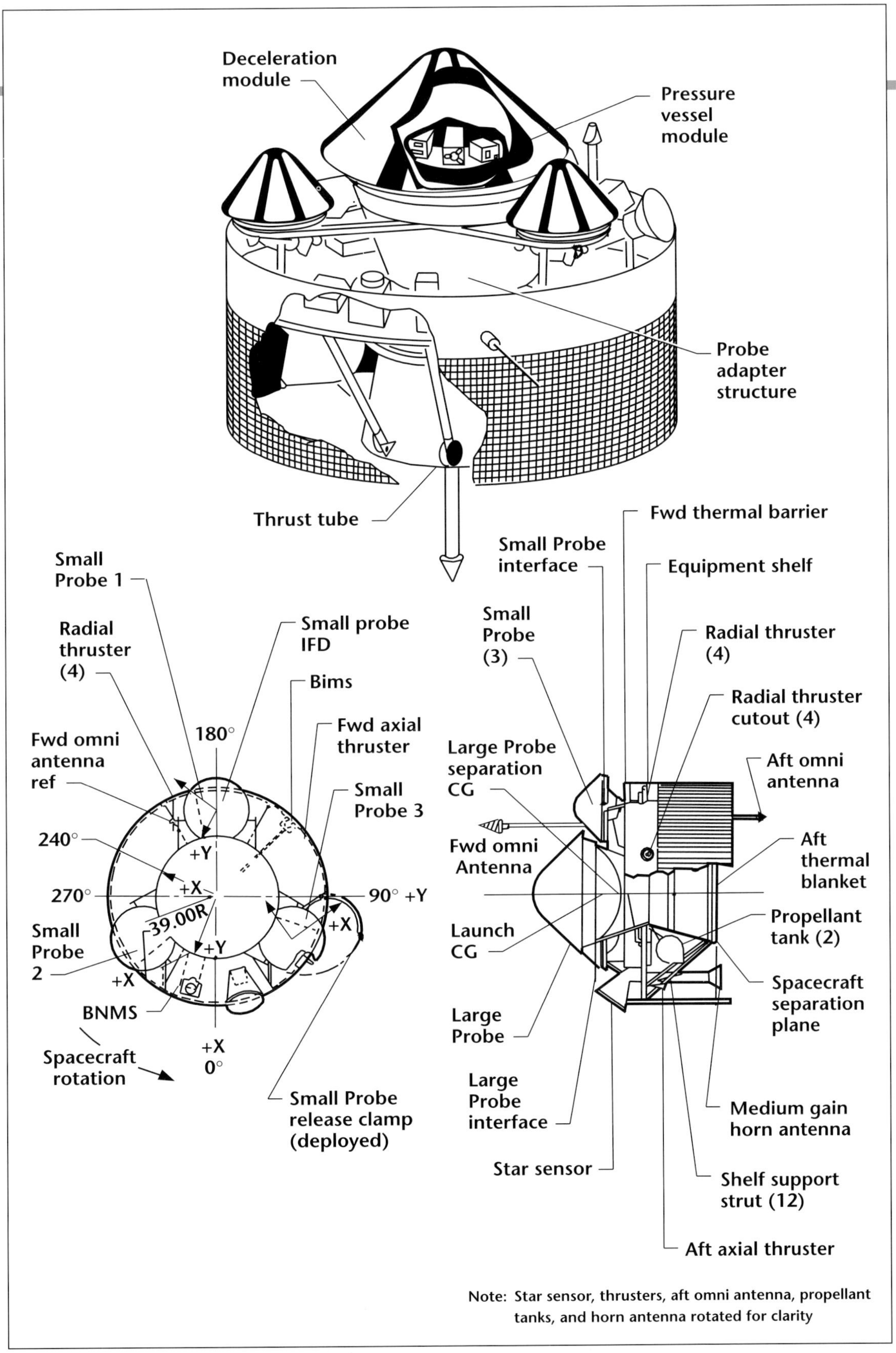

Deceleration module
Pressure vessel module
Probe adapter structure
Thrust tube

Small Probe 1
Radial thruster (4)
Fwd omni antenna ref
180°
Small probe IFD
Bims
Fwd axial thruster
Small Probe 3
240°
270°
+Y
+X
39.00R
+Y
+X
90° +Y
Small Probe 2
+X
BNMS
Spacecraft rotation
+X
0°
Small Probe release clamp (deployed)

Fwd thermal barrier
Small Probe interface
Equipment shelf
Small Probe (3)
Radial thruster (4)
Radial thruster cutout (4)
Large Probe separation CG
Aft omni antenna
Fwd omni Antenna
Aft thermal blanket
Launch CG
Propellant tank (2)
Spacecraft separation plane
Large Probe
Large Probe interface
Medium gain horn antenna
Star sensor
Shelf support strut (12)
Aft axial thruster

Note: Star sensor, thrusters, aft omni antenna, propellant tanks, and horn antenna rotated for clarity

Figure 3-14. The Multiprobe spacecraft. (Top) General view showing major parts. (Bottom) Detailed cross section and side view identifying major components.

diameter was 2.5 m (8.3 ft). From the bottom of the Bus to the top of the Large Probe mounted on it, the Multiprobe measured 2.9 m (9.5 ft).

During their flight to Venus, the four probes were carried on a large inverted cone structure and three equally spaced circular clamps surrounded the cone (Figure 3-15). Bolts held these attachment structures to the control thrust tube. This thrust tube formed the structural link to the launch vehicle. The Large Probe was centered on the spin axis. A pyrotechnic-spring separation system launched the probe from the Bus toward Venus. The ring support clamps that attached the Small Probes were hinged. To launch the Small Probes, the Multiprobe first spun up to 45 rpm. Then explosive nuts fired to open the clamps on their hinges. This sequence allowed the probes to spin off the Bus tangentially.

Figure 3-15. During the flight to Venus, the Multiprobe Bus carried the four probes in this configuration.

▼

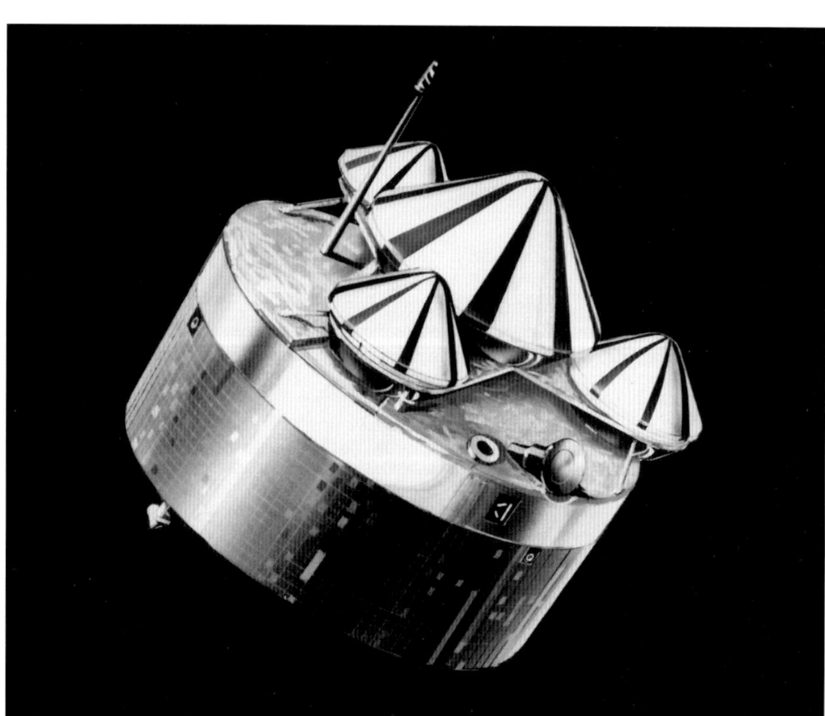

The forward omnidirectional antenna of the Multiprobe extended above the top of the Bus cylinder. An aft omni antenna extended below it. Both these antennas had hemispherical radiation patterns. A medium-gain horn antenna was on the instrument shelf and radiated aft of the spacecraft. Ground control used it during critical maneuvers when the aft of the spacecraft pointed toward Earth as the probes separated from the Bus.

The instrument-equipment compartment, as in the Orbiter, carried the scientific experiments and electronics for the spacecraft subsystems. The solar array provided electrical power from solar radiation. It contained the batteries and a power distribution system, Sun and star sensors, propellant storage tanks, and thrusters for maneuvering and stabilization. The Bus also carried radio transmitters and receivers, data processors, and a command and data handling system.

The thermal design was essentially the same as the Orbiter's. In addition, the Bus required protective surfaces near the Small Probes. These surfaces kept the probes at the required temperature during the cruise. They also protected the Bus itself from heating after the probes had separated from it.

Except for not having to position a high-gain antenna, orientation controls for the Multiprobe were the same as the Orbiter's. The propulsion system also was identical to the Orbiter's with one exception: the Multiprobe only had one aft axial thruster. The spacecraft did not, of course, carry a retrorocket.

Data-Handling System

The Multiprobe's data-handling system was virtually identical to Orbiter's. The only difference was its lack of data memory.

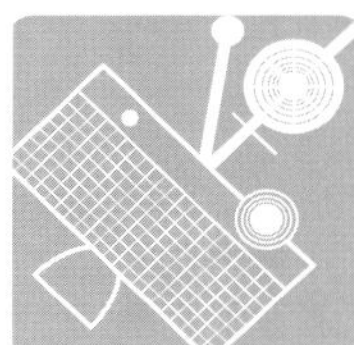

Data formats were organized for the Multiprobe's special mission requirements. Before the probes separated from the Bus, the Multiprobe handled data for the Bus and all probes. After separation, the probes used their own data systems.

The Multiprobe's data system accepted engineering and selected information that mission operations required and information from the four probes. It also accepted data from the Multiprobe Bus itself and from the experiments on the Bus. It converted analog data to digital form and prepared all information for transmission to Earth. Each telemetry major frame contained 64 minor frames composed of 64 eight-bit words. The system arranged these words in several formats. Each minor frame contained high-rate science or engineering data, plus subcommutated data, spacecraft data, and frame synchronization data. One subcommutated format carried low bit-rate science and science housekeeping information; two were for low bit-rate information from the spacecraft subsystems. The system used 12 real-time data transmission rates between 8 and 2048 bits/sec. Like the Orbiter, the Multiprobe also had high bit-rate formats for attitude control during maneuvers, for engineering data, and for reading out the contents of the command memory. A single format for use during entry into the Venus atmosphere transmitted science data at 1024 bits/sec.

Command, Communications, and Power

The Multiprobe's command and communications subsystems were similar to the Orbiter's. The command subsystem decoded all commands received via the Multiprobe's communications subsystem at a fixed rate of 4 bits/sec. The subsystem could either store these commands for later execution or route the commands as they reached their destination within the spacecraft and the probes where they were implemented. The communications subsystem provided reception and transmission for radio communications from and to Earth (Figure 3-16).

Also, the Multiprobe's power system was essentially the same as the Orbiter's. One difference, however, was a power interface unit that could send power to the probe heaters and the probe checkout buses, and to relay drivers for each of the probes. This system allowed the probes to receive power from the Bus without depleting their own batteries during the interplanetary cruise to Venus. The Multiprobe's solar array, consisting of 6.9 m^2 (74 ft^2) of 2 cm × 2 cm cells, provided 214 W near Earth and 241 W at Venus.

The Probes

The probes' designers faced a number of tremendous challenges. Among them—the high pressure in the lower regions of Venus' atmosphere, which was about 100 times greater than Earth's atmospheric pressure at sea level; the high temperature of about 480°C (900°F) at the surface; and, the corrosive constituents of the clouds, such as sulfuric acid. Moreover, these probes had to enter the atmosphere at a speed of about 41,600 km/hr (25,850 mph), or 43 times the speed of a typical commercial jet.

The Large and Small Probes were similar in shape. The main component of each probe was a spherical pressure vessel. Machined from titanium, the vessels were sealed against the vacuum of space and the high pressure of Venus' atmosphere. Within this pressure vessel were scientific instruments and various subsystems for the probe's operation.

An outer structure surrounded each spherical pressure vessel. This structure consisted of a

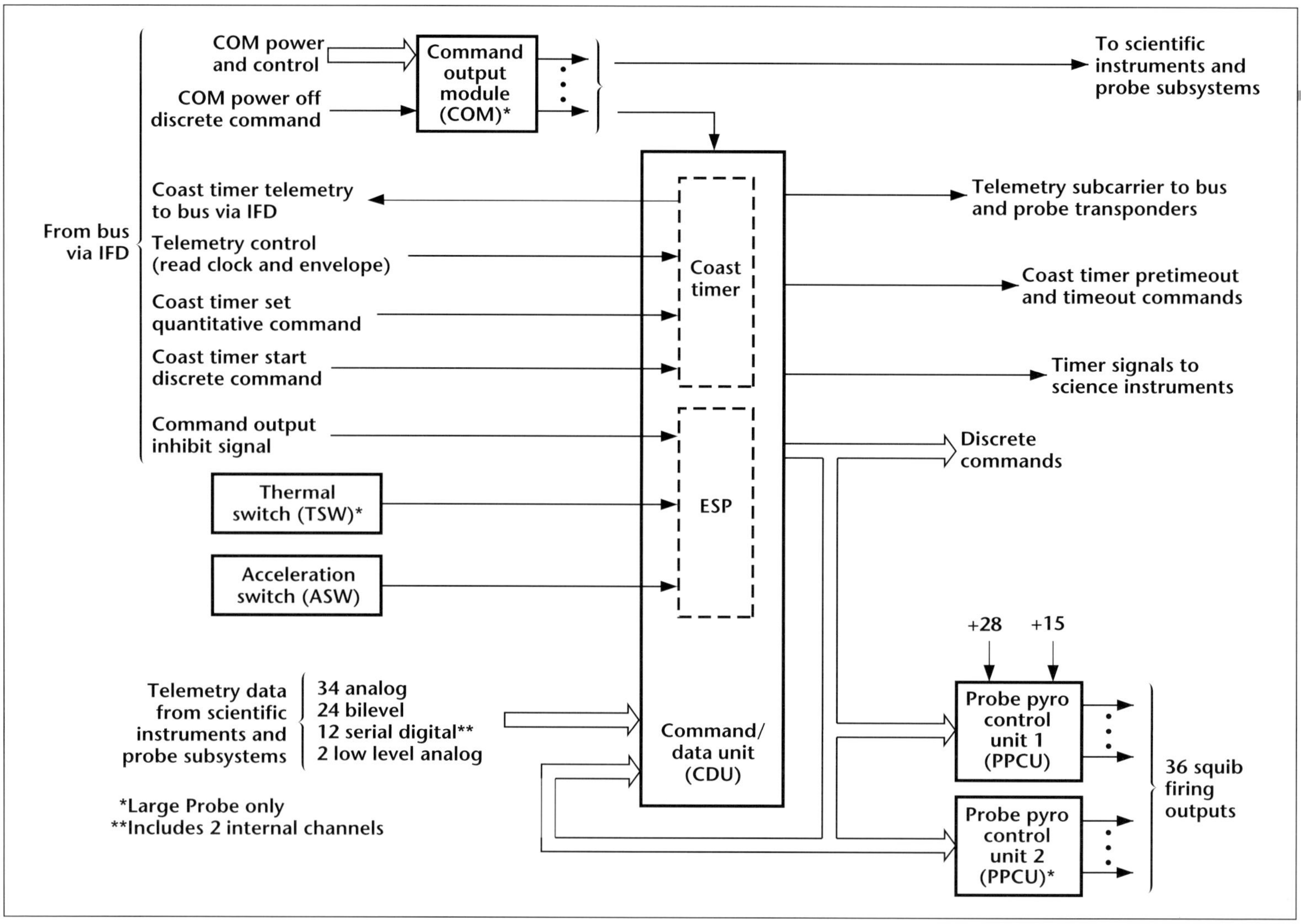

Figure 3-16. The main components of the command data subsystem for the Multiprobe spacecraft are related in this block diagram.

conical aeroshell and an aft shield. The aeroshell, shaped as a 45° cone with a hemispherical blunt top, was a one-piece aluminum structure with integrally machined stiffening rings. The aeroshell's heat shield protected the probe from the heat of a high-speed atmospheric entry. The aeroshell also acted aerodynamically to keep the probe stable on its flight into the atmosphere. The aft cover of fiberglass honeycomb had a Teflon flat section transparent to radio waves. It protected the aft hemisphere of the pressure vessel during entry into the Venusian atmosphere. Spin vanes kept the probes spinning during descent to maintain stability.

All instruments within the probes' pressure vessels required either observations or direct sampling of Venus' hostile atmosphere. Providing such access was a major design problem. The Large Probe had to have 14 sealed penetrations through the walls of its pressure vessel: one for the antenna, four for electrical cables, two for access hatches, and seven for scientific instruments. Each Small Probe required seven such penetrations: one for the antenna, three for electrical cables, one for an access hatch, and two for scientific instruments. Special diamond and sapphire windows admitted light or heat at wavelengths required for several of the science experiments.

The Large Probe

The Large Probe (Figure 3-17) weighed about 315 kg (695 lb) and was about 1.5 m (5 ft) in

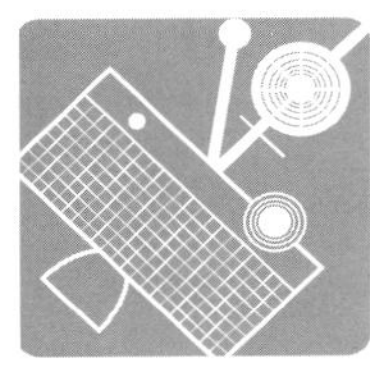

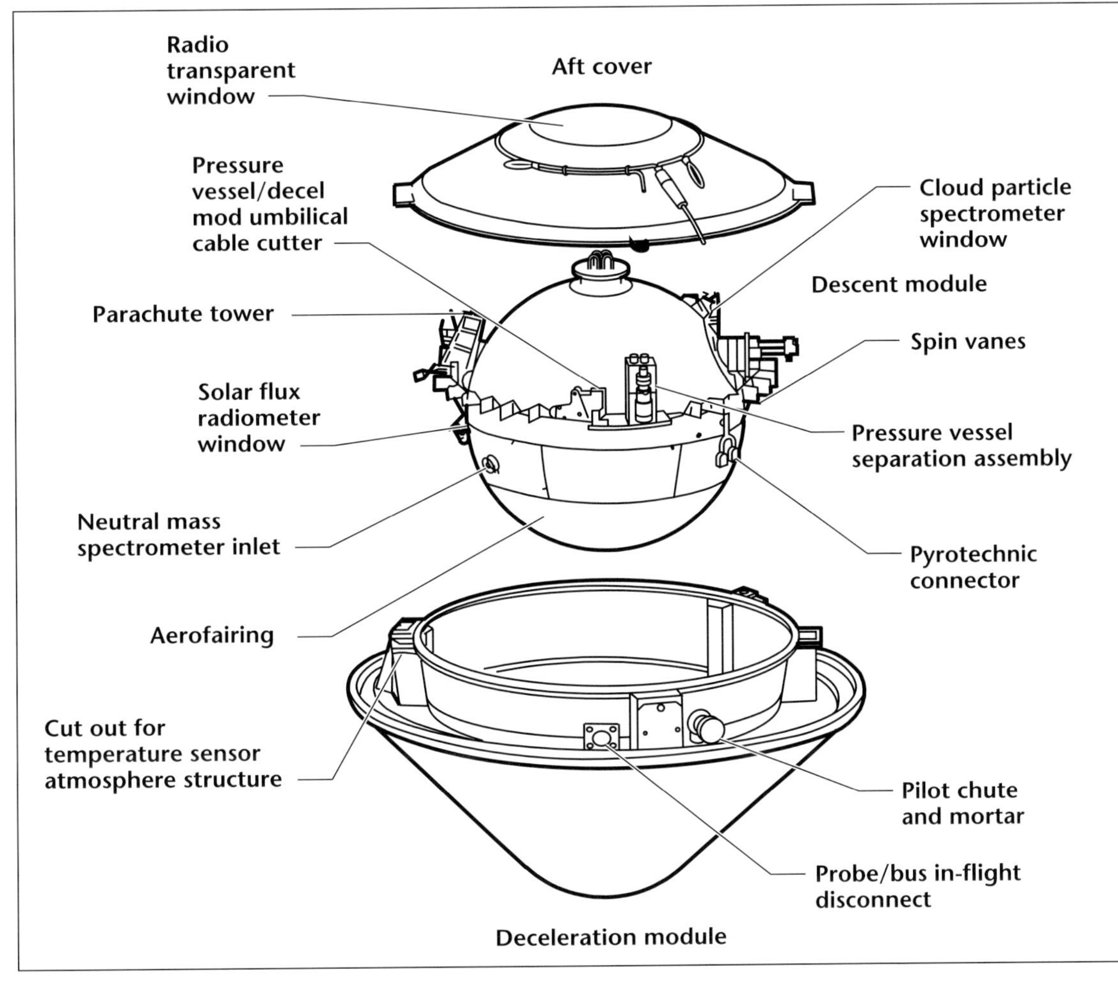

Figure 3-17. Detail of the Large Probe, including pressure vessel, protective nose cone, aft shield, and major components.

diameter. It consisted of a forward aeroshell heat shield, a pressure vessel, and an aft cover. Precisely machined from titanium to achieve high strength at high temperatures and still be lightweight, the pressure vessel (Figure 3-18) was 73.2 cm (28.8 in.) in diameter. It was made in three flanged pieces: an aft hemisphere, a flat ring section, and a forward cap. These were bolted together with seals between the flanges. The seals were a combination of two elements. The first were O-rings to prevent leakage of the probe's 102 kPa (15 psia) nitrogen atmosphere during transit to Venus. The second were Graphoil flat gaskets to prevent inward leakage of Venus' hot atmosphere during descent to the surface. A pressure bottle was on the forward shelf of the Large Probe. A stored command fired the bottle, which increased the probe's internal pressure by 41 kPa (6 psi). Inside the pressure vessel, two parallel beryllium shelves served as supports and heat absorbers for the instruments and spacecraft systems mounted on them. A 2.5-cm (1-in.) thick blanket of multilayered Kapton, which completely lined the interior, further protected equipment inside the pressure shell from heat encountered at Venus.

Four scientific instruments used nine observation windows through four of the pressure

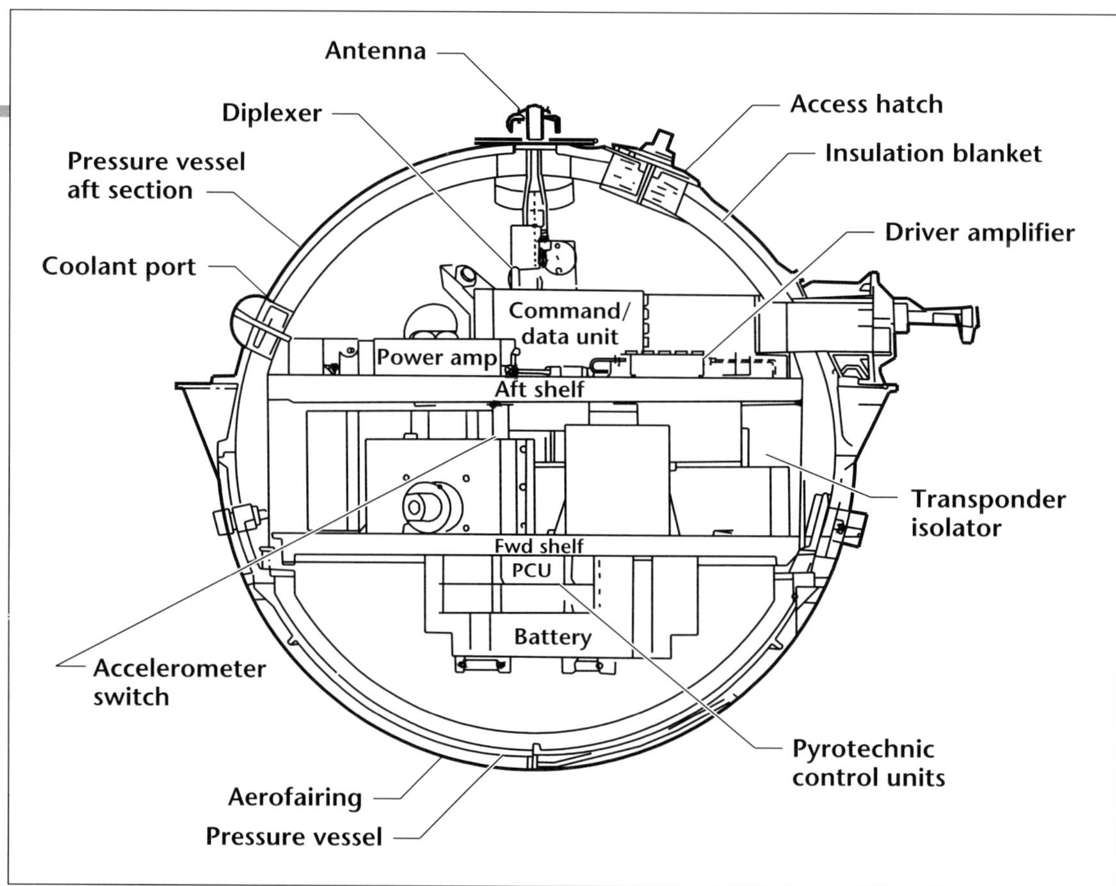

Figure 3-18. Arrangement of scientific instruments and other spacecraft components in the Large Probe's pressure vessel.

vessel penetrations. Eight windows were sapphire and one was diamond. Three vessel penetrations were inlets for direct atmospheric sampling by a mass spectrometer, a gas chromatograph, and an atmospheric structure experiment. At the aft pole of the pressure vessel, an antenna had a hemispherical radiation pattern. This provided communications with Earth when the probe had separated from its Bus. Extending 10 cm (4 in.) on one side of the pressure vessel, two arms held a reflecting prism for cloud particle observations. On the opposite side of the pressure vessel, a single arm carried a temperature sensor on its tip.

Three parachute-shroud towers were mounted above aerodynamic drag plates. These plates were at equal distances around the spherical vessel's equator. Of two access ports, one was for electronic checkout of the system before launch. The other provided a cooling port that scientists also used during ground tests.

During high-speed entry into Venus' atmosphere, a carbon phenolic ablative heat shield protected the Large Probe from overheating. This shield was bonded to and covered the outer surface of the forward-facing aeroshell. All other surfaces of the aeroshell and the aft cover were coated with a heat-resisting, low-density, elastomeric material.

The Large Probe performed a fixed sequence of operations when it reached Venus. Communications with Earth started 22 minutes before entry into Venus' atmosphere. A peak deceleration of 280 g occurred soon after entry, and the spacecraft jettisoned the aft cover to deploy a parachute. A mortar fired a pilot chute from a small compartment in the side of the aeroshell. Lines attached this parachute to the aft cover, which an explosive bolt separated so that it could then pull free. The cover, in turn, was attached to the main parachute. The pilot chute then pulled the main chute from its compartment within the conical

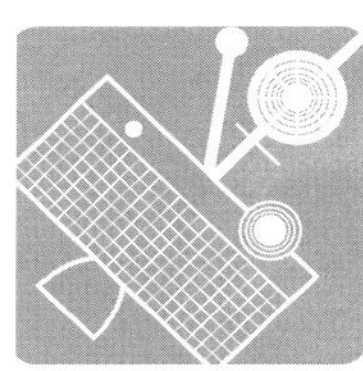

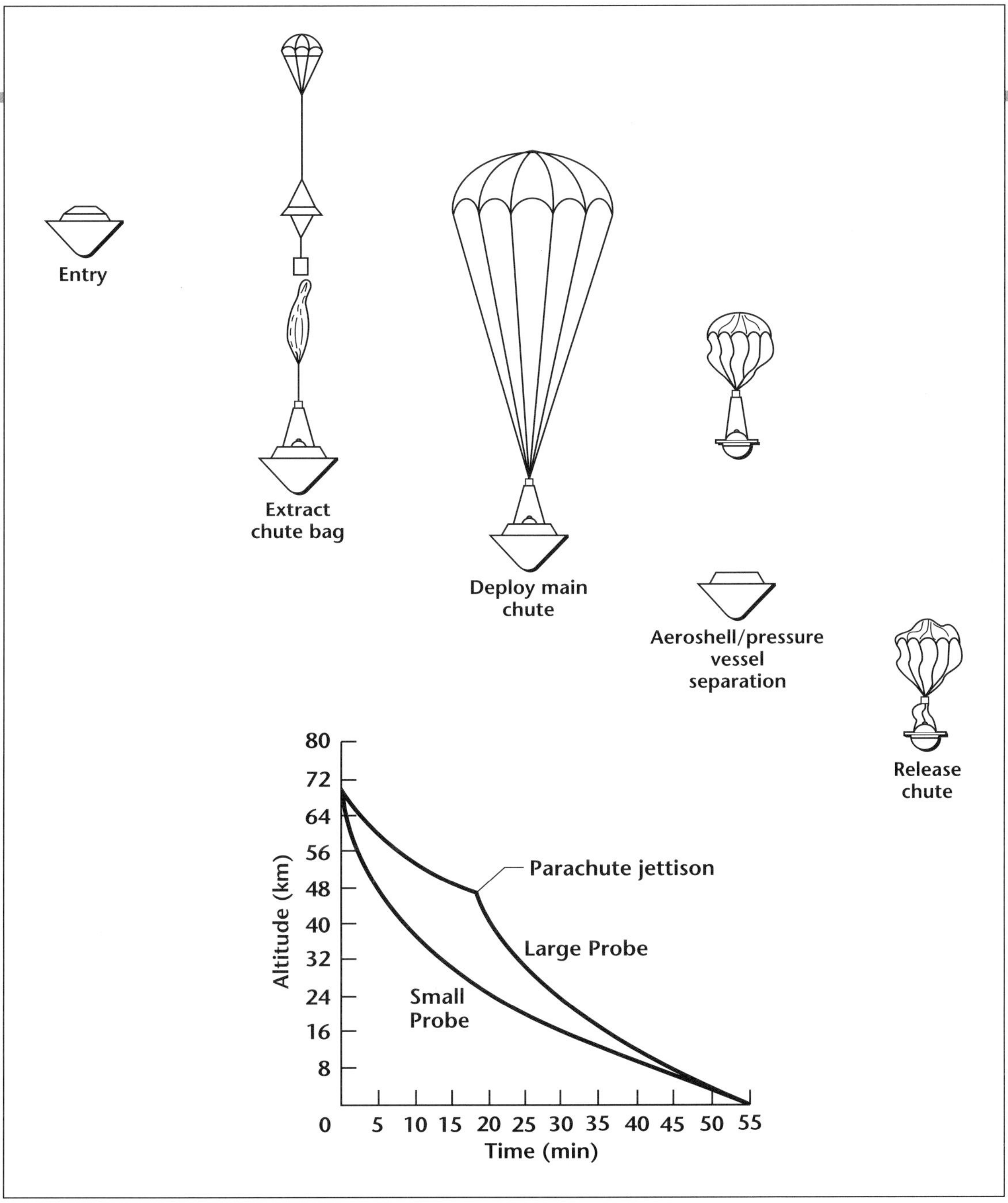

Figure 3-19. (Top) The release sequence for the Large Probe's parachute appears in these drawings. (Bottom) This graph displays altitude plotted against time for the Large and Small Probes. Both Large and Small Probes took about the same time to reach Venus' surface.

aeroshell. As soon as the spacecraft was stable, explosive nuts or cable cutters severed mechanical and electrical ties to the aeroshell. The main chute then pulled the pressure vessel free from the aeroshell (Figure 3-19).

The heat shield jettisoned about 67 km (42 miles) above the surface. About 47 km (29 miles) above the surface, the parachute released. The probe fell freely so that it reached the surface about 55 minutes after first entering the atmosphere. Spin vanes around the pressure vessel spun it at less than 1 rpm during its descent. A forward-facing aerofairing, a conical skirt, and sectional drag plates kept the spacecraft stable.

Communications Subsystem

The Large Probe's communications subsystem (Figure 3-20) had a solid state transmitter to return a data stream directly to Earth at 256 bits/sec. Four 10-W amplifiers provided a transmitter power of 40 W. A transponder received an S-band carrier from Earth at 2.1 GHz. It set the probe's transmitter to send at 2.3 GHz from the crossed dipole antenna on the aft hemisphere. The mission used the transponder receiver for two-way Doppler tracking only. The incoming signal carried no information, and the Large Probe did not receive commands from Earth.

Power Subsystem

A 40 A hr, silver-zinc battery powered the probe. During descent, the spacecraft maintained output at 28 V direct current. The power system consisted of the battery, a power interface unit, and a current sensor. Before the probe separated from the Bus, it received power from the Bus for checking and heating the probe. During this time, the internal battery was open-circuited by switches in the probe's power interface unit.

Command Subsystem

Once the Large Probe had separated from the Bus, its internal electronics provided all commands for its operation. The command

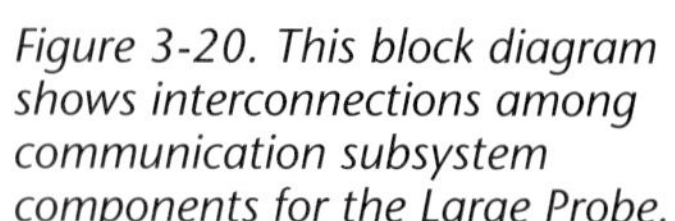

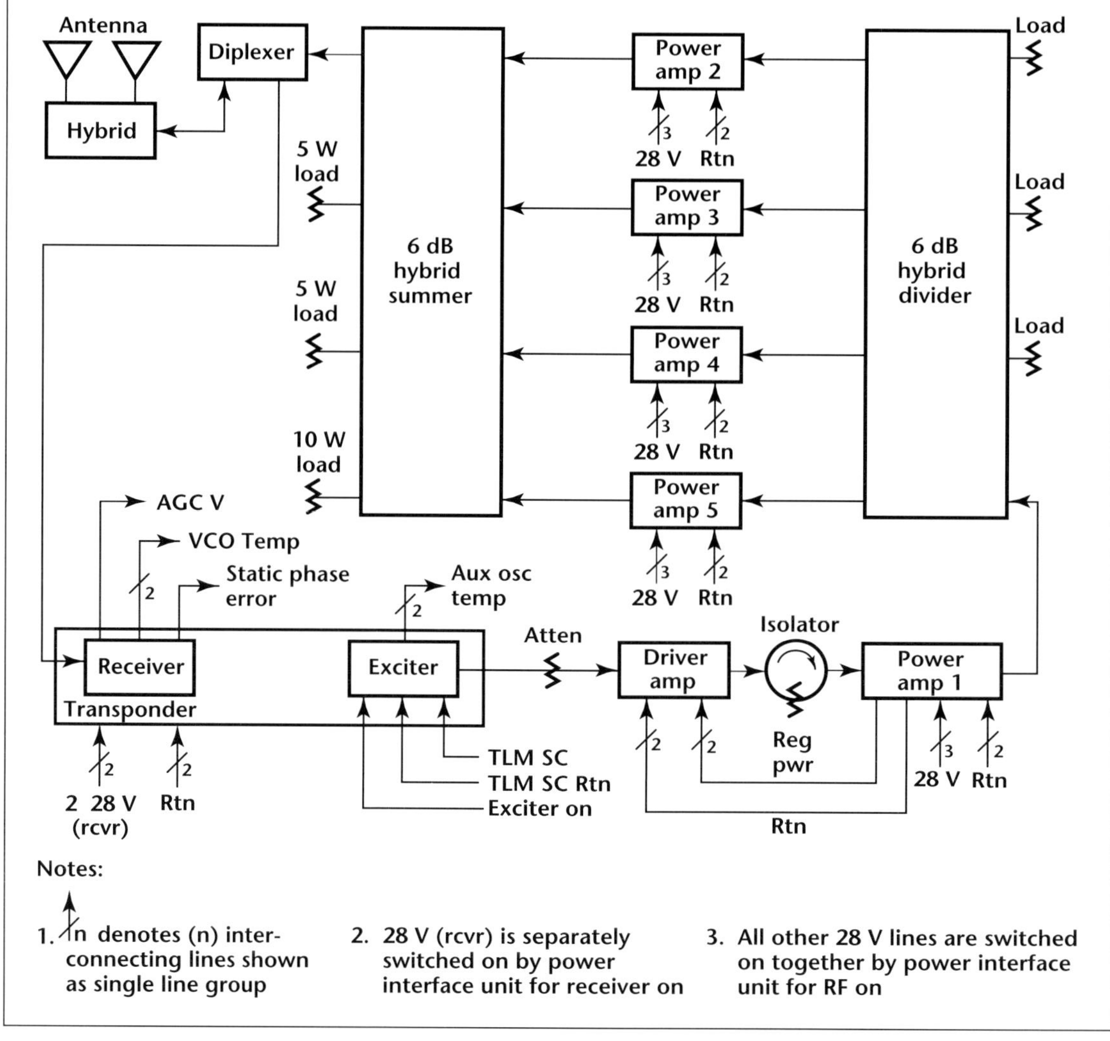

Figure 3-20. This block diagram shows interconnections among communication subsystem components for the Large Probe.

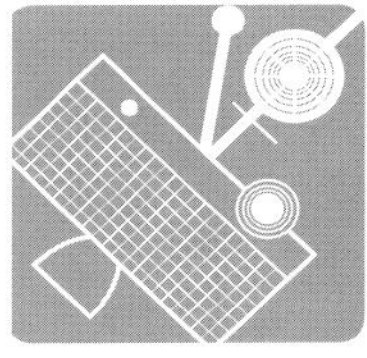

subsystem consisted of a command unit, a pyrotechnic control unit, and sensors to service the command unit, including sensors to measure the probe's deceleration. The internal command subsystem had 64 separate commands for the spacecraft itself and for its payload of scientific instruments. It contained a coast timer that was the spacecraft's only instrument working during separation from the Bus to entry into Venus' atmosphere. During this period, all other probe subsystems were off. There also was an entry sequence programmer and a command decoder. The entry sequence programmer transmitted 53 discrete commands in a fixed sequence. Two instruments could start these commands: the coast timer or an accelerometer switch that sensed the deceleration of entry. A temperature switch acted as back-up for the timer when the parachute jettisoned.

Pyrotechnic Control Unit

The pyrotechnic control unit consisted of 12 squib drivers. These drivers provided current to fire explosive nuts to release the aeroshell, the aft cover, and the parachute. There also were actuators for the cable cutter, the pilot chute mortar, and the instrument that released the protective cover of the mass spectrometer inlet port.

Data-Handling Subsystem

The Large Probe's data subsystem handled 36 analog, 12 serial digital, and 24 bilevel status channels from scientific instruments and subsystems within the probe. The unit converted all data into major telemetry frames. These frames consisted of 16 minor frames for time-multiplexed transmission to Earth. Each minor frame, in turn, consisted of a series of 64 8-bit words for a total of 512 data bits per minor frame.

The data handling subsystem provided two data formats: one for use during radio blackout caused by the plasma sheath during entry, the other for use during normal descent after the probe slowed down. There was a solid-state memory with a storage capacity of 3072 bits. This storage allowed the probe to store data during communications blackout and transmit it afterwards. The system stored data in memory at 128 bits/sec but read it out at 256 bits/sec. This was the normal bit-rate for data transmission to Earth during the descent. For 5 minutes before entry to 30 seconds after entry, the transmission bit-rate was only 128 bits/sec. The full bit-rate was allocated among the seven experiments at 16 to 44 bits/sec for each experiment. The nephelometer and atmospheric structure experiments, however, were able to use the blackout storage format of 4 and 72 bits/sec, respectively. Two subcommutated formats for low bit-rate phenomena also provided data for housekeeping and for the atmospheric structure, nephelometer, cloud particle spectrometer, and solar flux radiometer experiments.

The Small Probes

The three Small Probes (Figure 3-21) were identical. In contrast to the Large Probe, they did not carry parachutes. Aerodynamic braking slowed them down. Like the Large Probe, each Small Probe consisted of a forward heat shield, a pressure vessel, and an afterbody. The heat shield and the afterbody remained attached to the pressure vessel all the way to the surface. Each probe was 0.8 m (30 in.) in diameter and weighed 90 kg (200 lb).

Engineers precisely machined the pressure vessel (Figure 3-22) from titanium in two flanged hemispheres. Bolts joined the hemispheres with seals between the flanges. The vessel nested within the aeroshell and was permanently attached to it. The Small Probes used two types of seals similar to the Large

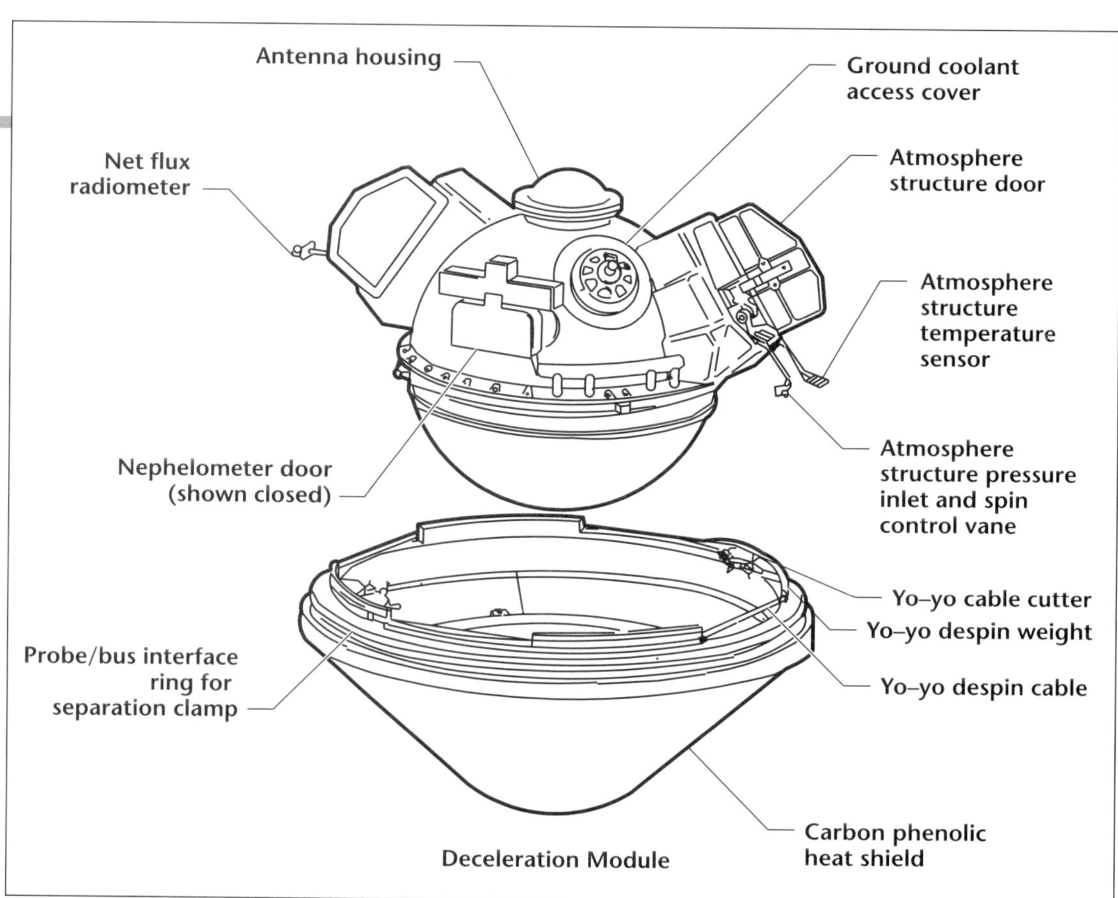

Figure 3-21. The three Small Probes were identical. Major components are identified.

Probe's: (1) O-rings to maintain internal pressure and (2) Graphoil flat gaskets to prevent Venus' hot atmosphere from leaking in. The afterbody also was a permanent part of the pressure vessel, its shape closely matching the pressure vessel's. Xenon filled each Small Probe's interior at a pressure of approximately 102 kPa (15 psia). Engineers used xenon instead of nitrogen—they used nitrogen in the Large Probe—to reduce heat flow from the pressure vessel walls to instruments and probe spacecraft systems. As in the Large Probe, a protective blanket lining of Kapton further slowed this flow. Instruments and spacecraft subsystems were on two beryllium shelves that absorbed heat.

The aeroshell had the same basic 45°, blunt cone design as the Large Probe. It, too, used a bonded carbon phenolic ablative coating as a heat shield. Because the shield had to protect the pressure shell down to the surface, engineers made the aeroshell from titanium (the Large Probe had an aluminum aeroshell). The shell had a stressed skin, or monocoque (one piece), construction.

The Small Probes' entry sequence started with communications 22 minutes before entry. About 5 minutes before entry, a pyrotechnic cable cutter cut two weights loose. The weights were now free to swing out like yo-yos on 2.4-m (8-ft) cables. As a result, each probe's spin rate slowed from about 48 rpm to 17 rpm. The probes then jettisoned the weights and cables. This reduction in spin rate allowed aerodynamic forces to line up the probes. Now their heat shields could protect them from the heating of entry. All probes entered the atmosphere at a speed of about 42,000 km/hr (26,099 mph). The probe making the steepest entry underwent a peak deceleration of 458 g, the others somewhat less. The probe making the shallowest entry decelerated the least at about 223 g. Three doors on the afterbody then opened at an altitude of about 70 km (43 miles) to give

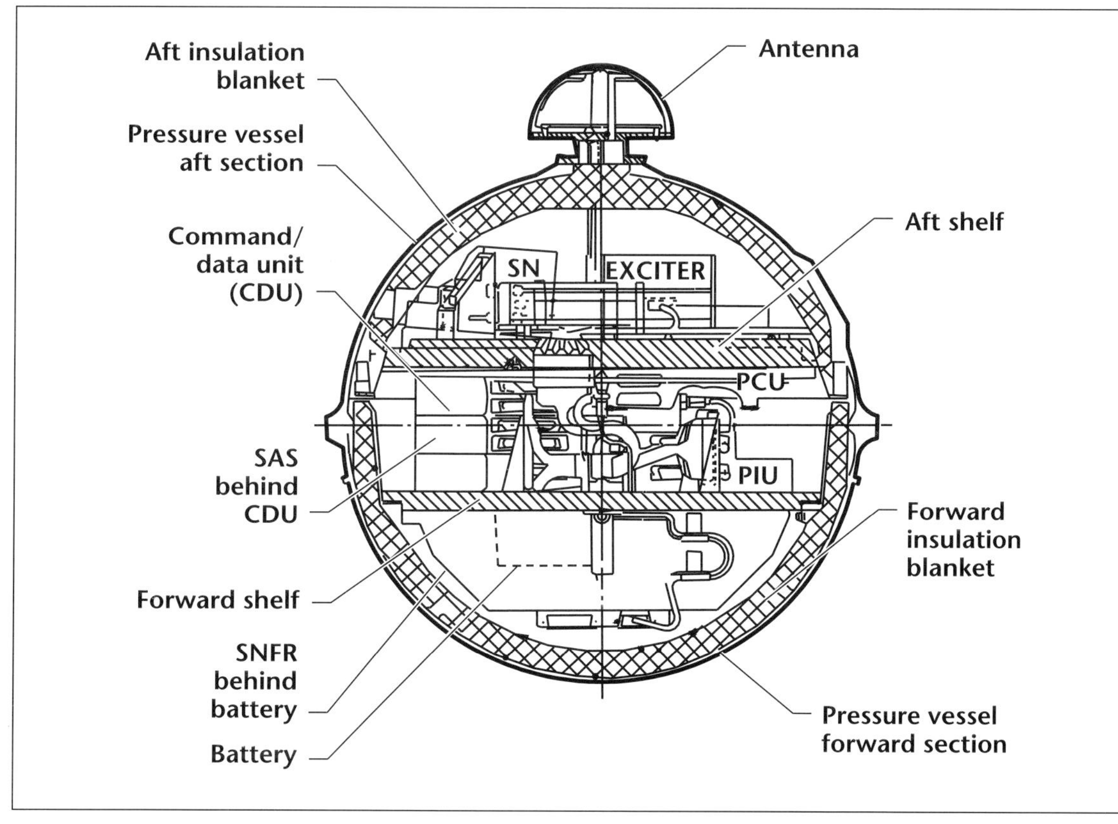

Figure 3-22. A titanium pressure vessel carried scientific instruments and spacecraft systems for each Small Probe. The Large Probe used the same type of pressure vessel.

three instruments access to the atmosphere. Two of the doors opened from each of two protective housings. One was for the atmospheric structure experiment and the other for the net flux radiometer experiment. The housings projected like ears from each side of the pressure vessel's sphere. The temperature sensor and atmospheric pressure inlet for the atmospheric structure instrument extended 10 cm (4 in.) from the door of one housing. The net flux radiometer sensor extended similarly on the opposite side.

When the housing doors opened after atmospheric entry, they slowed the spacecrafts' spin rate because they did not jettison. However, a small vane on the pressure sensor inlet kept the spacecraft spinning throughout its descent. This spin allowed the instruments to scan around the probe. A cover over the nephelometer folded down after it opened. Each Small Probe fell freely for about 53 to 55 minutes until it reached Venus' surface.

Communications equipment for each Small Probe (Figure 3-23) consisted of a solid state transmitter and a hemispherical-coverage antenna, similar to the Large Probe. The antenna was on the aft pole of the pressure vessel sphere and radiated through a Teflon window. Each transmitter had one 10-W amplifier, which was one-quarter the power of the Large Probe's transmitter. Until the probes penetrated to roughly 30 km (19 miles) above Venus' surface, the large 64-m (210-ft) antennas of the Deep Space Network could receive data at 64 bits/sec. From there on, they received data at 16 bits/sec only. The Small Probes did not carry a receiver for two-way Doppler tracking. Instead, each probe carried a stable

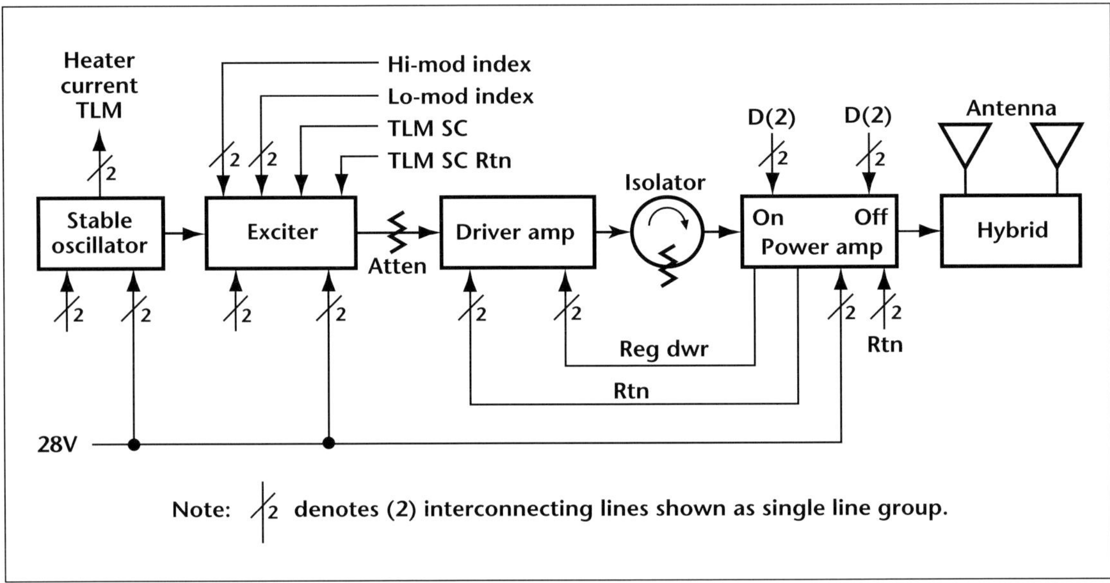

Figure 3-23. Components of each Small Probe's communication subsystem appear in this block diagram. The subsystem had only one power amplifier (the Large Probe had four). It did not have a receiver.

oscillator. This provided the reference frequency for the Doppler measurements that ground control used in its computations.

Each probe carried an 11 A hr silver-zinc battery. This provided 28 V direct current during the descent. As with the Large Probe, the power system had a power interface unit and a current sensor.

The command subsystem was identical to the Large Probe's. No uplink (Earth-to-probe) command capability existed. After separation, all probe commands originated from their respective coast timers, programmers, and acceleration switches. The coast timer maintained control. It started the entry sequence programmer, which transmitted all commands in a fixed sequence. The transmission lasted until each probe impacted with Venus' surface.

The Small Probe's data-handling subsystem components were the same as the Large Probe's. Each probe used three major data formats: upper descent, blackout, and lower descent. These formats contained 16 minor frames of 64 eight-bit words. As on the Large Probe, a 3072-bit solid state memory stored data when communications with Earth were blacked out by a plasma sheath on entry. These data were transmitted after the probes had slowed down and the plasma sheath had dissipated.

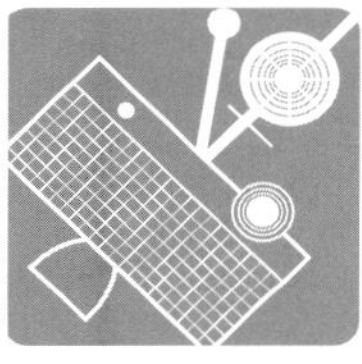

CHAPTER 4

SCIENTIFIC INVESTIGATIONS

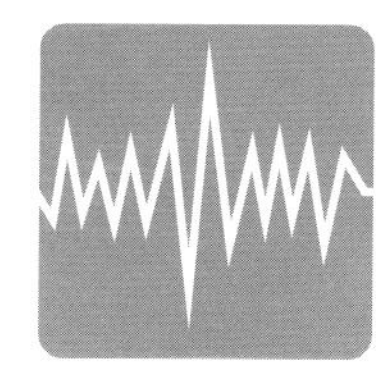

HAEC IMMATURA A ME JAM FRUSTRA LEGUNTUR; O.Y.

The Pioneer Venus project changed forever the way we looked at Venus. The project's foundations were carefully designed investigations and engineered instruments carried by two spacecraft, an Orbiter and a Multiprobe. The Orbiter had four scientific objectives: study Venus' upper atmosphere and ionosphere, clouds, surface, and gravitational field. The Multiprobe's experiments focused on the atmosphere to investigate the components, composition, structure, thermal balance, circulation around Venus, and interaction with solar wind. In this chapter, you learn how mission scientists achieved these objectives with specific instruments and investigations.

With this anagram in 1610, Galileo reported his first scientific observation of Venus. His observation broke centuries of failure to see what, in retrospect, is obvious: Earth is not the center of the Universe. Galileo's message, when unscrambled and translated into English, said, *The mother of the loves emulates the phases of Cynthia*. That is, Venus exhibits phases like the Moon.

In the centuries that followed Galileo's observation, scientists made many more discoveries about the cloud-shrouded planet. Yet, there were equally as many speculations about Venus' true nature. These speculations ranged from a dust-ridden world to a world of swamps to one in a sea of hydrocarbons. Modern Earth-based observations disproved many of the earlier theories. Such observations used highly sophisticated new instruments and data reduction techniques. New data also came from several Venus flybys and some Russian probes that had landed on the surface. Despite this work, many unknowns about Earth's sister planet still remained.

To resolve these unknowns, scientists designed the scientific payloads of the Pioneer Venus Orbiter and Multiprobe spacecraft to obtain new information about Venus. Six spacecraft carried advanced scientific instruments that revised our notions about Venus. These spacecraft altered our understanding as drastically as Galileo's observations changed many of his contemporaries' beliefs. With Pioneer Venus, scientists were the first to look globally through the thick cloud layers. They sampled the constituents of Venus' dense atmosphere. Also they made long-term observations of changes within that atmosphere and of its ultraviolet cloud markings.

New viewpoints resulted from the Pioneer mission to Venus. These viewpoints influenced comparative planetologists and other scientists as they worked to refine theories to explain the evolution of the Solar System and its planets.

Orbiter Scientific Objectives

The Orbiter explored Venus in four important ways. First, it investigated the clouds globally. To do this, it used advanced technology sensors aboard the spacecraft. It also observed how Venus' atmosphere affected radio signals from the spacecraft to Earth as the spacecraft was occulted by Venus. Second, it measured the upper atmosphere and ionosphere's features over the entire planet and detected how the solar wind interacts with the ionosphere. Third, it used a radar instrument to penetrate the Venusian cloud layers and obtain information about the planet's surface. Finally, it determined the general shape of Venus' gravitational field and detected local anomalies in it by measuring how the field affected the spacecraft's orbit.

To achieve its science objectives, the spacecraft carried a complement of 12 scientific instruments. Three instruments provided information to answer basic questions about how Venus interacts with the solar wind. A magnetometer measured magnetic fields. A plasma analyzer measured solar wind. An electric field detector measured electric fields.

An ultraviolet spectrometer measured the intensity of ultraviolet radiation at various wavelengths. Its aim was to check how sunlight reflects and scatters off clouds and haze layers in Venus' atmosphere. This instrument also detected day and night glows in the upper atmosphere. They are caused by solar radiation acting on the gases there and recombination of molecules when solar radiation is absent at night. The instrument also investigated a hydrogen gas corona surrounding the planet.

An infrared radiometer measured radiation at selected wavelengths within the infrared part of the electromagnetic spectrum. It was sensitive to the atmosphere's emitting temperature at several levels. The instrument also detected and mapped both water vapor distribution in the atmosphere and reflected solar radiation.

A radar mapper penetrated the cloud layers to determine surface topography and surface scattering properties. This instrument revealed surface details that cloud layers obscured. By using side-looking mapping, it also provided information on the radar brightness of the surface.

An ultraviolet spin-scan imager mapped the Venusian clouds. To build a picture, this instrument made a series of narrow scans across Venus. The process is similar to the way a television creates a picture by scanning a series of lines across the tube face. The spacecraft's rotation swept the viewpoint of the instrument across the planet. While this was happening, the spacecraft's motion along its orbit placed the scan paths side by side to build images. The spin-scan imager also measured intensity and polarization of light reflected from Venus' clouds. When operating in a polarimetry mode, it provided information about size, shape, and types of particles making up clouds and haze layers.

When the Orbiter was closest to Venus, at orbit periapsis, it passed briefly through the ionosphere and upper atmosphere. During those periods, it used several instruments. One identified the atmosphere's neutral, uncharged particles. Another measured composition and concentration of positively charged thermal ions. A retarding potential analyzer and electron temperature probe also were aboard the Orbiter. These instruments measured the abundances of charged particles in the ionosphere and in layers between the ionosphere and the region of the solar wind. It determined ion composition and electron and ion energy.

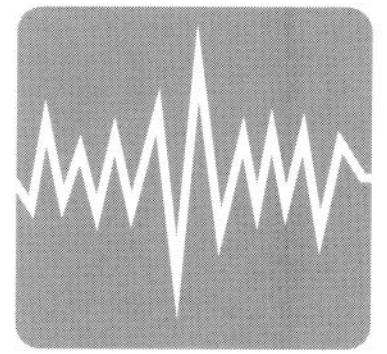

The Orbiter also carried an instrument that was not connected with Venus exploration. The instrument measured gamma ray bursts coming from space. Before a special research spacecraft probed space in the 1980s, scientists could not determine the source of these rays. The observation platform that Pioneer Venus provided complemented experiments that scientists were conducting near Earth. While en route to and in orbit around Venus, it presented an opportunity to obtain another set of data. Researchers used these data to triangulate with Earth-Orbiter observations and find each source's direction.

Multiprobe Scientific Objectives

The Multiprobe spacecraft consisted of a Bus and probes to investigate Venus' atmosphere in four major ways. First, its instruments sampled gases and particles within the clouds to establish their nature and composition. Second, its science experiments determined composition, structure, and thermal balance of the planet's atmosphere, by direct sampling and measurements of radiation from high altitudes down to the surface. Third, observations of the atmospheric probes' paths checked how the atmosphere circulates about the planet. Fourth, the spacecraft gathered data to further investigate how the planet interacts with solar wind.

To achieve these science objectives, the Multiprobe spacecraft carried 18 scientific experiments. These included two aboard the Bus, three on each of the three identical Small Probes, and seven on the Large Probe.

One instrument on the Bus was a neutral mass spectrometer. This sophisticated instrument measured density and analyzed gas composition in the upper atmosphere. The other instrument was an ion mass spectrometer that was identical to the Orbiter's. It determined the composition of thermal ions in the upper atmosphere and measured their concentration and temperature.

Each Small Probe carried an instrument to detect and measure optical properties of particles at various levels in Venus' atmosphere. It also carried an instrument complex to measure atmospheric temperature and pressure. These sensors had two main functions. First, they defined the properties of the atmosphere and clouds from an altitude of about 65 km (40 miles). Second, they enabled investigators to establish the probe's altitude during each measurement. A third device monitored the amount of sunlight penetrating to different atmospheric levels. The instrument also measured the amount of planetary infrared radiation emitted back to space.

The Large Probe carried the first two experiments, which were described above, to determine atmospheric and cloud structure. In addition, it carried a neutral mass spectrometer to measure the composition of the neutral atmospheric components. It took measurements from an altitude of about 65 km (40 miles) to the surface. This instrument identified vapors that condense to form Venus' clouds. It also measured the number of rare gas isotopes in the atmosphere. This isotopic measurement was important in tracing the planet's history and atmospheric evolution. Another instrument, the gas chromatograph, measured the abundances of atmospheric gases.

To find out which solar radiation penetrates the atmosphere and reaches ground level, the Large Probe included yet another instrument. Such measurements are important to our understanding of why Venus is so much hotter than Earth. A separate instrument measured the infrared part of the solar radiation flux at all

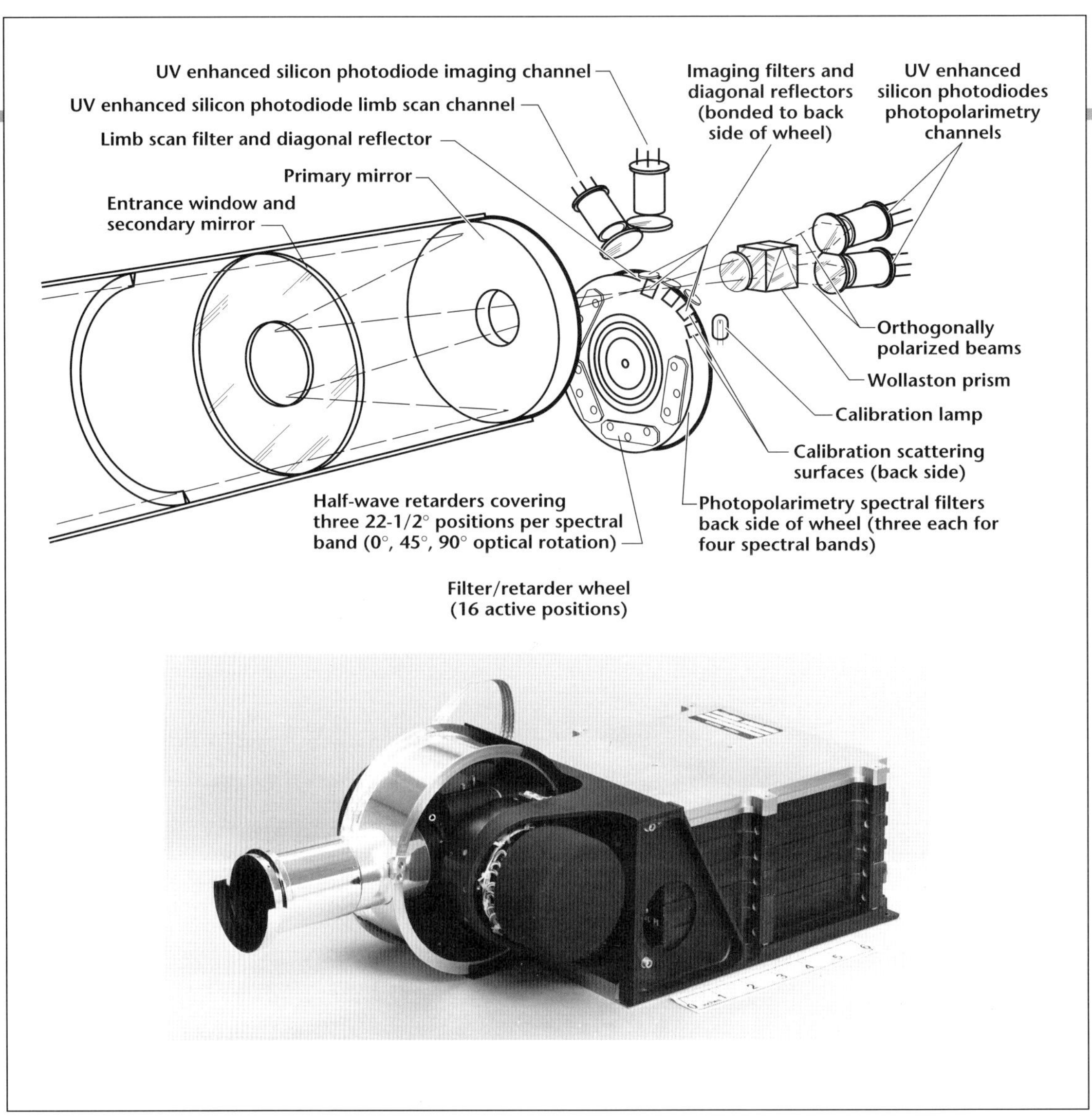

Figure 4-1. Orbiter cloud photopolarimeter (OCPP). (Top) Diagram identifying components of instrument's optical system, including its telescope, filter/retarder wheel, and photodiodes. (Bottom) Photograph of complete instrument.

levels in the atmosphere. It also detected the presence of clouds and water vapor. Another instrument measured particle sizes in clouds and in the lower atmosphere, and determined particle concentrations at various levels.

Earth stations received radio signals from all probes and the Multiprobe Bus. Science investigators used these signals to make extremely accurate measurements of the various probes' velocities. From these measurements, researchers calculated wind speeds and circulation patterns in Venus' atmosphere.

Orbiter Instruments and Experiments

Cloud photopolarimeter

The photopolarimeter measured distribution of cloud and haze particles and detected ultraviolet markings and cloud circulations. The ultraviolet images from this instrument provided visual references for data from other Orbiter experiments and for its own polarization readings. Principal investigator for this instrument was L. D. Travis, NASA Goddard Institute for Space Studies.

The photopolarimeter (Figure 4-1) weighed 5 kg (11 lb) and required 5.4 W of electrical power. It consisted of a 3.7-cm (1.5-in.) aperture telescope with a rotating filter wheel.

There were 16 active positions on the filter wheel, three filters for each of four spectral bands (255-285, 355-380, 540-555, and 930-945 nm), limb-scan filters, and imaging filters. A Wollaston prism directed the light beams for the photopolarimetry channels to two silicon photodiodes enhanced to detect ultraviolet light. Diagonal reflectors at two positions on the back of the filter wheel sent the beams to two other silicon photodiodes. One was for the imaging channel and the other for the limb-scan channel.

This telescope observed Venus at fixed angles. It used the Orbiter's rotation to lay scans across the planet. To set these scans side by side, it used the motion along the spacecraft's trajectory. Ground control could set the angle of the telescope's axis to the spacecraft's spin axis. By this means, investigators could direct the telescope to observe the planet from any point along the Orbiter's elliptical orbit.

In the imaging mode of operation, when the spacecraft measured only the intensity of received radiation, the polarimeter's field of view was about 0.5 mrad. This corresponds to a resolution of about 30 km (19 miles) directly below the Orbiter. In this mode, approximately 3.5 hours were required to record an image of Venus' full disk. An ultraviolet filter revealed the fast moving cloud markings that appear only in ultraviolet pictures of Venus. A maximum of five full-disk planetary images were possible during each spacecraft's orbit.

In the photopolarimetry mode, the instrument's field of view was approximately 0.5°. This corresponds to a resolution of about 500 km (310 miles) directly below the Orbiter. In this mode, the photopolarimeter used four pass-bands. The instrument measured polarization of scattered sunlight, the characteristics of which depend on particle size, shape, and density in clouds and hazes. These data yielded vertical distribution of cloud and haze particles relative to atmospheric pressure.

When the Orbiter neared periapsis, the instrument could observe in visible light the atmosphere's high haze layers. This was done by programming the telescope to scan across the limb of the planet. In this mode, the field of view was about 0.25 mrad, or an altitude resolution of about 0.5 to 1.0 km (0.3 to 0.6 miles). Such observations provided information about layers above Venus' main cloud deck.

In Phase II of the Orbiter's mission, the instrument provided a detailed record of the long-term evolution of significant haze effects. This is important for understanding photochemical and aerosol processes and the atmosphere's mechanisms of meridional transport. From observations of many years during Phase II, the instrument showed build-up and dissipation of midlatitude jet streams and provided insight into zonal circulation.

Surface Radar Mapper

The radar mapping instrument (Figure 4-2) weighed 9.7 kg (21.3 lb) and required 18 W of electrical power. The radar team leader was G. H. Pettengill, Massachusetts Institute of Technology. The experiment produced the first maps of large areas of Venus unobservable by Earth-based radar. From radar echoes, experimenters derived surface heights along the spacecraft's suborbital trajectory to an accuracy of 150 m (492 ft). They were able to make a good estimate of global topography and shape. The team also derived surface electrical conductivity and meter-scale roughness from the radar data.

A low-power (20 W peak pulse power), S-band (1.757 GHz) radar system observed the surface. Ground controllers mechanically moved the

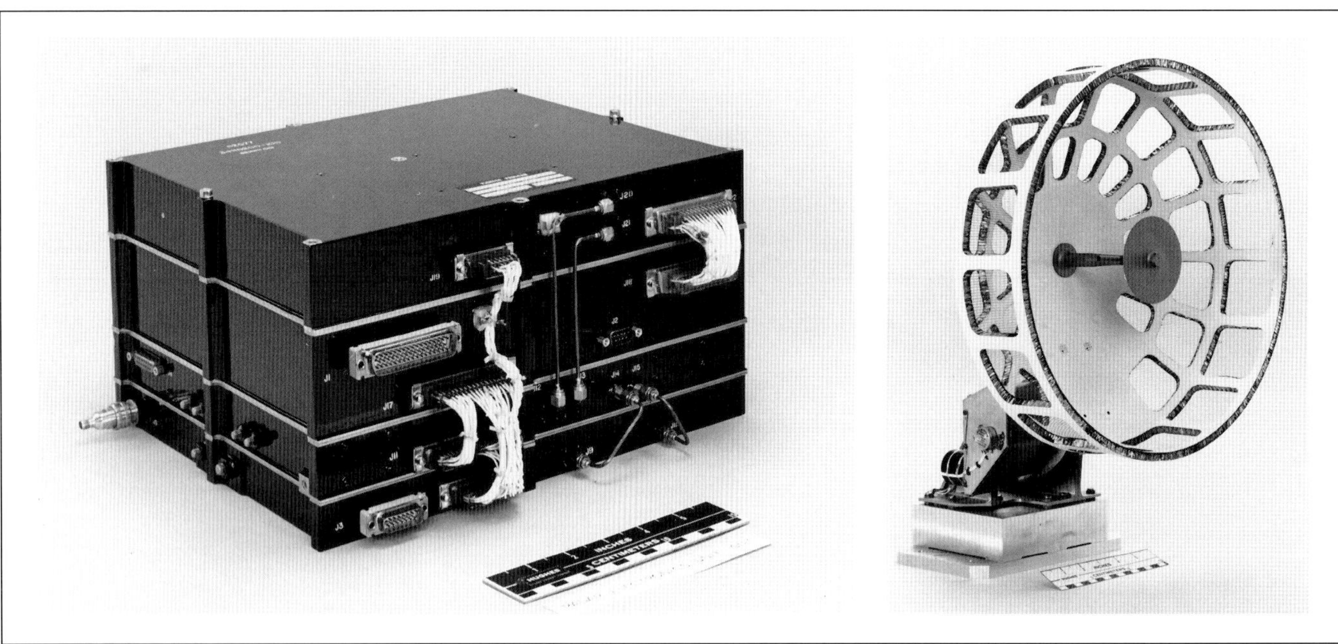

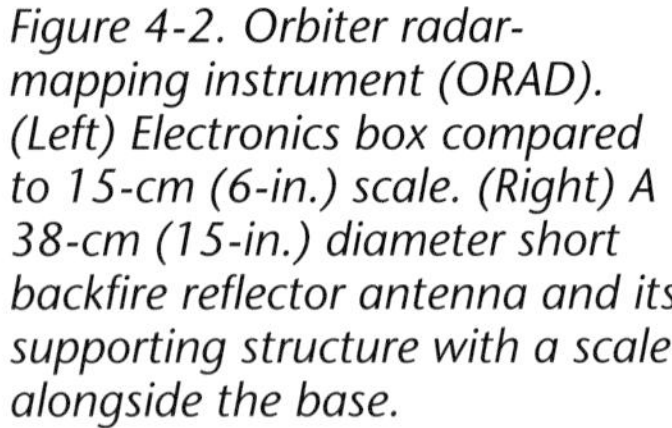

Figure 4-2. Orbiter radar-mapping instrument (ORAD). (Left) Electronics box compared to 15-cm (6-in.) scale. (Right) A 38-cm (15-in.) diameter short backfire reflector antenna and its supporting structure with a scale alongside the base.

antenna in a plane containing the spacecraft's spin axis. The controllers did this to view the suborbital point on the planet's surface once during each spacecraft roll. The Orbiter took measurements whenever it was below 4700 km (2920 miles). These measurements were subject to constraints that the spinning spacecraft set. They also had to compete with other experiments for the Orbiter's limited telemetry capacity. To minimize telemetry requirements, echoes were processed on board the spacecraft. The spacecraft spun at a rate of about five revolutions per minute. During this period, radar observations occupied about 1 second out of the total rotation period of 12 seconds. The instrument automatically compensated for Doppler shift. (Radial motion of the Orbiter toward and away from the planet during each elliptical orbit caused the shift.) When the spacecraft was closer than 700 km (435 miles) to the surface, the received frequency was stepped to make range measurements of the areas lying just ahead and behind the spacecraft's path.

Investigators wanted to find absolute topographical elevations. To do this, they subtracted the observed distance between the Orbiter and the surface from the spacecraft's orbital radius. This radius was obtained from Deep Space Network (DSN) tracking of the spacecraft. Surface resolution was best at periapsis. It was then 23 km (14 miles) along the track and 7 km (4.3 miles) across the track. Relatively long pulses were used to obtain a good signal-to-noise ratio from each pulse.

When the radar operated in its other mode (namely, side-looking radar imaging at altitudes below 550 km, or 342 miles), the functional parameters for altimetry measurements changed. This mode relied upon uncoded pulses at a pulse repetition frequency of 200 Hz to avoid ambiguities in range and surface mapping. The antenna, pointing to one or both sides of the ground track, made a sequence of surface brightness measurements. Commands from Earth determined which way the antenna pointed. The illuminated surface area was divided into 64 picture elements (pixels). When the spacecraft was close to periapsis, each pixel covered an area about 23 km (14.3 miles) square on the surface.

The radar mapper operated during Phase I and part of Phase II of the mission when the periapsis altitude was low enough for useful radar images. This period of low periapsis ended March 31, 1981, and the mapper experiment

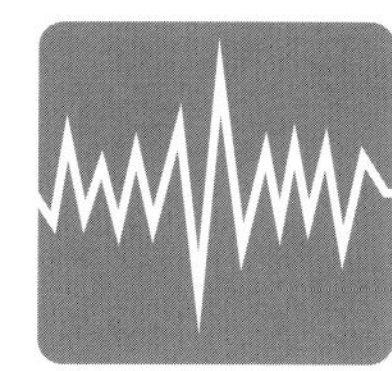

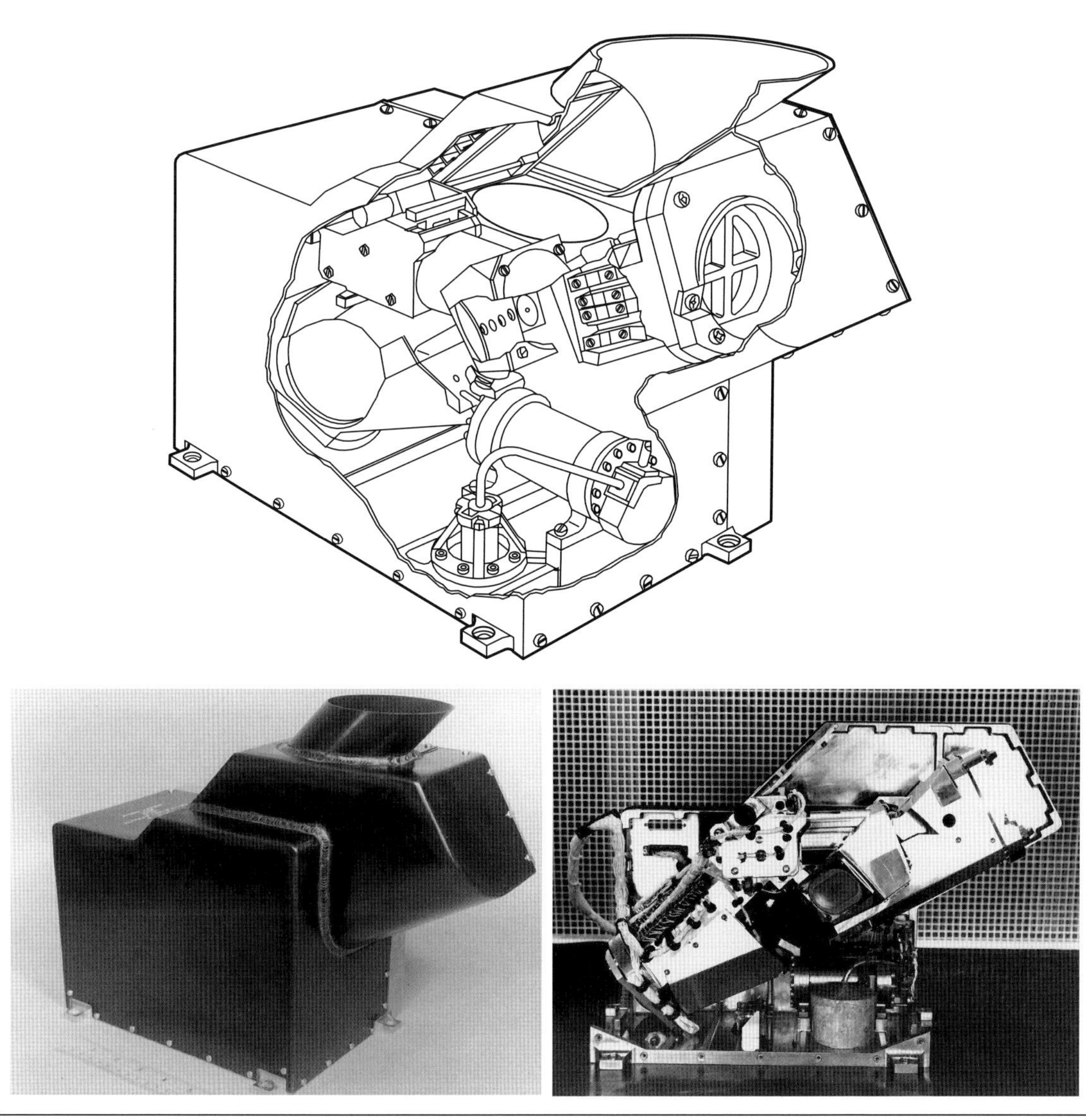

Figure 4-3. Orbiter infrared radiometer (OIR). (Top) Cutaway drawing of instrument related to outline of its housing. (Bottom left) Packaged instrument. (Bottom right) Instrument without its housing.

ended. Toward the end of Phase III of the mission, the periapsis altitude was again suitable for radar mapping. However, because the Magellan spacecraft was by that time producing higher resolution radar imaging, project management did not reactivate the Orbiter's instrument.

Infrared Radiometer

The infrared radiometer (Figure 4-3) weighed 5.9 kg (13 lb) and required 5.2 W of electrical power. The principal investigator was F. W. Taylor, Oxford University, England. The university developed and constructed a pressure modulation unit and molecular sieve for one channel of the instrument to make measurements over a wide range of temperatures and pressures. The radiometer measured infrared radiation that Venus' atmosphere emitted at various altitudes. These altitudes ranged from 60 km (37 miles) at the top of the cloud deck, where the atmospheric pressure was 250 mbars, to 150 km (93 miles), where the pressure was 10^{-6} mbars. This region includes those parts of Venus' atmosphere where the four-day circulation takes place, where there is maximum cooling by radiation, and where there is maximum deposition of solar energy. The instrument searched for water vapor above the cloud layers and measured the extent of the heat-trapping cloud layers, and measured the albedo. Data from the radiometer yielded about 800,000 vertical profiles of upper atmosphere temperatures. By keeping sample time short, scientists were able to obtain a temperature sensitivity of more than 0.5 K at 240 K. Such information was important for discovering both the extent and the driving forces of the upper atmosphere's four-day circulation.

The radiometer had eight detectors, each sensitive to a different part of the spectrum. Because the instrument covered such a wide spectrum range, it needed several different measurement techniques. Five detectors measured infrared emissions at five selected wavelengths of the absorption band for carbon dioxide near 15 microns. Each wavelength sampled a specific altitude region in the atmosphere, depending on the heat-absorbing traits of the carbon dioxide molecule and the temperature variation with altitude. One detector exclusively detected and mapped water vapor distribution in the upper atmosphere. This device centered on the strongest part of the pure rotational band of water vapor at 40 to 50 microns. Another instrument, operating in the 2.0-micron band of carbon dioxide, measured cloud layer size and shape. The wide-band albedo channel from 0.2 to 4.5 microns measured total solar reflectance.

A 48-mm (1.9-in.) aperture parabolic mirror gathered radiation for all eight channels of the instrument. The instrument's axis was at 45° to the Orbiter's spin axis. This position allowed rotation of the spacecraft to scan the instrument's field of view across the planet. When looking at the limb of the planet, the instrument provided a vertical resolution of 5 km (3 miles) at periapsis.

Unfortunately, on February 4, 1979, the radiometer malfunctioned after 72 orbits, and could no longer be operated.

Airglow Ultraviolet Spectrometer

The airglow ultraviolet spectrometer mapped and made spectroscopic analyses of ultraviolet light that Venus' clouds and gases scattered or emitted. The instrument (Figure 4-4) weighed 3.1 kg (6.8 lb) and required 1.7 W of electrical power. The principal investigator was A. I. Stewart, University of Colorado.

How the planet's clouds and atmosphere reflect ultraviolet sunlight depends on the details of the size and makeup of cloud aerosols. It also depends on distribution of ultraviolet-absorbing gases. Both spectral intensity (how the brightness of the light varies with its wavelength) and maps, or images, carry the "finger-print" of these factors. Analysis of such information reveals three-dimensional details of the distribution of clouds, hazes, and gases. From images they made on successive days, investigators traced the variations and movement of gas bodies and cloud markings that can be seen only in ultraviolet light.

Absorption of extreme ultraviolet radiation from the Sun by the upper atmosphere's gases causes a fluorescence known as "airglow." Each gas has its own special emissions, and each of the many physical and chemical processes involved in airglow has its own characteristics, too. By measuring emissions, experimenters sought to learn how the Sun's radiation modifies the upper atmosphere's composition and temperature.

One of Venus' big mysteries is why it lacks water. The ultraviolet spectrometer helped solve this problem. It measured the emission of Lyman-alpha radiation from hydrogen atoms that form a corona around Venus. Scientists used this measurement to derive the amount of hydrogen that must be escaping from the top of the planet's atmosphere. The information is important because escaping

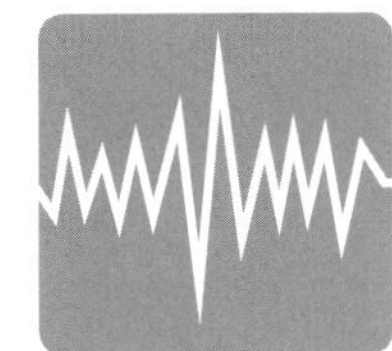

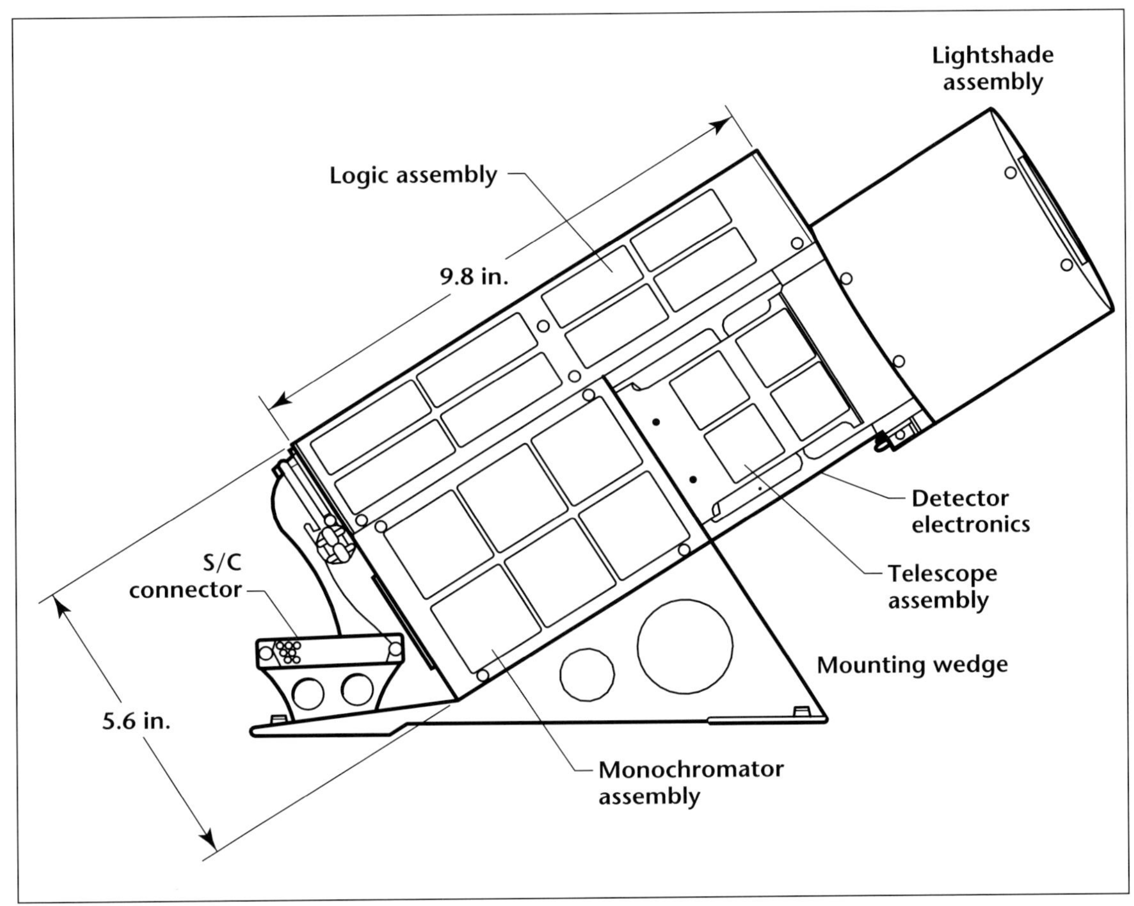

Figure 4-4. Orbiter ultraviolet spectrometer (OUVS). The diagram identifies the instrument's major components.

atomic hydrogen is the last step before a planet loses water. Incoming solar radiation breaks water into hydrogen and oxygen by photolytic processes. The oxygen is too heavy to escape from a planet the size of Venus, but hydrogen can escape into space from the top of the atmosphere. Yet, this process does not account for the extreme dryness of Venus today. Scientists needed much more information about conditions on Venus.

The spectrometer featured a 5-cm (2-in.) aperture f/5 Cassegrain telescope, protected by a light shade. It had an f/5, 12.5-cm (5-in.) focal length monochromator of Ebert-Fastie design. The monochromator used a diffraction grating with 3600 grooves per millimeter. A programmable step motor commanded from Earth was used to select the desired wavelength for each observation. The spectral resolution was 13 angstroms, and each grating step was 4.4 angstroms. (An angstrom is a unit of wavelength equal to 10^{-8} cm; this is approximately the diameter of a hydrogen atom.) Two exit slits passed the dispersed light from the monochromator to two photomultiplier tubes. They converted the light from Venus into electrical impulses that the spacecraft then telemetered back to Earth.

One photomultiplier had a cesium iodide cathode with a lithium fluoride window. It was sensitive to the wavelength range from 1100 to 1900 angstroms. The other had a cesium telluride cathode and a quartz window and was sensitive from 1800 to 3400 angstroms.

The instrument could operate in several modes. In the spectral mode, it scanned the complete spectrum in four 256-word sections. Each section was acquired in 1 second and required one or more complete spins of the spacecraft to transmit it to Earth.

The spectrometer performed mapping and imaging in the wavelength mode. In that mode, commands from Earth selected the grating position to choose the wavelength, the detector tube, and the length and location of the data arc. If the instrument command system or data memory failed, backup modes with lesser capabilities were available to ensure data collection.

On a typical orbit, the ultraviolet spectrometer viewed the planet from 150 to 35 minutes before periapsis. It viewed the planet again 15 minutes before periapsis to 10 minutes after. The first period gathered airglow and cloud images. The second obtained data for studying limb airglow profiles and limb hazes. For the rest of the orbit, the instrument observed bright, hot stars for calibration purposes. Measurements could be made of Lyman-alpha radiation emitted by hydrogen atoms throughout the orbit.

The principal investigator's objectives for the extended mission included continued measurements of Venus and measurements of selected comets. The spectrometer mapped and monitored the distribution of two components of the dayside Venusian thermosphere. One was the horizontal distribution of atomic oxygen. The other was the horizontal and vertical distribution of carbon monoxide. The aim was to characterize circulation properties and the role of vertical eddy mixing in this region. The instrument also determined dependence on solar activity of the dayside and nightside circulation patterns within the thermosphere. Another aim was to determine long-term behavior of sulfur dioxide in the cloud tops. Additionally, Phase II operations showed how Venus' hydrogen corona responded to changes in solar activity during an entire solar cycle.

Comets were a target of opportunity for this instrument. As a result, it had a number of research objectives that were finding out how comets lose water, the ratios of carbon, oxygen, and hydrogen, the rotation rate of the nucleus, and the extent and nature of the ultraviolet emitting coma.

Neutral Mass Spectrometer

The neutral mass spectrometer (Figure 4-5) was one of two mass spectrometers that the Orbiter carried. It weighed 3.8 kg (8.4 lb) and required an average of 12 W of electrical power. It measured the densities of neutral atoms and molecules in an upper atmosphere range. That range extended from near periapsis to a maximum altitude of 500 km (311 miles). Principal investigator was H. B. Niemann, NASA Goddard Space Flight Center. Information about vertical and horizontal distributions of neutral gas molecules was important. Scientists could use it to define the chemical, dynamical, and thermal state of Venus' upper atmosphere. Researchers also were able to determine the height above the planet's surface at which atmospheric mixing ends. (This region is the turbopause.) They did this by comparing inert gas densities at altitudes accessible to Orbiter with densities that the Large Probe and the Multiprobe Bus measured below 150 km (93 miles).

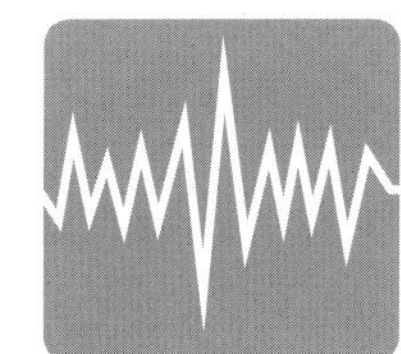

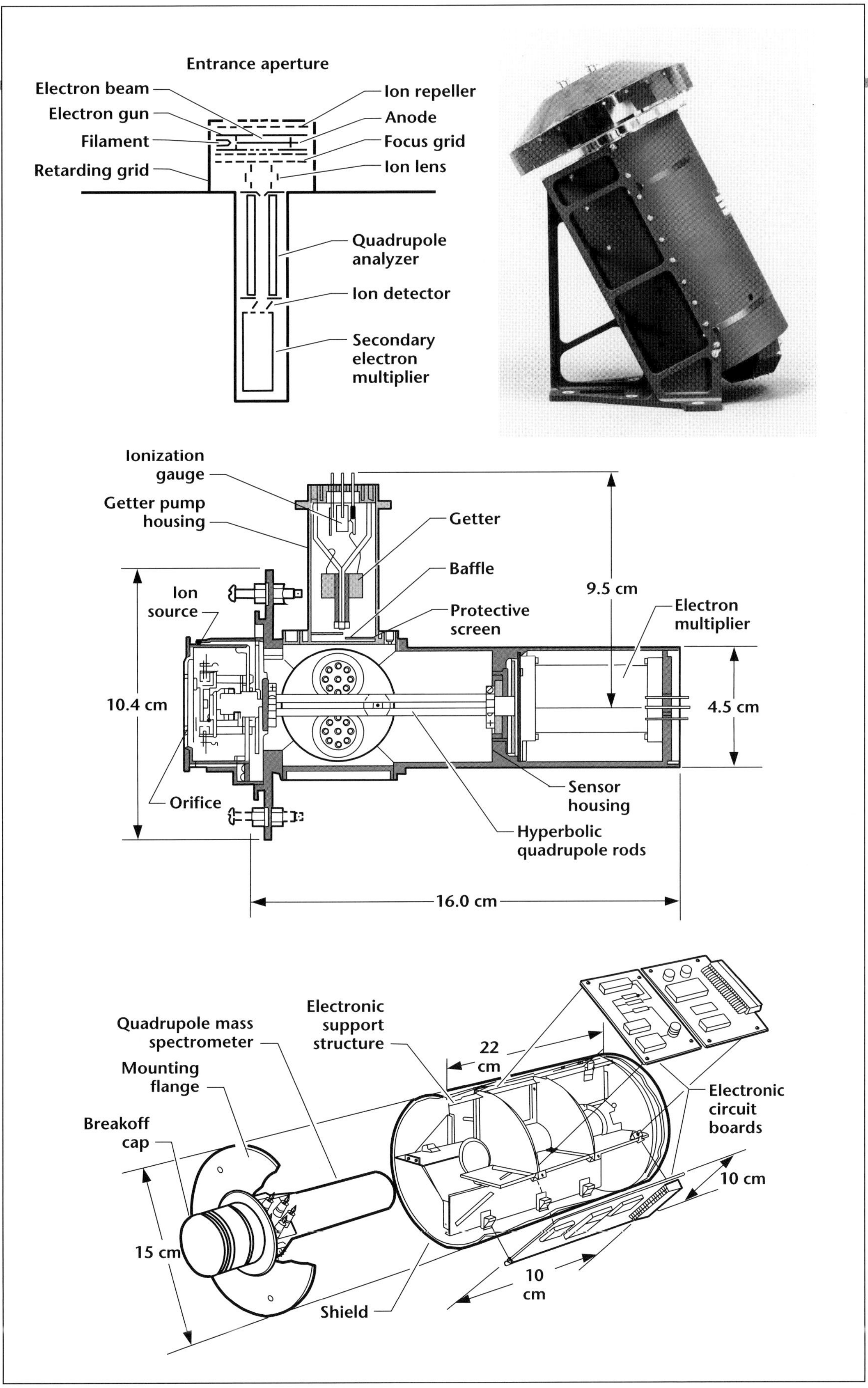

Figure 4-5. Orbiter neutral mass spectrometer (ONMS). (Top left) A simplified diagram of the instrument with entrance orifice at top. (Top right) Photograph of assembled instrument. (Middle) A more detailed diagram showing whole instrument in longitudinal cross section. (Bottom) A cutaway perspective of the instrument with spectrometer separated from its housing. The breakoff cap covering the inlet during cruise to Venus is to the left of this drawing.

Investigators identified and measured noble gases, other nonreactive gases, and chemically active gases of up to 46 atomic mass units. They used a quadrapole mass spectrometer with an electron-impact ion source and a secondary electron multiplier ion detector. The instrument first ionized gas molecules. Then a quadrapole mass filter separated them according to their mass. The ion source was inside a chamber that connected to the outside atmosphere via a knife-edged orifice. It operated in two modes alternately: an open-source mode and a closed-source mode.

In the open-source mode, the device analyzed only those ions that came from ionization of free-streaming particles. Such particles had a large kinetic energy with respect to the Orbiter since it was moving through the atmosphere at nearly 10 km/sec (6.2 miles/sec) at periapsis. For atomic oxygen, this kinetic energy was about 8 eV. By contrast, it was about 0.025 eV for surface-reflected particles. A retarding potential analysis discriminated between surface-reflected and free-streaming particles after the electron beam had ionized them. To be effective near periapsis, the mass spectrometer's axis had to point in the general direction of the Orbiter's motion once per spin period. Researchers accomplished this by mounting the device on the spacecraft's instrument platform so its axis was 27° from the spacecraft's spin axis. In this mode, the instrument measured concentrations of chemically active gases, such as atomic oxygen.

In the closed-source mode of operation, almost all particles the instrument analyzed were surface-reflected particles. The gas density in the ion source was significantly enhanced because inflowing gas stagnated in the source chamber. This mode was suitable for determining concentrations of noble gases, such as helium, and of nonreactive gases, such as carbon dioxide and molecular nitrogen. Surface-reflected particles adjusted to the surface temperature before making multiple passes through the ionization region. As a result, this mode had enhanced sensitivity. This permitted measurements to much lower concentrations than was possible in the open-source mode.

To keep internal surfaces clean and allow instrument testing during launch preparations and cruise, a metal-ceramic breakoff cap covered the ion source. It maintained the internal pressure below 10^{-4} Pa (10^{-6} torr). A pyrotechnic actuator removed the cap after the spacecraft entered orbit.

Ground commands could program the mass spectrometer to scan continuously from 1 to 46 atomic mass units, or to scan any combination of eight masses within that range. The kinetic energy of the ionizing electrons could be chosen by ground command to be 70 or 27 eV, so constituents of equal mass could be discriminated during analysis.

In Phase I, the instrument made measurements within the neutral atmosphere below 250 km (155 miles) in the planet's northern hemisphere. In Phase III, the orbit was oriented so measurements could be made in the southern hemisphere.

Solar Wind Plasma Analyzer

The solar-wind plasma analyzer (Figure 4-6) weighed 3.9 kg (8.6 lb) and required 5 W of electrical power. It measured the velocity, density, flow direction, and temperature of the solar wind, and its interactions with Venus' ionosphere and upper atmosphere. Principal investigator for the solar-wind plasma experiment was initially J. H. Wolfe, NASA Ames Research Center. A. Barnes, also of Ames Research Center, succeeded him.

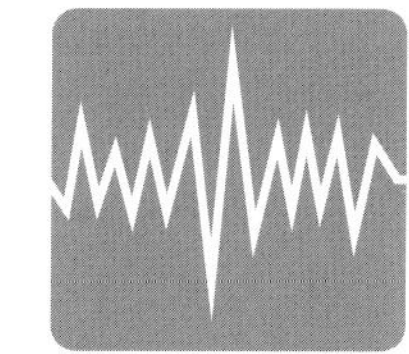

The plasma analyzer was an electrostatic, energy-per-unit-charge spectrometer. Mounted near the outer edge of the equipment shelf, the instrument had a field of view normal to the spacecraft's spin axis. The field of view rotated with the spacecraft. The rate of flow (flux) of the solar wind was measured by the deflection of incoming particles subjected to an electrostatic field between two metal plates. If the particles were within the range of energy and incidence determined by the aperture's orientation and the voltage between the plates, they exited to hit one of five detectors. Which target detector a solar-wind particle hit depended on the wind's direction. By varying the voltage between plates, scientists could measure a complete solar-wind particle velocity distribution.

The instrument's analyzer section was a nested pair of quadrispherical plates with a mean radius of 12 cm (4.72 in.). These plates were 1.0 cm (0.39 in.) apart. Charged particles, such as protons and electrons, that passed through the instrument's entrance aperture entered the region between the charged plates. There the electrostatic field deflected them into a curved path. Following this, an array of five current collectors—located at the curved plate's exit end—collected them. Each target was connected to an electrometer amplifier.

The instrument had two modes of operation that scientists could command from Earth: a scan mode and a step mode.

The scan mode first found the maximum flux over one spacecraft rotation for each voltage step. It then identified the collector and spacecraft azimuth of this maximum flow. The energy/charge range was normally 32 logarithmically equal steps over the range of 50 to 8000 V for high-energy positive ions, or 15 steps from 3 to 250 V plus a zero step at 0.25 V for electrons and low-energy positive

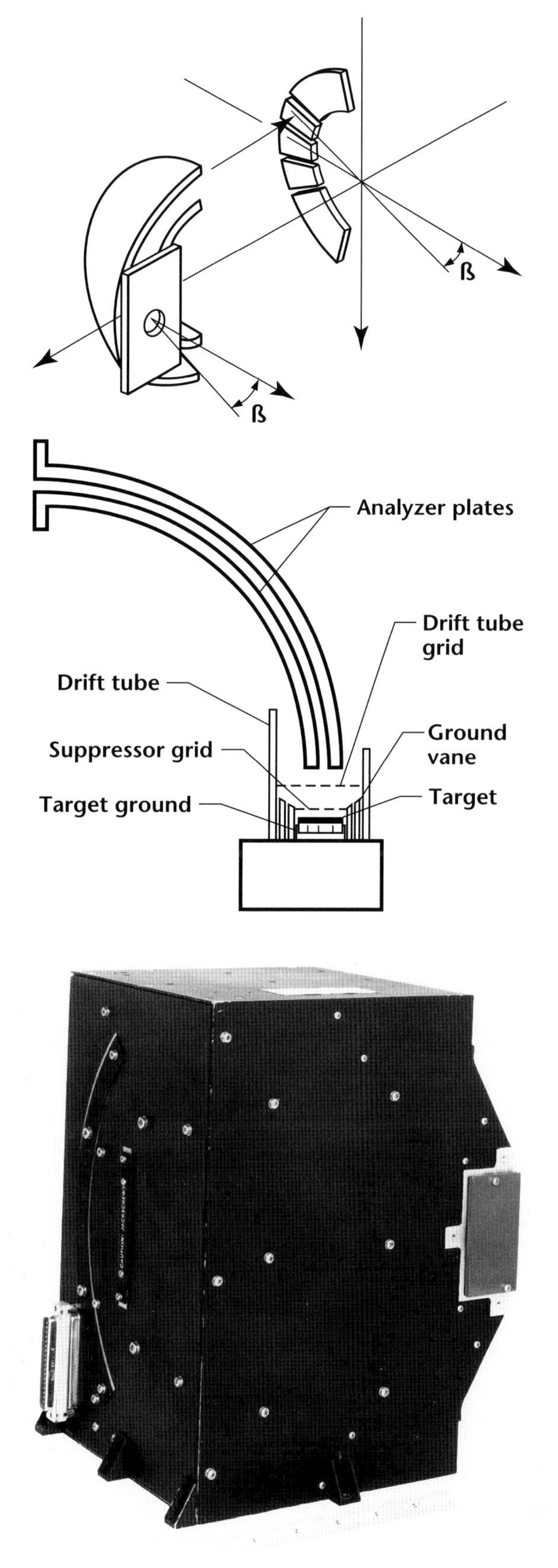

Figure 4-6. Orbiter solar-wind plasma analyzer (OPA). (Top) Diagram showing arrangement of the curved, electrostatically charged plates with respect to the five detectors that recorded velocity of solar-wind particles. (Bottom) Photograph of assembled instrument in its housing.

ions. Next, the instrument made a polar and an azimuthal scan. It did this at the four consecutive steps beginning with the step before the one in which it measured the peak flux. Each polar scan measured the flux at all five collectors at each step. All azimuth scans measured the flux in 12 sectors centered on the peak flux direction.

In the step mode, only maximum flux scan occurred, with about 1 second allocated to each voltage.

During the early part of Phase II, this experiment increased our understanding of conditions within the ionosheath. A more detailed knowledge of bow shock allowed calculation of how much solar wind the planet's ionosphere absorbs. In Phase III, the instrument gathered data similar to that in Phase I but at a different part of the solar cycle.

Magnetometer

A flux-gate magnetometer recorded Venus' extremely weak magnetic field. The instrument weighed 2 kg (4.44 lb) and required 2.2 W of electrical power. The principal investigator was C. T. Russell, University of California, Los Angeles. The magnetometer searched for surface-correlated magnetic features. These included regions of Venusian crust that might have been magnetized in the past. If present, the features would have shown that Venus once had a field more like Earth's. Although Venus' magnetic field is extremely weak, scientists thought that it might play an important part in the interactions between solar wind and the planet. Their aim was to clarify whether solar wind was deflected by a field intrinsic to Venus, by an induced field, or by the ionosphere itself.

The magnetometer consisted of three sensors mounted on a 4.7-m (15.4-ft) boom. The long boom isolated the sensors from the spacecraft's magnetic field. This feature allowed it to measure weak fields in the nanotesla (nT) or gamma range. (The field of Earth at its surface is about 50,000 nT.) Two sensors were at the end of the boom: one parallel to the spacecraft's spin axis and the other perpendicular to it. An inboard sensor, one-third of the way down the boom, tilted 45° to the spin axis. This inner sensor measured the Orbiter's magnetic field. This value was subtracted from the readings of the outboard sensors to correct them for the spacecraft's presence. Each sensor consisted of a ring, around which was wrapped a ribbon of permeable metal to form the sensor's core. It was surrounded with drive, sense, and feedback coils. Any external field caused the core to produce an electrical signal. A feedback signal then canceled the external field so the magnetometer always operated in a zero-field condition. The strength of the feedback signal needed to produce the zero-field condition was a measure of the external magnetic field.

Engineers designed the magnetometer so that it did not need gain changes when it moved to and from low- or high-field regions. The instrument's range remained fixed at 128 nT. The resolution, however, changed from 0.0625 nT to plus or minus 0.5 nT in response to field changes.

During Phase II, investigators had a number of objectives. Among them were gathering new information about solar-wind interaction with the equatorial ionosphere and determining how much material is lost into space from Venus' atmosphere. They also were interested in learning how energy moves from solar wind to ionosphere and about conditions in Venus' wake and tail. Phase II repeated the geometry

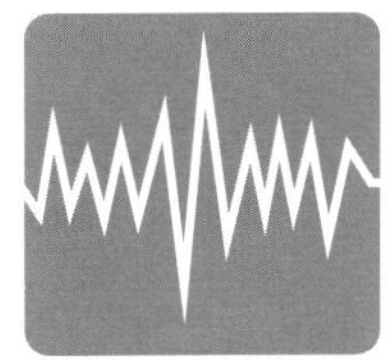

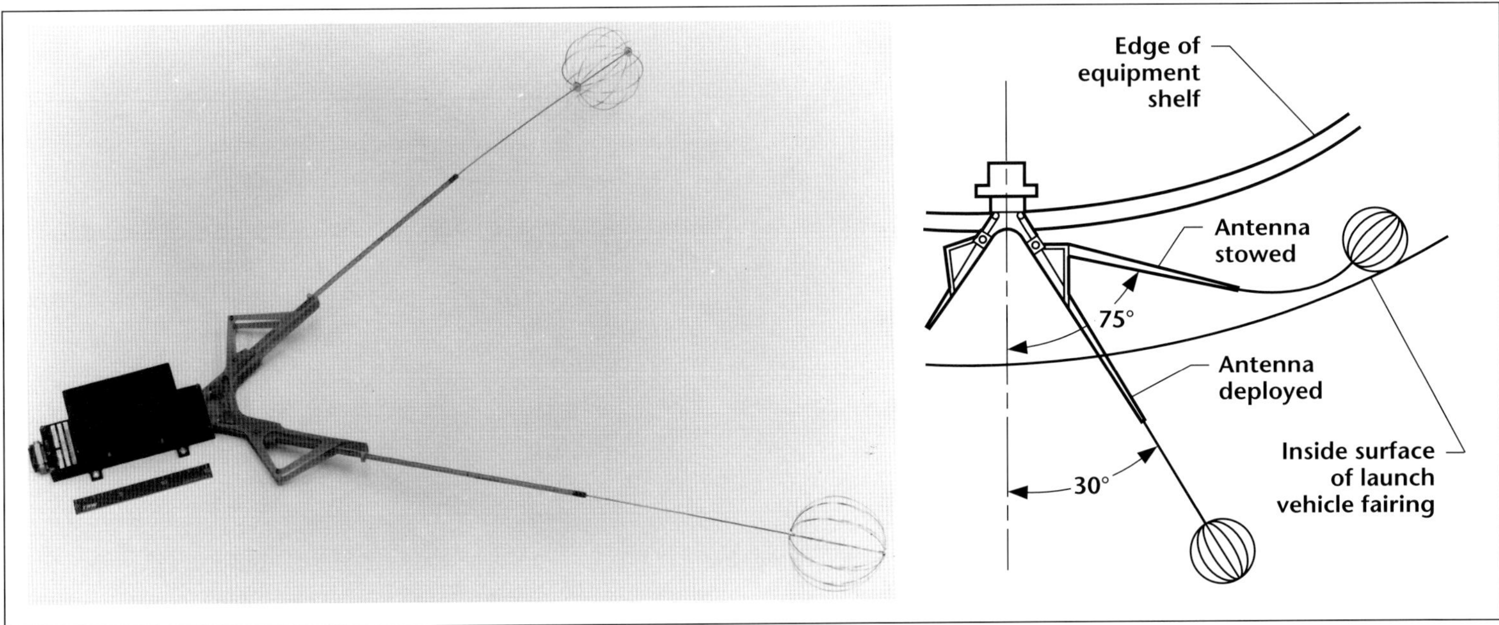

of Phase I but made measurements in a different part of the solar cycle. Further, at the end of Phase III, the Orbiter carried the magnetometer rapidly through the ionosphere. This allowed the instrument to obtain data, at least for several periapsis passages, at altitudes lower than in Phase I.

Electric Field Detector

Investigators designed the electric field detector (Figure 4-7) to answer questions about interactions between Venus and the solar wind. The instrument weighed 0.8 kg (1.76 lb) and required 0.7 W of electrical power. The principal investigator was initially F. L. Scarf, TRW Systems. R. J. Strangeway, University of California at Los Angeles, later replaced him. The instrument provided information about how Venus deflected solar wind around the planet and how much the solar wind heated the ionosphere. It also provided data about the extent of ionization that the exosphere-solar-wind interaction caused. It gave information about solar-wind turbulence, too. Additionally, it allowed scientists to measure variable locations of the bow shock, the ionopause, and the wake-cavity boundary.

The electric field detector measured electric components of plasma waves and radio emissions in the frequency region from 50 to 50,000 Hz. Currents were induced in a 66-cm (26-in.) long V-type electric dipole antenna, and they were amplified to relay information to Earth. Four 30% bandwidth channels, centered at 100, 730, 5400, and 30,000 Hz, were used. Each was needed at different points along the spacecraft's orbit when the Orbiter passed through varying densities of solar wind.

The instrument also searched for "whistlers," or electromagnetic disturbances that travel along a magnetic field line. Scientists designed the device so that it could detect electron whistler mode signals in the 100-Hz channel at all orbital locations.

Observations during Phase II paid valuable dividends. They provided an extended database to assess time variations and solar cycle

Figure 4-7. Orbiter electric field detector (OEFD). (Left) Photograph of V-type antenna and its mounting. (Right) Diagram to illustrate how antenna was stowed within the launch vehicle and antenna's position when deployed.

Figure 4-8. Orbiter electron temperature probe (OETP). Two Langmuir probes mounted outside the spacecraft and the electronics package.

effects on the instrument's measurements, and their implications.

Electron Temperature Probe

The electron temperature probe measured thermal properties of Venus' ionosphere. Measurements included electron temperature, electron concentration, ion concentration, and the spacecraft's own electrical potential. Scientists needed such measurements to help them understand how the ionosphere obtains heat. The principal investigator for the electron temperature experiment was L. H. Brace, NASA Goddard Space Flight Center.

The probe (Figure 4-8) weighed 2.2 kg (4.76 lb) and required 4.8 W of electrical power. It consisted of two cylindrical Langmuir probes: an axial probe and a radial probe. The former was mounted parallel to the spacecraft's spin axis at the end of a boom that was 40 cm (15.75 in.) long. The latter was mounted at the end of a 1-m (39.37-in.) boom that extended radially from the spacecraft's periphery. Each probe was 7 cm (2.8 in.) long and 0.25 cm (0.1 in.) in diameter. Both probes had their own power generator but shared in-flight data analysis circuitry.

A sawtooth voltage swept each probe twice a second. The voltage was electronically adapted to match the existing electron density and temperature being measured. The sweep amplitude varied automatically over the range 0.5 to 10 V, to suit the electron temperature being measured. Appropriate bias voltages were added to compensate for the spacecraft's potential. At the beginning of each sweep, automatic current-ranging circuits sampled this ion current. They adjusted the electrometer gain to suit the variations in ion concentration. The instrument's design included such adaptive functions so the resolution could be as large as possible over a wide range of electron concentrations and temperatures.

A commandable mode permitted sampling of one probe instead of alternating between two probes. This allowed experimenters to take advantage of having two probes that, because of their orientation, responded differently to changes in electron concentration while maintaining high spatial resolution.

During Phase II, this instrument investigated two important regions of Venus' ionosphere. The first included the ionopause, ionosheath, and bow shock in the front stagnation region.

The second included ionosphere, ionopause, and wake in the region immediately downwind from Venus. In Phase III, the southward drift of periapsis allowed the spacecraft to examine different ionospheric regions.

Ion Mass Spectrometer

The ion mass spectrometer (Figure 4-9) weighed 3 kg (6.6 lb) and required 1.5 W of electrical power. It measured the distribution and concentration of positively charged ions in Venus' atmosphere above 150 km (93 miles). The spectrometer was similar to the instrument the Multiprobe Bus carried. The principal investigator for both ion mass spectrometers was initially H. A. Taylor, Jr., NASA Goddard Space Flight Center (1974-1988). P. R. Cloutier, Rice University, replaced him. The instrument directly measured ions in a mass range from 1 (protons or hydrogen ions) to 56 atomic mass units. Scientists wanted the data for a greater understanding of Venus' ionosphere and its solar-wind interactions.

The basic measurement cycle was 6.3 seconds. The instrument first made an exploratory sweep of 1.8 seconds. This explore mode searched for up to 16 different ions. Then the instrument entered its adapt mode and made a series of sweeps for 4.5 seconds. The device repeated the sampling of the eight most prominent ions that it identified during the exploratory sweep. The instrument the Orbiter used had commandable modes to regulate its explore-adapt logic circuit. This allowed the number of prominent ions for adaptive repeats to be reduced from 8 to 4 or 2. A commandable option also allowed the spectrometer to remain in the explore mode.

In flight, the sensor—a Bennett-type radio-frequency ion mass spectrometer tube—encountered a stream of atmospheric ions. They flowed into an aluminum cylinder enclosing a series of parallel wire grids. Next, a variable negative sweep potential accelerated each ion species along the spectrometer's axis. Engineers programmed this process to step and then dwell at voltage levels needed to detect particular ions. In this way, ions that passed through the radio-frequency analyzer stages in phase with the applied voltage gained sufficient energy to penetrate a retarding direct-current field and impinge on a collector at the rear of the sensor cylinder. The ion stream's accelerating voltage yielded the identity of the ions and its amplitude revealed their concentration. A dual collector system that consisted of a low-gain grid collector and a high-gain solid disk collector detected ion currents.

In Phase II and Phase III, the instrument investigated the superthermal ion concentrations and flow properties in the upper altitude regions of Venus' wake. It also gathered data about structural details of superthermal ion distribution in the ionopause, ionosheath, and bow shock regions.

Charged-Particle Retarding Potential Analyzer

The charged-particle retarding potential analyzer measured temperature, concentrations, and velocity of the most abundant ions in the ionosphere. It also measured concentration and energy distribution of photoelectrons in the ionosphere, temperature of thermal electrons, and the spacecraft's potential. The analyzer provided experimenters with important data on plasma quantities in the ionosphere, planetary tail, and boundary layers surrounding Venus.

The instrument weighed 2.8 kg (6.2 lb) and required 2.4 W of electrical power. The principal investigator was W. C. Knudsen, initially with Lockheed Missiles and Space Company and later with Knudsen Research. The Fraunhofer Institut für Physikalische

Figure 4-9. Orbiter and Multiprobe Bus ion mass spectrometer (OIMS/BIMS). (Top) Schematic diagram to show sensor components and equations used to derive results. (Bottom) Photograph of assembled instrument with inlet to the left.

(1) Sensor at rest relative to plasma

$$M = \frac{K\,|V_a|}{S^2 F^2}$$

(2) Sensor moving relative to plasma

$$M = \frac{K\,(|V_a| - 1/2\,mv^2 + \phi_{sc})}{S^2 F^2}$$

Where:

- M = Mass of ion (amu)
- V_a = Accelerating voltage
- m = Mass of ion
- v = Sum of spacecraft and ion velocities
- ϕ_{sc} = Spacecraft charge
- = Inter-grid spacing
- F = rf (radio frequency)
- K = Constant

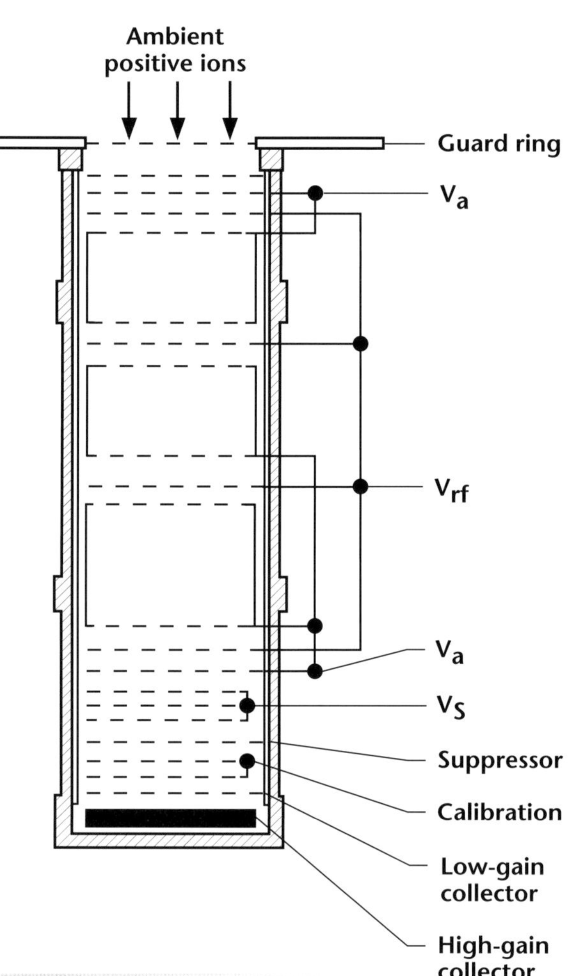

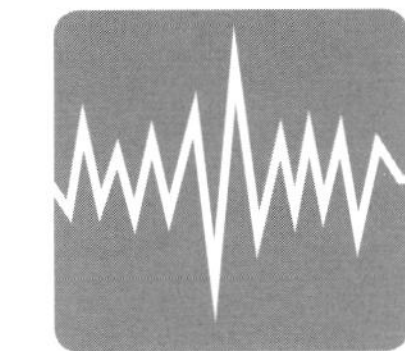

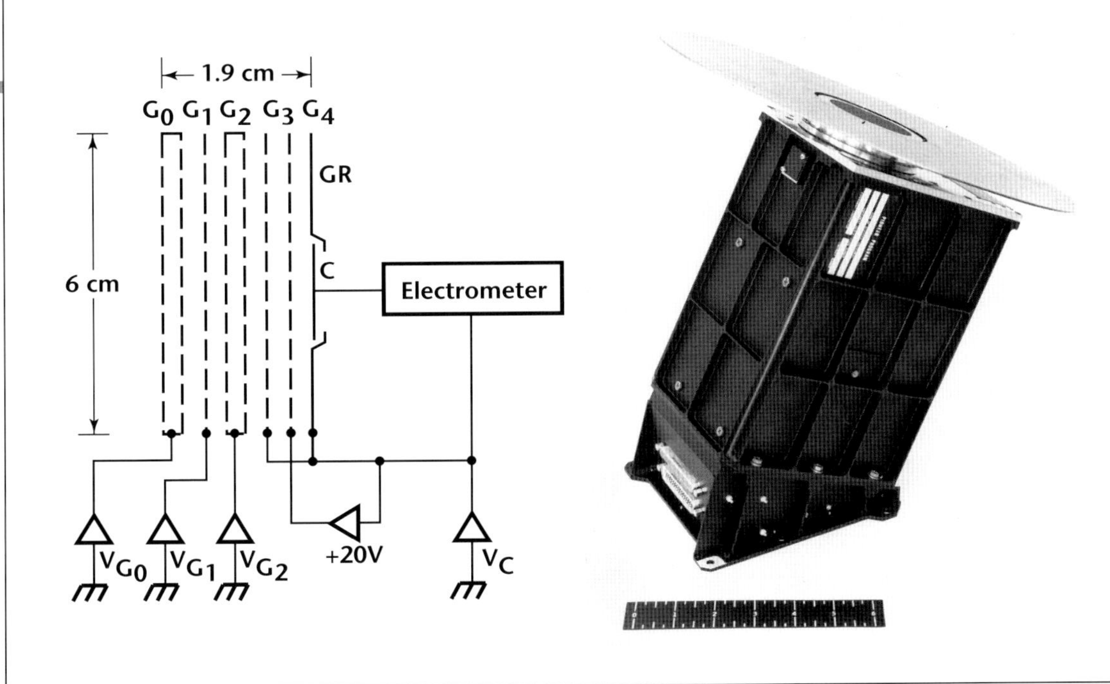

Figure 4-10. Orbiter charged particle retarding potential analyzer (ORPA). (Left) Simplified diagram to illustrate arrangement of grid system and detectors. (Right) Photograph of complete instrument with circular ground plane at top. The scale at bottom of the instrument is 15 cm (6 in.).

Weltraumforschung, West Germany, developed and fabricated the instrument's sensor.

The instrument (Figure 4-10) detected low-energy plasma particles in Venus' ionosphere, as opposed to the much more highly energized solar-wind particles. Nevertheless, the analyzer did provide data about the interaction between ionosphere and solar wind at an altitude of 400 to 500 km (249 to 311 miles). This is the level where solar-wind streams into the ionosphere.

Because of their varying electrical potentials, 6-cm (2.4-in.) diameter collector grids selectively allowed various ionospheric particles to strike a detector. An electrometer-amplified current was induced in the detector. Large entrance grids and a collector guard ring provided a uniform flux radially from the instrument's axis. The collector sampled the central region of this flux. Multiple retarding grids, coated with colloidal graphite, kept systematic error low. Surrounding the entrance grid was a 30-cm (11.8-in.) diameter ground plane. This ensured that the plasma sheath remained planar even at low electron concentrations.

By applying control voltages and a special program, the investigator could operate the instrument in three modes. These modes were an electron Langmuir probe mode, an ion mode, and a photoelectron mode. Onboard data analysis by the instrument selected the optimum point in the spacecraft's rotation to sample the plasma. Each scan occupied a small fraction of a spin period. The device took scans repeatedly, sensing, storing, and transmitting to Earth scans for which it was optimally oriented. Scans were typically spaced at 120-km (75-mile) intervals along the orbital path.

By recording three scans as it pointed to three different celestial longitudes in three successive spin cycles, the instrument measured vector ion velocity. The investigator could command a special operation mode to measure total ion concentration at 20-m (66-ft) intervals.

During Phase II, changing orbital properties allowed several series of observations. Sampling at higher altitudes aided a search for the source of ion heating in the nightside ionosphere. Scientists also investigated how the mantle region developed downstream from the planet. They looked for the source of nightside ionization and superthermal electrons. Another source they sought was for ion heating on the dayside at altitudes between 150 and 170 km (93 and 105 miles). Additional information was needed about the nature of the mantle at the subsolar point, how ions are accelerated across the terminator, and characteristics of the plasma within flux ropes.

During Phase III, this experiment provided information about the relative roles of solar protons and solar wind in several Venusian phenomena. This was possible because scientists could compare these measurements with those obtained earlier at a different part of the solar cycle.

Gamma Ray Burst Detector

Its designers did not intend the gamma ray burst detector to obtain information about Venus. Onboard the Pioneer spacecraft in orbit about the planet, the detector provided another set of important data concerning the intense short-duration bursts of high-energy photons from beyond the Solar System. Lasting from one tenth to a few tenths of a second, these bursts were first observed by scientists in 1973. They occurred randomly, roughly 18 bursts each year, and their source was a mystery. The Orbiter provided a means to obtain a direction for the bursts. It achieved this by correlating observations from Venus with simultaneous observations from Earth-orbiting satellites. Several years of high quality observations from the Pioneer Orbiter contributed much to the early stages of these astronomical observations.

The instrument (Figure 4-11) weighed 2.8 kg (6.17 lb) and required 1.3 W of electrical power. It consisted of two sodium-iodide photomultiplier detector units to provide a near uniform sensitivity over a wide field of view. These detectors were sensitive to photons with energies between 0.2 and 2.0 MeV. To accommodate high data rates that occurred during intense gamma ray bursts, the experiment included a 20-kilobit buffer memory for storing the data until they could be telemetered to Earth at a lower rate.

The principal investigator for the gamma ray experiment was initially W. D. Evans, Los Alamos Scientific Laboratory (1974-1982). R. W. Klebesadel, also of Los Alamos Scientific Laboratory, replaced him.

Orbiter Radio Science Experiments

The experiments connected with instruments on the spacecraft were not the only experiments. There were several investigations that involved radio signals exchanged between the Orbiter and Earth. The team leader for these investigations was G. H. Pettengill, Massachusetts Institute of Technology.

Radio science experiments included the following: occultation studies by A. J. Kliore, Jet Propulsion Laboratory, and T. A. Croft, SRI International; internal density distribution of Venus by R. J. Phillips, Jet Propulsion Laboratory; celestial mechanics by R. Reasenberg, Massachusetts Institute of Technology; atmospheric and solar-wind turbulence by R. Woo, Jet Propulsion Laboratory; solar corona by T. A. Croft, SRI International; and atmospheric drag by G. M. Keating, NASA Langley Research Center.

Radio science experiments used the spacecraft's Doppler tracking system. An antenna of the DSN transmitted a microwave signal at a frequency of about 2.1 GHz. When the spacecraft received the signal, it phase-coherently multiplied it by 240/241 and then retrans-

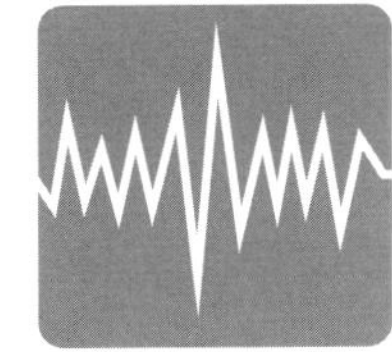

mitted the signal. This frequency multiplication allowed the spacecraft receiver to detect the incoming signal while its transmitter was operating. It was able to discriminate between the two signals. The frequency multiplication also served a similar purpose for the ground station.

When the DSN received the signal, it mixed it with another locally generated signal. This process produced a video signal, offset by a known frequency from that resulting from the Doppler effects. The Doppler shift was then reconstructed from this biased Doppler video signal. The ground station counted the biased Doppler signal cycles. The differences between uniformly spaced samples of the cycle count divided by the count interval and corrected for the effects of the known frequency offset, provided the primary Doppler data. These data approximated the average rate of change for the range between the ground station and the spacecraft, and thus contained information about the spacecraft's acceleration.

Most of the observed Doppler shift was due to the relative motions of Earth and Venus. The mean elliptical trajectory of the Orbiter accounted for the greater part of the remaining Doppler shift. Scientists attributed a significant part of this Doppler shift to perturbations in the spacecraft's trajectory. They attributed a smaller part to direct effects of the propagation media. Several factors caused the trajectory perturbations. These factors included other planets and the Sun, atmospheric drag effects, and irregularities in Venus' gravitational potential. Analysis of Doppler data provided a model of these irregularities.

The Doppler shift that the propagation media caused had several components. Each component originated from a different location: Earth's troposphere and ionosphere, the solar corona and plasma that flows from it, the interplanetary medium, and, for some geometries, Venus' neutral atmosphere and ionosphere. Some of the radio science experiments concerned the characterization of components of the propagation media.

In addition to transmitting an S-band signal at 2.293 GHz, the spacecraft could transmit an X-band signal at 8.407 GHz. This latter signal also was phase coherent with the S-band signal. This X-band signal was received and processed on the ground in the same way as the S-band signal. The propagation delay at a given frequency caused by charged particles (plasma) was inversely proportional to the square of the frequency. This allowed investigators to use the dual-band spacecraft transmissions to measure the change of the total charged-particle content of the path from the spacecraft to the ground station.

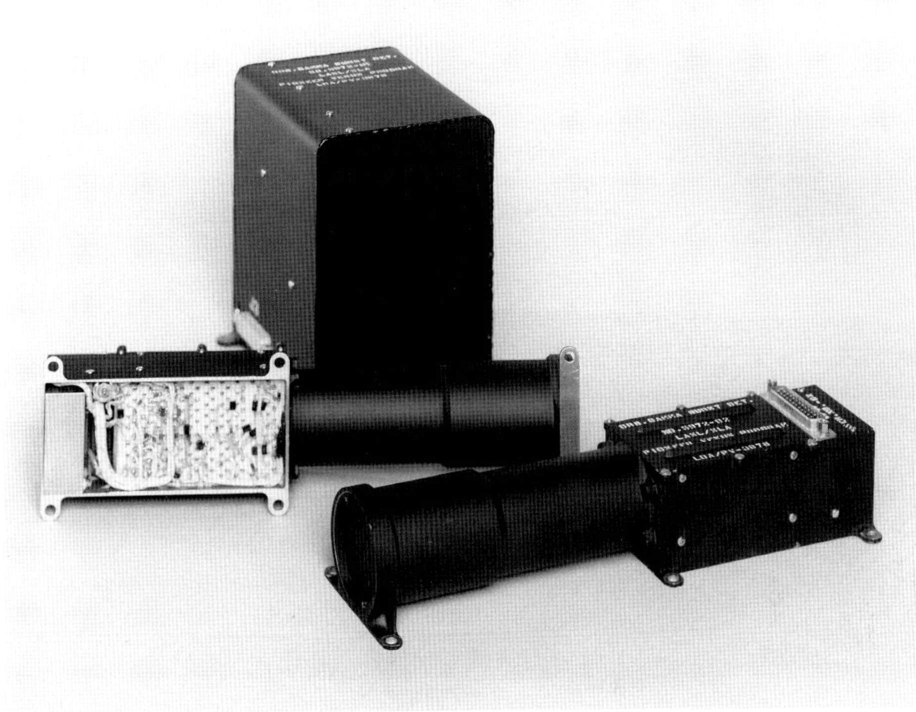

▲ *Figure 4-11. Orbiter gamma ray burst detector (OGBD). The photograph shows two photomultiplier detector units, one without cover to reveal associated electronics. Also, in the background is the electronic data processing package.*

Internal Density Distribution Experiment

In the internal density distribution experiment, researchers studied the relationship between Venus' surface features and internal densities. In their study, they used data about Venus' shape and the spacecraft's gravitational perturbations. The Orbiter's two-way Doppler tracking data allowed these researchers to infer the planet's gravity field. When used with topographic data obtained from the radar experiment, the gravity data provided a constraint on the internal density distribution. Geophysicists also used these data. It allowed them to investigate whether there were any continuing physical processes taking place within Venus similar to those moving Earth's crustal plates. This experiment was practical only during Phase I and Phase III when periapsis occurred at low altitudes.

Celestial Mechanics Experiment

The celestial mechanics experiment used the spacecraft's radio tracking system and its onboard radar system. Doppler tracking produced data to develop a high-resolution map of Venus' gravitational potential. This map, which showed the irregularities in gravity's vertical component at Venus' surface, correlated with topography from the onboard radar. Researchers could compare the topography and gravity in the spatial frequency domain. This comparison, in turn, yielded the spectral admittance, which provided a constraint on Venus' near surface structure. Investigators also used the Doppler tracking data to study the time-variable structure of Venus' upper atmosphere.

Simultaneous radio tracking of the Orbiter with extragalactic radio sources allowed precise determination of Earth and Venus' orbits with respect to these sources. This experiment occurred only during Phase I when periapsis occurred at low altitudes. It was not repeated during Phase III. During Phase II, the increasing altitude of periapsis, coupled with lack of propulsive maneuvers and atmospheric drag, allowed researchers to measure the shape of the gravitational field globally.

Dual-Frequency Radio Occultation Experiment

The dual-frequency radio occultation experiment provided information about Venus' atmosphere. This was done by observing how the Orbiter's S-band and X-band radio signals penetrated the planet's ionosphere and neutral atmosphere just before and after occultations. Observation data from multiple occultations with Pioneer were very rewarding. There was far more data than from earlier observations of a single spacecraft passing behind a planet during a flyby. Each occultation recorded Doppler frequency shifts and changes in signal strength caused by refraction and absorption by the planet's atmosphere.

The Orbiter's repetitive path was practically unchanging in orientation to Venus. However, motions of Venus and Earth around the Sun precessed the occultation points around the limb of the planet. During the nominal mission, 80 occultations sampled the atmosphere and ionosphere. Samples were over all latitudes from the North Pole to about 60° south latitude. Nearly all observations were, however, in Venus' night hemisphere. Observations not in the night hemisphere were at polar latitudes. It was during the extended mission that investigators acquired data on the day side.

During occultations, ground control aimed the Orbiter's high-gain antenna precisely to ensure that radio signals traveled to Earth after they had been refracted by Venus' atmosphere. In this way, there was maximum penetration of the atmosphere by the signals. From this deep penetration, scientists could identify and define microwave absorbing cloud layers.

Analyzing the Doppler frequency variations in the radio signals revealed much. For example, investigators determined the structure, the index of refraction, temperature, pressure, and density of the atmosphere above 34 km (21 miles). Radio signal refraction at Venus is so strong that any level ray that penetrated below 33 km (20.5 miles) curved down to hit the surface and became useless for this study.

Atmospheric and Solar-Wind Turbulence Experiment

The atmospheric and solar-wind turbulence experiment observed turbulence of scale sizes smaller than 10 km (6 miles) in Venus' atmosphere above 34 km (21 miles). Experimenters sought the global distribution of this turbulence. Their experiment also revealed fluctuations in the ionosphere's electron density.

Detailed information about the atmosphere was obtained just before and after occultation. At these times, the radio signal passed through deep regions of the atmosphere on its way from the Orbiter to Earth. Signal scintillations, akin to the twinkling of stars for Earth-bound observers, occurred during the passage. These scintillations revealed variations in the atmosphere's density and the presence of atmospheric layers.

For this experiment, the ground station made a wide-band linear recording in the frequency interval known to contain the signal. Subsequently, the signal was detected by a digital computer simulation of the phase-lock loop in a receiver acting on a digitized record of that wide-band signal plus noise. The digital approach was superior to ordinary, analog radio signal detection in many respects. This was particularly true when it involved critical scientific applications.

Scientists applied advances in phase scintillations and spectral-broadening measurements to study solar wind. They made these measurements after they had completed the nominal mission. At that time, Venus, with the Orbiter, approached superior conjunction. The spacecraft's radio waves then passed close to the Sun on their way to Earth. This was an ideal time to investigate solar wind near the Sun. Because the wind is variable, repeated observations provide information about its density, turbulence, and velocity. Two DSN stations simultaneously recorded fluctuations in the S-band and X-band signals as the signals passed through the solar wind.

Scientists compared Pioneer Venus data from the inner Solar System with data from Voyagers 1 and 2 and Pioneer Saturn spacecraft in the outer Solar System. Their comparisons formed the basis for a special period of international collaborative solar corona observations. This was the first scheduled event of the Solar Maximum Year.

Atmospheric Drag Experiment

The atmospheric drag experiment used drag measurements made for the first time within another planet's atmosphere. The aim was to model the upper atmosphere's mean behavior. It also included searching for variations in atmospheric density that correlated with solar-wind activity and changes in ultraviolet radiation. In addition, experimenters sought evidence that the four-day rotation extended into the upper atmosphere.

Investigators extracted drag effects from the spacecraft's estimated orbital parameters. The navigation team obtained these parameters from the S-band tracking data. By use of an *ad hoc* model, experimenters determined atmospheric density at each periapsis. This was where drag was greatest. Scientists evaluating the atmospheric density model relied on the periodic variation of the spacecraft's periapsis altitude. They determined the drag coefficient in free molecular flow from two observations. The first was the spacecraft's orientation relative to the flightpath. The second was an estimate of the atmosphere's composition. Scientists inferred temperature and composition variation with altitude and time from several factors. These were density, density scale height, and knowledge of dominant atmospheric components. Further analysis yielded models of pressure gradients and flow patterns. Since it required that periapsis should be at low altitude, this experiment was useful during Phase I and Phase III only.

Phase III provided good atmospheric drag measurements near the terminator on the nightside. Scientists compared this information with Phase I to show how the neutral upper atmosphere and the helium-rich regime above 200 km (124 miles) changes. They also used it to show what happens to the lower cryosphere's vertical structure and variability.

Multiprobe Bus Experiments

After its four probes separated 20 days before reaching Venus, the Multiprobe Bus also became a probe. It provided important information on the density and composition of Venus' high atmosphere, in particular for the altitude range from 150 to 130 km (93 to 81 miles). For this experiment, the Multiprobe Bus carried two mass spectrometer instruments. Each instrument was on the equipment shelf with its inlet projecting over the flat top of the spacecraft cylinder.

Neutral Mass Spectrometer

Between about 700 km (435 miles) and 130-km (81 miles) altitude, the neutral mass spectrometer measured the components (atoms and molecules) of Venus' high atmosphere. The Bus did not have protective thermal shields, so there was no way to prevent or delay its destruction by atmospheric heating as it plunged at high speed into Venus' upper atmosphere. It could not penetrate much below 130 km (81 miles).

The spectrometer weighed 6.5 kg (14 lb) and used 5 W of electrical power. The principal investigator for this experiment was Ulf von Zahn, University of Bonn, Germany.

From information gathered by this instrument, the investigator derived the height of the turbopause, or homopause. Above this region, the atmospheric gases do not mix, but become stratified as the lightest gases congregate toward the top of the atmosphere. The data also revealed chemical composition of the ionospheric region where density is greatest. An additional discovery was the temperature of the exosphere, the atmosphere's outer fringe.

The neutral mass spectrometer (Figure 4-12) ionized atmospheric components by bombarding them with electrons. By deflecting them magnetically, the device separated the ions according to their masses, up to 46 atomic mass units. The spectrometer featured a fast data sampling and telemetering capability. This feature allowed it to cope with the 3 km/sec (6700 mph) speed of the Bus' vertical descent (at an altitude of 150 km, or 93 miles). The Bus traveled much faster when it first entered the atmosphere. And, because it made a very shallow entry, most of its speed was in a horizontal direction.

One day before the Bus encountered Venus, a small glass vial released a known amount of gas into the spectrometer to calibrate it. The gas provided a reference to determine the instrument's sensitivity after its cruise through interplanetary space.

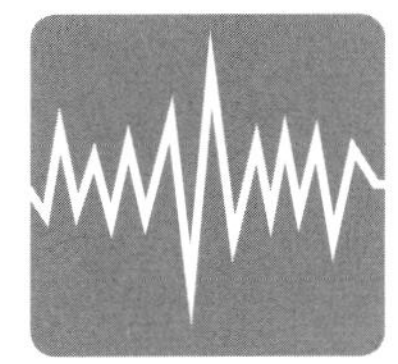

The instrument was a double-focusing Mattauch-Herzog electric and magnetic deflection mass spectrometer. Small and compact, it provided constant sensitivity at high pressures. The design also permitted use of a dual collector system for a large dynamic signal range.

The spectrometer had several major parts. One was an ion source, where electron bombardment ionized atmospheric particles. Another was an electric analyzer for mass separation of ions. The spectrometer also had a collector system, consisting of multiple elements. These elements enabled the system to collect ions of more than one mass at the same time according to their mass. Also, two detectors were Spiraltron electron multipliers. One detected ions between 1 and 8 atomic mass units, and the other detected ions between 12 and 46 atomic mass units. In addition, there was a titanium sublimation pump and an ion getter pump. These devices maintained a pressure differential of more than 1000 to 1 between the ion source and the mass analyzer.

The instrument first operated in a peak stepping mode. It sampled only tops of selected mass peaks and required zero levels. However, below altitudes of about 215 km (134 miles), the instrument operated for about 25% of the time in a fly-through mode. In this mode, it sampled only high-energy ions.

Ion Mass Spectrometer

The Bus' ion mass spectrometer was identical to the Orbiter's ion mass spectrometer. It measured the distribution and concentration of positively charged ions in the planet's upper atmosphere above 120 km (75 miles). The principal investigator was H. A. Taylor, NASA Goddard Space Flight Center. He also was principal investigator for the Orbiter's ion mass spectrometer experiment.

Large Probe Experiments

The Large Probe carried seven scientific instruments. A gas chromatograph and a mass spectrometer measured the composition of the atmosphere directly. A group of pressure sensors measured pressure directly, with inlet ports penetrating the probe's shell. The other five instruments observed through the probe's windows and sensed the probe's motion. They also measured temperature through externally mounted sensors.

An infrared radiometer required a diamond window because diamond was the only material transparent to the wavelengths of interest. It also was the only material capable of withstanding the high temperatures and pressures within Venus' lower atmosphere. The window was about 1.9 cm (0.75 in.) in diameter and 0.32 cm (0.125 in.) thick, or about the size of a quarter. It weighed 13.5 carats. Diamond cutters in the Netherlands shaped it from a 205-carat industrial-grade, rough diamond from South Africa.

A nephelometer used two sapphire windows. A cloud particle instrument also used a sapphire window, directing a laser beam through it. The beam traveled to an outside reflecting prism and then back to its sensor. A solar flux radiometer used five sapphire windows.

Neutral Mass Spectrometer

The neutral mass spectrometer (Figure 4-13) measured the composition of the lower 62 km (38 miles) of Venus' atmosphere. This region was mostly below the cloud layers. Information on the relative abundance of gases in this region was important. With it, scientists could better understand the planet's evolution, structure, and heat balance.

The spectrometer, which weighed 10.9 kg (24 lb) and required 14 W of electrical power,

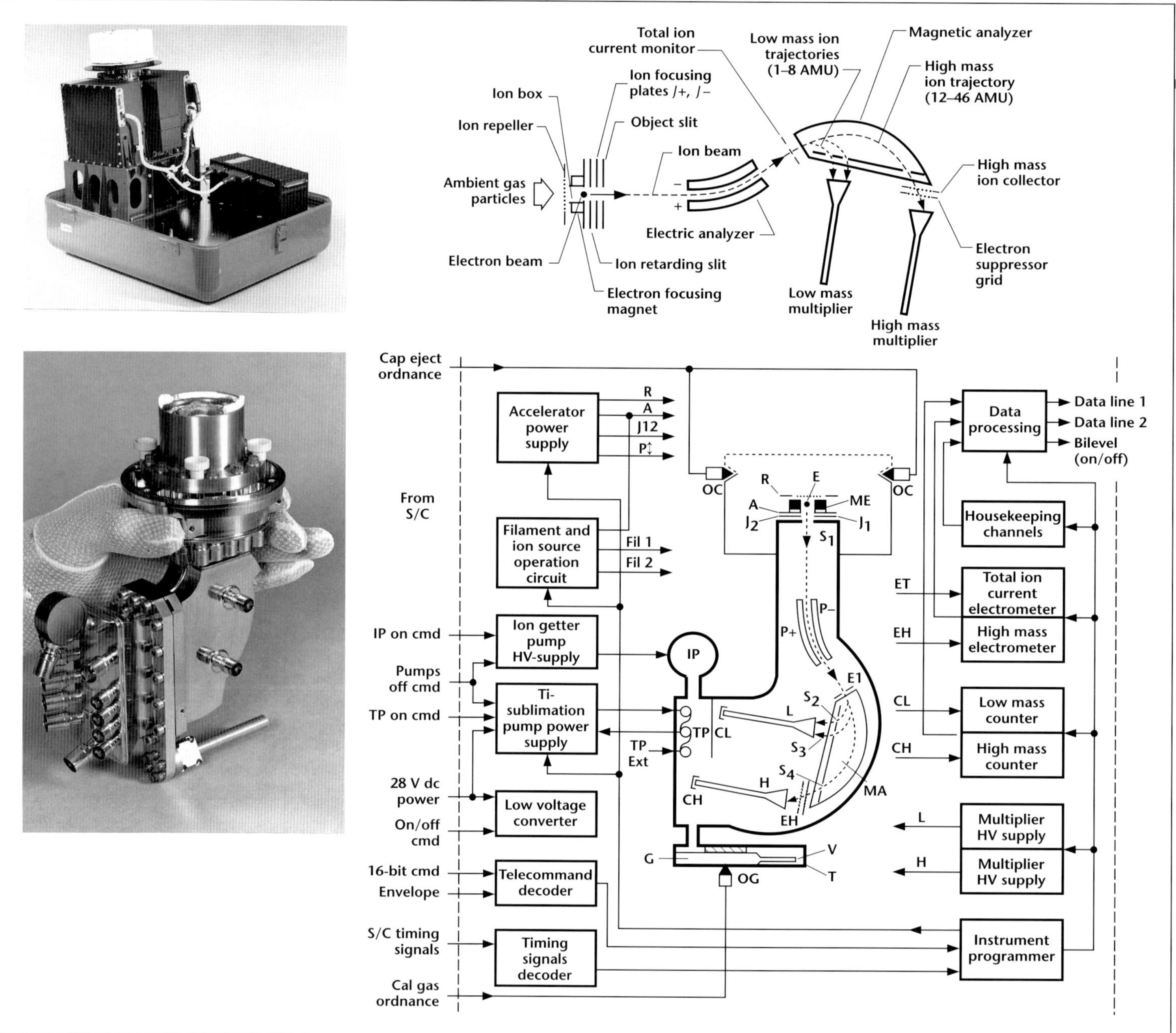

Figure 4-12. Multiprobe Bus neutral mass spectrometer (BNMS). (Top left) Mass spectrometer. (Bottom left) Electronics package. (Top right) Diagram to illustrate ionized gas path through the instrument to detectors. (Bottom right) Schematic of the instrument and its electronics.

consisted of two units. Both units were on a single baseplate on the probe's lower shelf. A mass analyzer, ion source, pumping system, isotope ratio measuring cell, and valves were in one unit. Electronics were in the other. The principal investigator was J. H. Hoffman, University of Texas, Dallas.

The instrument had wide dynamic and mass ranges to survey atmospheric gases and determine cloud composition. Its design made sure that the sampling process did not alter chemically active species. To prevent such alteration, it collected samples through a chemically passive inlet leak.

The inlet consisted of a pair of microleaks, each formed by compressing the tip of a tantalum tube into a slit. The tubes projected through the probe wall to beyond the

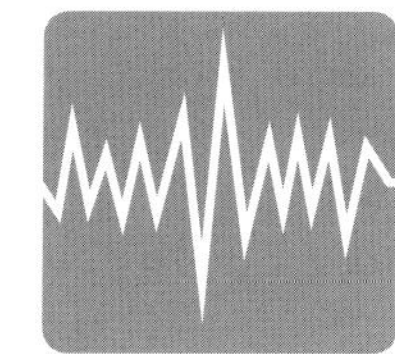

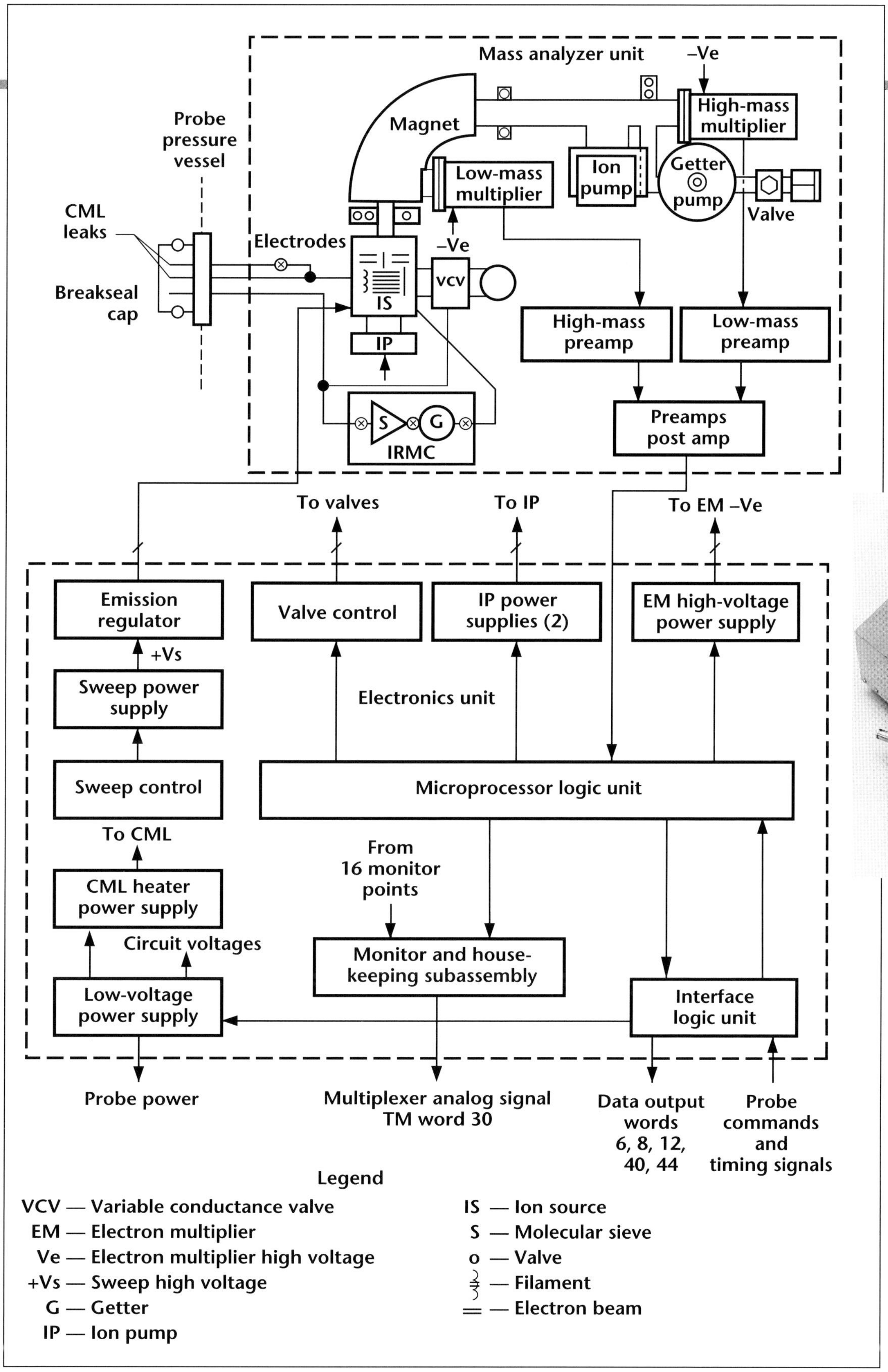

Legend

VCV — Variable conductance valve
EM — Electron multiplier
Ve — Electron multiplier high voltage
+Vs — Sweep high voltage
G — Getter
IP — Ion pump
IS — Ion source
S — Molecular sieve
o — Valve
(filament symbol) — Filament
═ — Electron beam

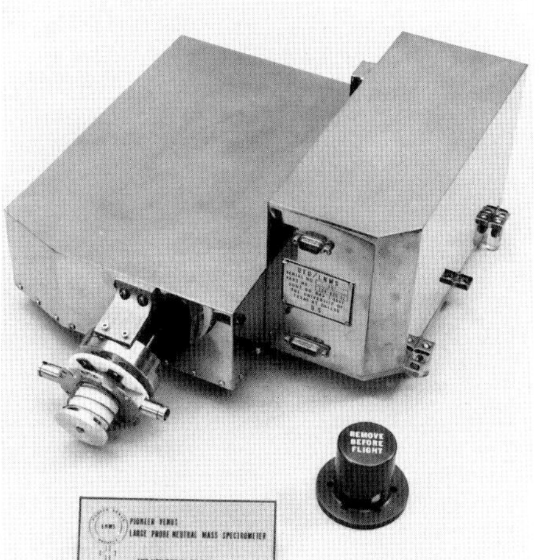

▲

Figure 4-13. Large Probe neutral mass spectrometer (LNMS). (Left) Schematic diagram of the instrument showing various functions in the operation of the spectrometer and its associated electronics. (Right) Photograph of assembled unit.

boundary layer. When the atmospheric pressure reached 1.5 bars, the tube with the larger conductance closed off. This prevented too large a sample deeper within the atmosphere when pressure increased rapidly. Atmospheric gases and vapors were pumped into an ion source through a variable conductance valve. During descent, the valve gradually opened to keep a constant pressure at the ion source. A magnetic sector field mass spectrometer analyzed the gas sample. Its range was 1 to 208 atomic mass units. The spectrometer detected minor constituents in 1-ppm concentration over the entire descent. To identify unknown substances and separate parent peaks from fragmentary ions, ionizing electron energy was stepped through three levels.

Each mass spectrum took 64 seconds to sample. A microprocessor controlled the mass scan mode, sequencing of ion source energy, and data accumulation and formatting. The instrument converted accumulated counts for each spectral peak into 10-bit, base-2, floating-point numbers. With a rate of only 40 bits/sec, the spacecraft successfully transmitted to Earth data from about 50 spectra obtained during the descent.

The instrument used an isotope ratio measuring cell to collect a sample shortly after the parachute's deployment. In this cell, the sample was purged of carbon dioxide and other active gases. After purging, an enriched sample of inert gases was left. Then the device pumped out the ion-source cavity and analyzed the sample to determine the isotope ratios of such inert gases as xenon, argon, and neon. All these gases are important for understanding how Venus' atmosphere evolved.

Gas Chromatograph

The gas chromatograph experiment also measured the gaseous composition of Venus' lower atmosphere. It was a modified version of the gas exchange experiment the Viking lander carried to Mars in 1976. It measured gases likely to be on Venus, with the aim of answering questions about Venus' evolution, structure, and thermal balance. The principal investigator was V. Oyama, NASA Ames Research Center.

The instrument (Figure 4-14) weighed 6.3 kg (13.9 lb) and required 42 W of electrical power. It sampled the lower atmosphere three times during the Large Probe's descent. During each sampling process, atmosphere flowed through a tube into a helium gas stream. This stream swept the sample into two chromatograph column assemblies. There the device identified atmospheric components by the time each took to flow through the columns.

A long column assembly consisted of a matched pair of 1585-cm (624-in.) packed columns bifilarly wound. Each column contained polystyrene (Porapak N) and operated at 18°C (64°F). A proportional heater surrounded by a shell of phase change material (n-hexadecane) controlled the temperature. The long columns were for gases with masses between those of neon and carbon dioxide.

A short column assembly also was part of the instrument. It consisted of similarly wound 244-cm (96-in.) columns. These columns contained a mixture of polymer spheres (80% polydivinyl benzene, 20% ethylvinyl-benzene). The materials remained at an operating temperature of 62°C (144°F). These short columns were for gases in the mass range from carbon dioxide to sulfur dioxide.

As the gases sequentially emerged from the columns, they passed to a thermal conductivity detector that generated data. These data remained in a buffer memory awaiting

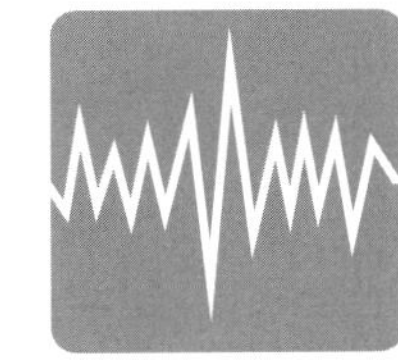

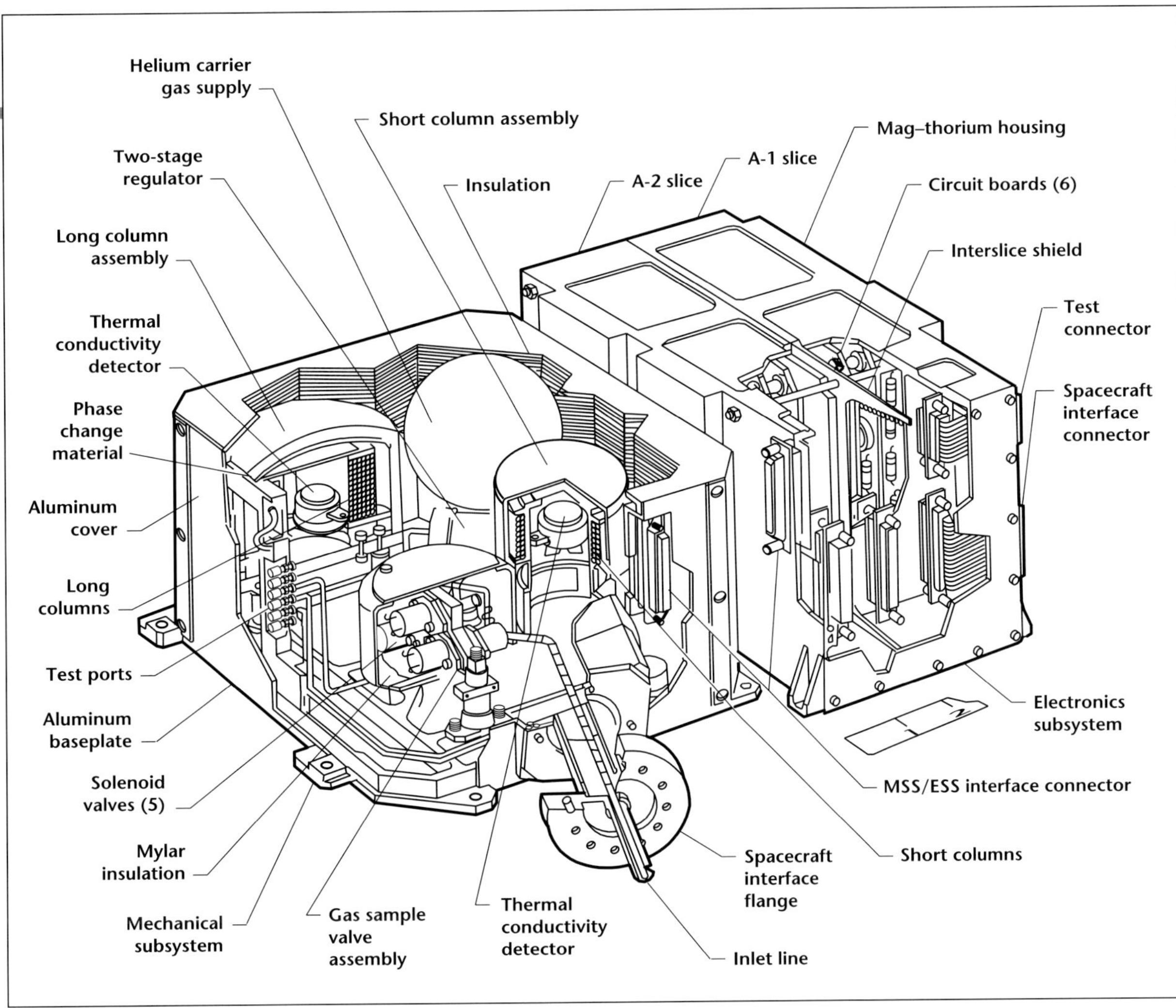

telemetry. As a calibration check, two samples of Freon (a gas not likely to be in Venus' atmosphere) were added to each third sample.

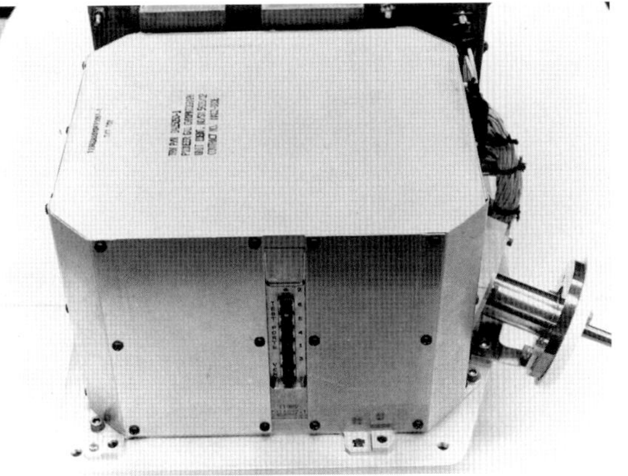

Figure 4-14. Large Probe gas chromatograph (LGC). (Top) Cutaway drawing of instrument showing its major parts. (Bottom) Photograph of assembled instrument.

Solar Flux Radiometer

The solar flux experiment measured the height of the region in Venus' atmosphere where solar energy is deposited to heat the atmosphere. The principal investigator was M. Tomasko, University of Arizona. The radiometer (Figure 4-15) weighed 1.6 kg (3.5 lb) and required 4 W of electrical power. It revealed how much sunlight clouds absorbed and how much reached the surface. This information was important for understanding Venus' heating mechanism. Does heat result from a greenhouse effect where the planet absorbs solar energy efficiently but reradiates it inefficiently?

Figure 4-15. Large Probe solar flux radiometer (LSFR). (Top) Detailed diagram of the instrument's detector head shows the location of quartz lenses, light pipes, and filters. (Bottom) Photograph of assembled unit: detector head on left and electronics package on right.

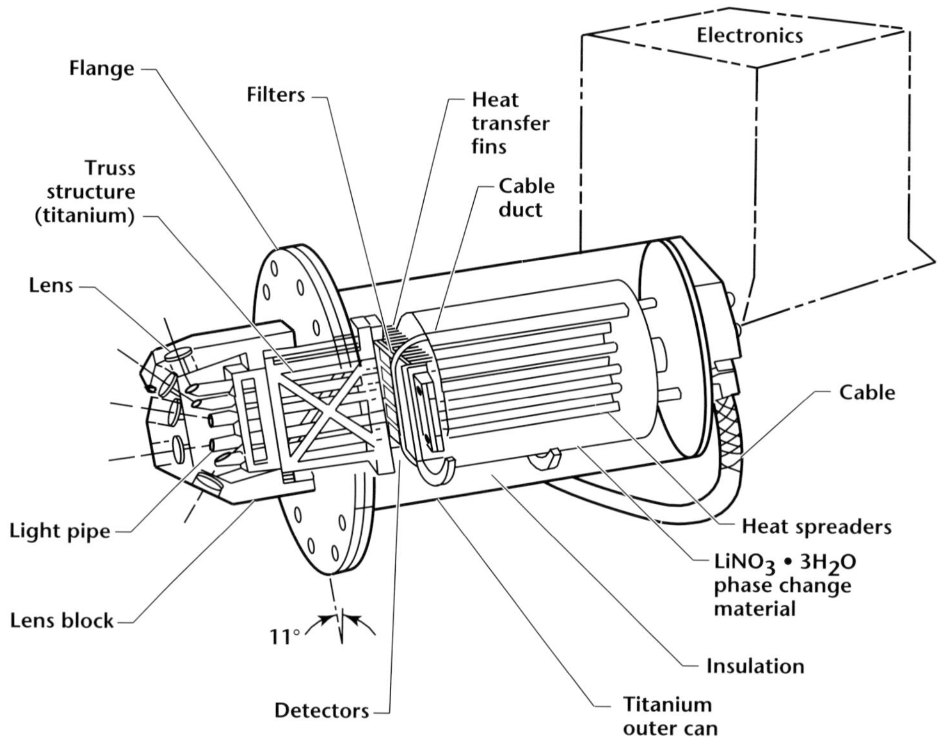

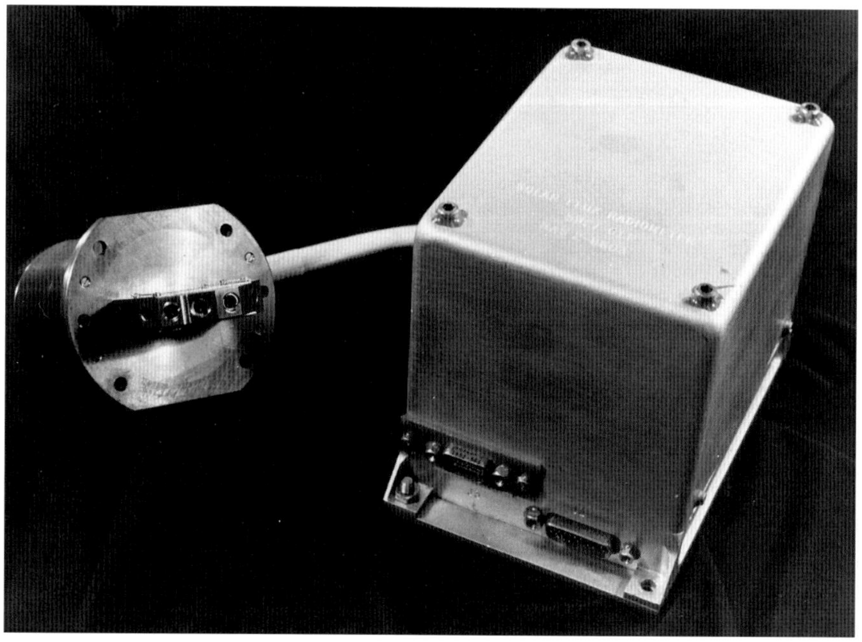

The instrument continually measured the difference in intensity of sunlight directly above and below the probe's horizon. Five quartz lenses, 3 mm (0.125 in.) in diameter, inside five flat sapphire windows collected the light and transmitted it along quartz rods to a detector array of 12 separate photovoltaic detectors. The intensity of sunlight was detected over the spectral range of 0.4 to 1.8 microns. This is where 83% of solar energy is concentrated. Two broad and flat spectral channels were included at each azimuth and zenith sample. One filtered a channel from 0.4 to 1.0 microns, the other a channel from 1.0 to

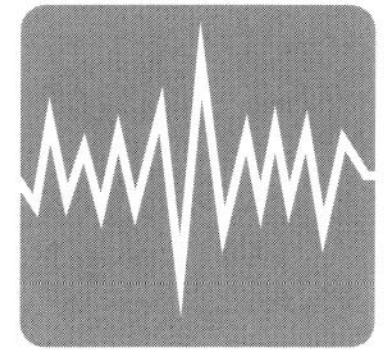

1.8 microns. The instrument also used a narrow filter from 0.6 to 0.65 microns at one of the upward-looking zenith samples and one of the downward-looking samples. This channel provided information about the single scattering albedo and the clouds' optical depth along the descent path.

A mass of phase-change lithium salt, which absorbed heat as it melted, cooled the detector array. The detector head consisted of lenses, quartz rods, filters, detectors, and their supporting structure. It had 12 electronic channels, and the electronics package contained 12 logarithmic amplifiers for these channels.

Mission scientists were concerned that either the probe or the parachute might affect the measurements. To avoid this, engineers restricted the instrument's field of view to a narrow 5° over a carefully selected set of azimuth and zenith angles.

The instrument operated in two modes. At the start, it detected the intensity peak at the solar azimuth. It used the time of successive peaks to control a mode-1 azimuth sampling according to preset values. If a period of 16 seconds passed without detecting a peak, the instrument then automatically switched to a second mode. In mode 2, it collected samples at each zenith angle as frequently as the telemetry rate allowed, which was every 8 seconds. This provided a vertical resolution of 300 m (984 ft), or 2.67 times better than mode-l resolutions. When the probe penetrated to an altitude of 54 km (34 miles), the instrument locked into mode 2 for the rest of the descent.

Infrared Radiometer

The infrared radiometer (Figure 4-16) measured vertical distribution of infrared radiation in the atmosphere. It took measurements from the time the Large Probe's parachute deployed until the probe reached the planet's surface. It also detected cloud layers and water vapor, both important traps for solar heat. The instrument weighed 2.6 kg (5.8 lb) and required 5.5 W of electrical power. The principal investigator for this experiment was R. Boese, NASA Ames Research Center.

The radiometer consisted of two sections: an optical head and an electronics box. On the aft side of the probe's forward shelf, it gathered information through a diamond window. The window was heated to prevent contamination during descent through the clouds. It provided an unobstructed conical field of view of 25° centered at 45° upward and downward from the horizontal.

Designers chose six pyroelectric infrared detectors. Because they required no special cooling equipment, they were well suited to Venus' high temperatures. Each detector viewed the atmosphere through rotating light pipes (to minimize stray light). They also used a different infrared filter between 3 and 50 microns. These detectors possessed uniform sensitivity throughout the infrared range. Although the detectors needed no protection from heating, preamplifiers, which were closely connected to them, did need protection. So phase-change material was put around the detector package to control temperature.

Filters for the 6 channels covered these ranges: 3 to 50 microns, 6 to 7 microns, 7 to 8 microns, 8 to 9 microns, 14.5 to 15.5 microns, and 4 to 5 microns. The first channel allowed measurement of the entire thermal flux. The next two channels searched for water vapor. The fourth channel provided information on cloud opacity. The fifth channel, centered in a strong band of carbon dioxide, revealed any obscurities of the outer window. The sixth band determined window temperature.

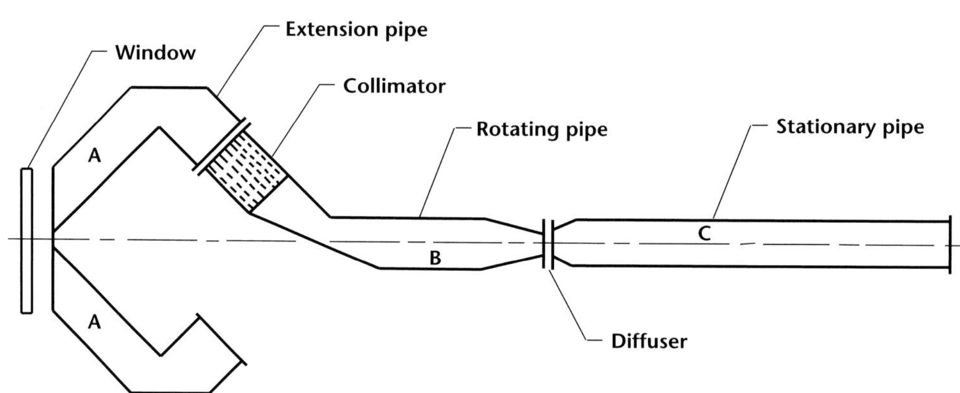

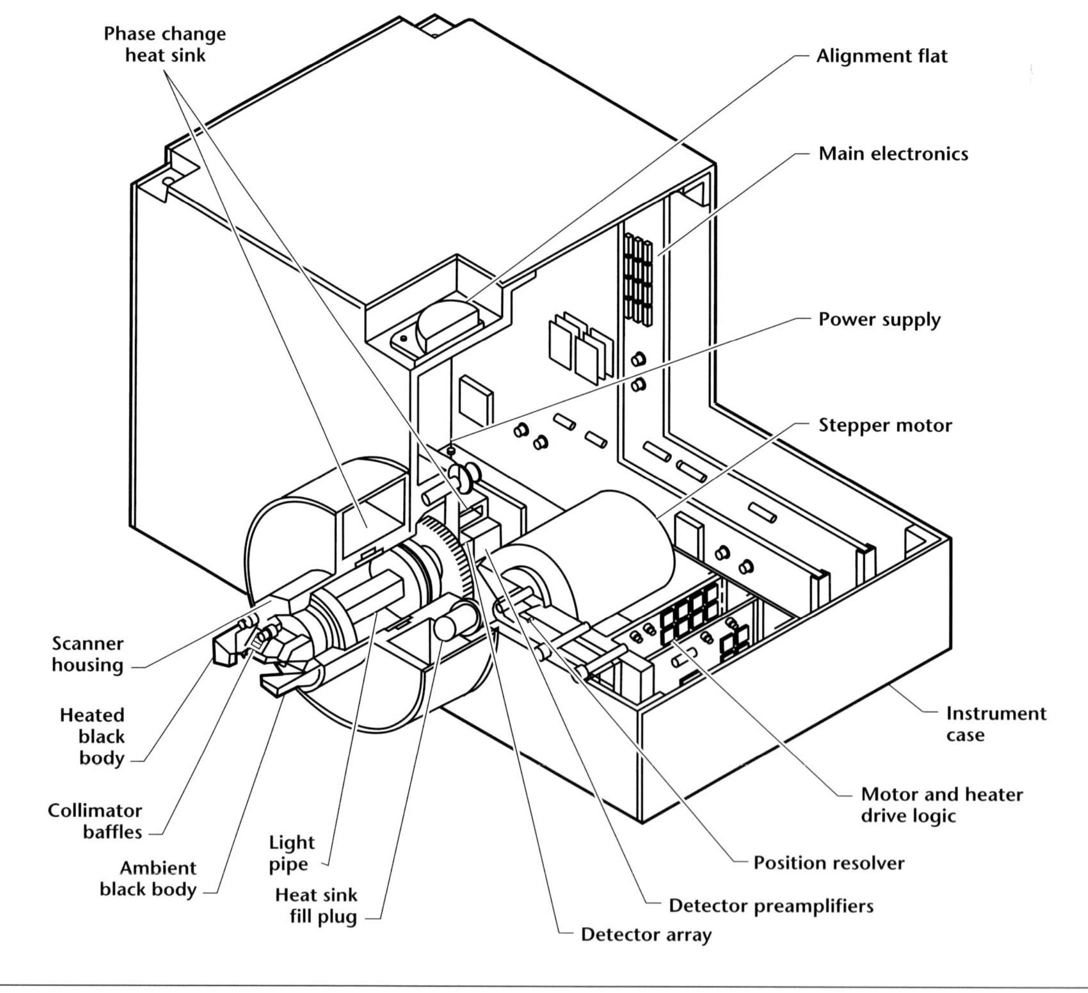

▲

Figure 4-16. Large Probe infrared radiometer (LIR). (Above) Photograph of assembled unit. (Top right) Simplified diagram to show instrument's major components. (Bottom right) Cutaway of detector and filter package identifying various parts.

Two black bodies within the instrument provided a calibration system. These remained at temperatures sufficiently different to generate a signal-to-noise ratio of at least 100:1. This happened in all the detector-filter channels. The instrument was commanded into this calibrate mode approximately 6% of the time during descent.

An electronics box conditioned power from the spacecraft's electrical system. This enabled it to provide closely regulated voltages that

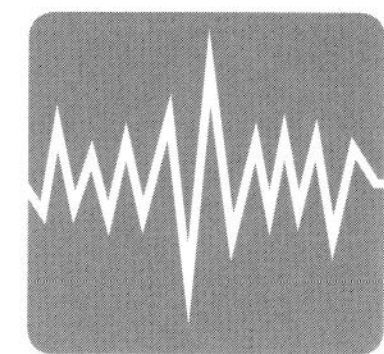

items within the instrument needed. It also conditioned the output signals from the detectors and prepared data for telemetry to Earth.

Vertical resolution within Venus' atmosphere varied from about 260 m (853 ft) at the top of the atmosphere to about 90 m (295 ft) near the surface. The telemetry bit rate assigned to the experiment governed the resolution, which allowed integration of data over a six-second period.

Cloud Particle Size Spectrometer

The cloud particle size spectrometer (Figure 4-17) measured sizes, shapes, and densities of particles within clouds and in the lower atmosphere. R. Knollenberg, Particle Measuring Systems, Inc., directed the investigation. By measuring particle size and mass, the investgation provided a vertical profile of particulate concentration for 34 different size classes. These categories ranged from 1 to 50 microns. Such measurements provided clues to basic cloud formation processes and interactions between clouds and sunlight. The spectrometer also determined if there were ice crystals. It did this by determining if particles had the typical ratio of particle thickness to size for ice. In this way, the instrument could tell them apart from other crystal-like particles.

With this instrument, investigators could resolve the heights of clouds to within 400 m (1312 ft). Its prime measuring technique was optical array spectrometry. This technique covered particle sizes in sequential ranges of 5 to 50 microns, 20 to 200 microns, and 50 to 500 microns. It used multiplexed photodiode arrays to achieve this. Each size range included 10 size classes of equal size width. Also, a scattering subrange used one of the light paths to measure particle sizes from 0.5 to 5 microns.

The instrument, weighing 4.4 kg (9.6 lb) and requiring 20 W of electrical power, directed a laser beam onto an external prism. The prism was supported 15 cm (6 in.) from the outer wall of the probe's pressure vessel. A metal flexible bellows mechanically decoupled it from the wall. The prism directed the laser beam back into the pressure vessel to a backscatter detector. There, a system of lenses and beam splitters generated three independent optical paths. When a particle entered the instrument's field of view, its shadow was cast onto a photodiode array detector. The instrument measured and recorded the shadow's size. Another way of measuring particle size used light scattered by single particles. This process resolved 5-micron particles. A third measurement of particle transit time gave the average thickness of the particle. (Particle transit time is the time a particle needs to pass through the beam.)

Experiments Common to Large and Small Probes

There were two experiments common to the three Small Probes and the Large Probe. These were the atmospheric structure experiment and the nephelometer experiment. Each of the four probes carried identical instruments for these experiments.

Atmospheric Structure Experiment

The atmospheric structure experiment was aimed at finding the structure of Venus' atmosphere from 200 km (124 miles) down to the surface. It involved four well-separated entry sites. Temperature, pressure, and acceleration sensors on all four probes yielded data. These data included location and intensity of atmospheric turbulence and temperature variation with pressure and altitude. The atmosphere's average molecular weight and the radial distance from the planet's center

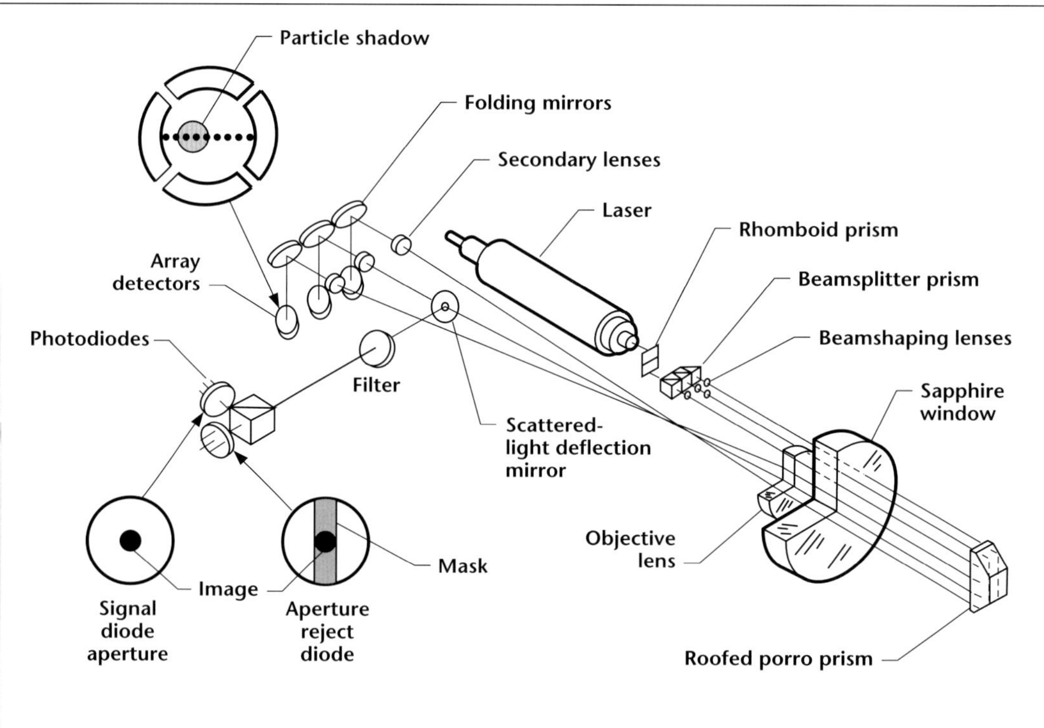

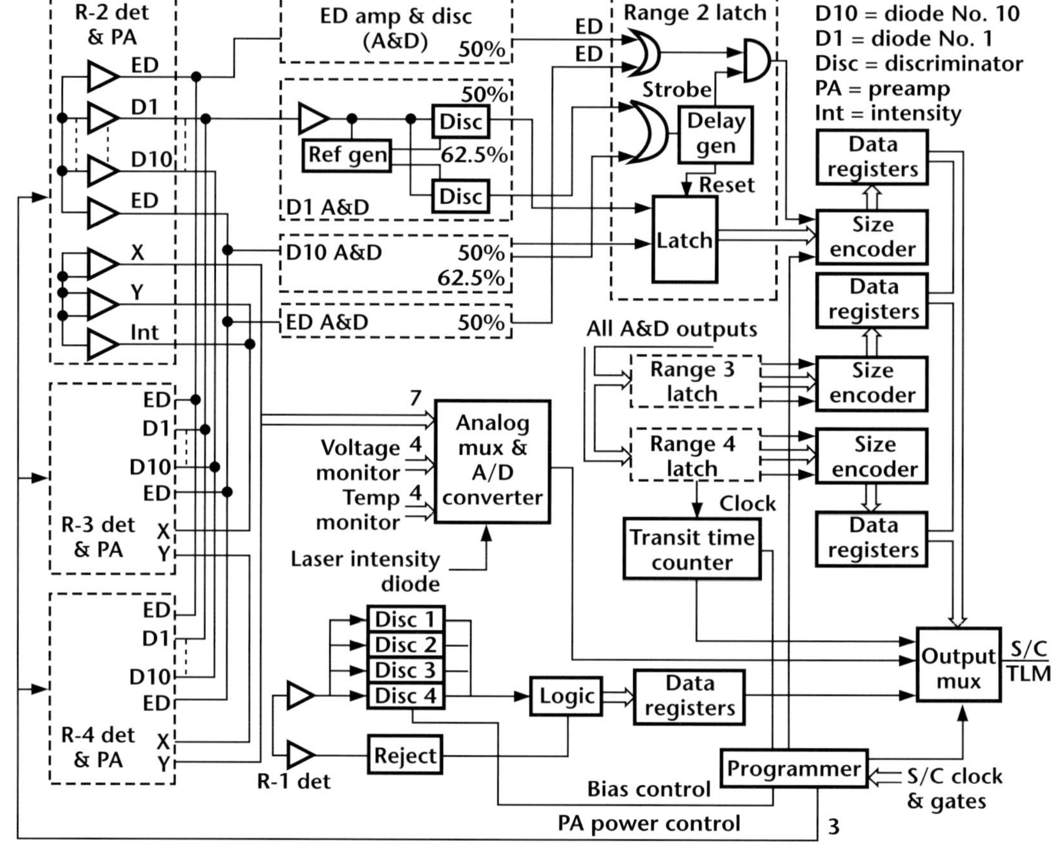

▲

Figure 4-17. Large Probe cloud particle size spectrometer (LCPS). (Above) Photograph of the assembled spectrometer. (Top right) Diagram of optical path. (Bottom right) Block schematic of instrument.

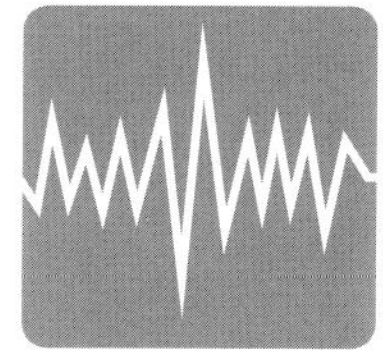

also were among the data. A. Seiff, NASA Ames Research Center, was the principal investigator.

The Large Probe's instruments for this experiment weighed 2.3 kg (5.1 lb) and required 4.9 W of electrical power. The instruments on each Small Probe weighed 1.2 kg (2.7 lb) and required 3.5 W of electrical power (Figure 4-18).

The temperature sensors were dual resistance thermometers. Each had one free wire element protruding into the atmosphere for maximum sensitivity. Another wire element was bonded to the support frame for maximum survivability. The sensors could record temperatures from –100°C (–148°F) to 525°C (977°F). A current source of 10 mA, constant to within 20 ppm, stimulated the sensor. The potential drop across the sensor measured temperature.

The pressure sensors were multiple-range, miniature, silicon-diaphragm sensors. They had to operate over a wide dynamic range from 30 mbars to 100 bars. To meet this requirement, the device used 12 sensors, each covering a small pressure range. These sensors were sampled in a way that preserved data even if one did not work properly. Each sensor had a strain element, diffusion-bonded onto the pressure side of the diaphragm. Engineers arranged the four resistors as a Wheatstone bridge. Two resistors could deform, two could not.

Engineers developed acceleration sensors (four on the Large Probe, one on each Small Probe) from highly accurate guidance accelerometers. They used a pendulous mass maintained in a null position. Interaction of a current in a coil inside the mass with a permanent magnetic field made this possible.

The amount of current needed to keep the mass in the null position was a measure of the acceleration. By changing load resistors and amplifier gain, the sensors could switch over a range from 0.4 microgravity to 600 gravities. The spacecraft used four ranges during entry and two during descent.

An electronics package distributed power to the sensors, sampled their output, and changed their ranges. It also stored their data, ready for telemetry. There were separate data formats for the high-speed entry phase, transition to the descent phase, the descent phase itself, and use on the surface if the probe survived.

Nephelometer

The nephelometer (Figure 4-19) searched for cloud particles. The objective was to find out if cloud layers vary from location to location, or if they were uniformly distributed around the planet. By providing all four probes with a nephelometer, investigators were able to resolve such questions. Each instrument weighed 1.1 kg (2.4 lb) and required 2.4 W of electrical power. The experiment's principal investigators were B. Ragent, NASA Ames Research Center, and J. Blamont, University of Paris, France.

To investigate cloud particles, the nephelometer used a solid-state, light emitting diode (LED) operating at 9000 angstroms. The LED illuminated the surrounding Venusian atmosphere near the probe (but beyond the aerodynamically disturbed region). The device measured the intensity of light backscattered by atmospheric particles. On those probes entering the sunlit hemisphere, the instrument also measured background solar light penetrating the atmosphere. It made measurements at two wavelengths: 3550 angstroms and 5200 angstroms. The LED illuminated the atmosphere through a window in the probe's pressure vessel. Through a second window, receivers measured intensity of backscattered light and background solar light. A plastic

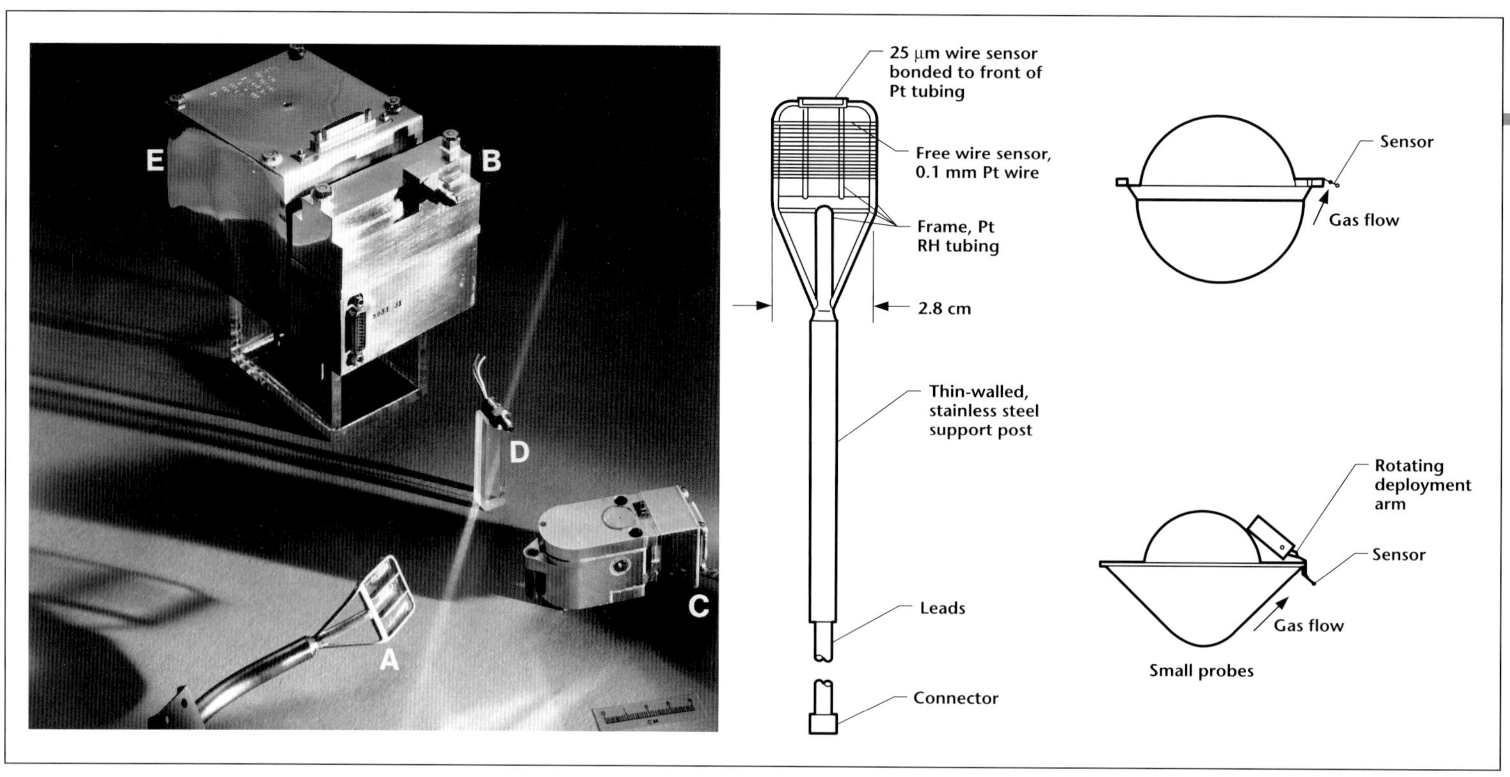

▲

Figure 4-18. Probe atmospheric structure experiment (LAS/SAS), on all probes. (Left) Photograph of instrument. The letters identify: A, multirange atmospheric pressure sensor; B, single-axis accelerometer; C, electronics package; D, one multirange sensor; E, electronics package. (Right) Diagram of one temperature sensor and its location on the probes.

Fresnel lens focused the beams. Investigators fixed calibration targets to the Small Probes' window covers and to the Large Probe's aeroshell.

The instrument consisted of an optical subsystem and an electronics subsystem. The former consisted of two major optical trains of elements: a transmitter, a receiver, and a lens barrel for each. A fiber optics light pipe, shielded from direct reflections, conducted some of the light reflected from the front surface of the window through which transmitted light passed from the probe. The system used this light pipe to monitor the state of the window and the condition of the light-emitting diode. Three solid-state photodiodes detected backscattered light, ultraviolet background, and visible background. The lens barrels for each channel gave some thermal insulation and also collimated the light. Borosilicate glass elements provided further thermal insulation.

The electronic subsystem converted electrical power for the instrument. It provided timing and logic control and conditioned the LED pulse power. It also compressed data and prepared it for telemetry. Digital data telemetered to Earth included measurements of backscattered light and calibration and monitoring data. These data included temperature, channel noise, and the window's condition. Investigators used the experiment to construct a vertical profile of particle distribution in the lower atmosphere. The two Small Probes descending on the planets' sunlit side also measured vertical distribution of scattered solar light in the ultraviolet and visible regions of the spectrum.

Small Probe Experiment

One experiment was exclusive to the Small Probes—the net flux radiometer experiment. It mapped planetary positions of sources and absorbers of radiative energy and their vertical distribution. This experiment enhanced our understanding of what powers Venus' atmospheric circulation. The principal investigator for this experiment was V. E. Suomi, University of Wisconsin.

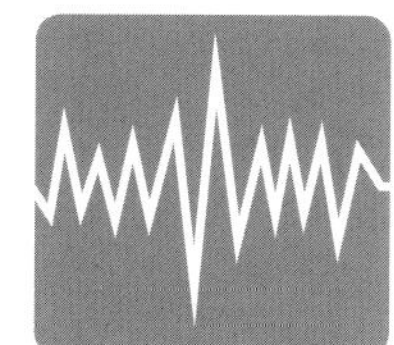

Figure 4-19. Probe nephelometer (LN/SN) on all probes. (Top) Photograph of nephelometer showing its compact form. (Middle) Instrument's various components. (Bottom) Optical path through the instrument.

The instrument (Figure 4-20) weighed 1.1 kg (2.4 lb) and required 3.8 W of electrical power. It consisted of a sensor assembly outside each Small Probe's pressure vessel. This assembly, inside a protective enclosure, was deployed only after the probe experienced its maximum deceleration during atmospheric entry. The sensor was a net flux detector on an extension shaft that could rotate periodically through 180°. This rotation canceled offsets of the instrument and reduced asymmetric heating effects. The detector also included a temperature sensor and a heater. The latter reduced condensation on the detector's diamond windows. The windows—two per detector—were cut from the same stone as the infrared radiometer window.

The flux plate was parallel to Venus' surface. A difference between upward and downward radiant energy falling on the two sides of the flux plate produced a temperature gradient through it. This induced an electric current, a measure of the flux difference. The plate was flipped through 180° every second.

An electronics module processed two flux parameters. These parameters were the integral, time-averaged flux and the maximum and minimum values of a periodic input. Internal timing controlled the system, which operated over four dynamic ranges. In addition to science measurements, the instrument performed other duties. For example, it transmitted detector housing temperature, amplifier temperature, and status of the detector and its heater.

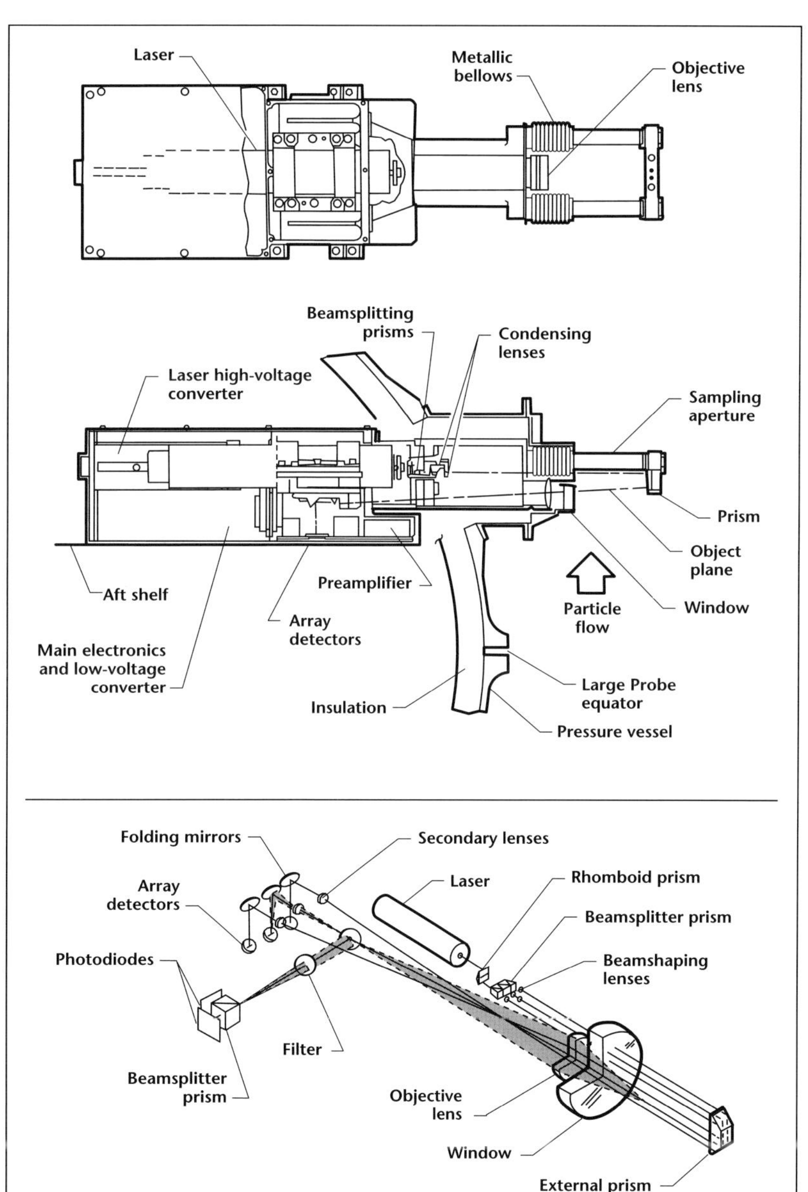

Multiprobe Radio Science

As with the Orbiter, scientists used radio signals from the Multiprobe mission (probes and Bus) for several experiments that did not require instruments onboard the spacecraft. These were a differential, long-baseline interferometry experiment, an atmospheric

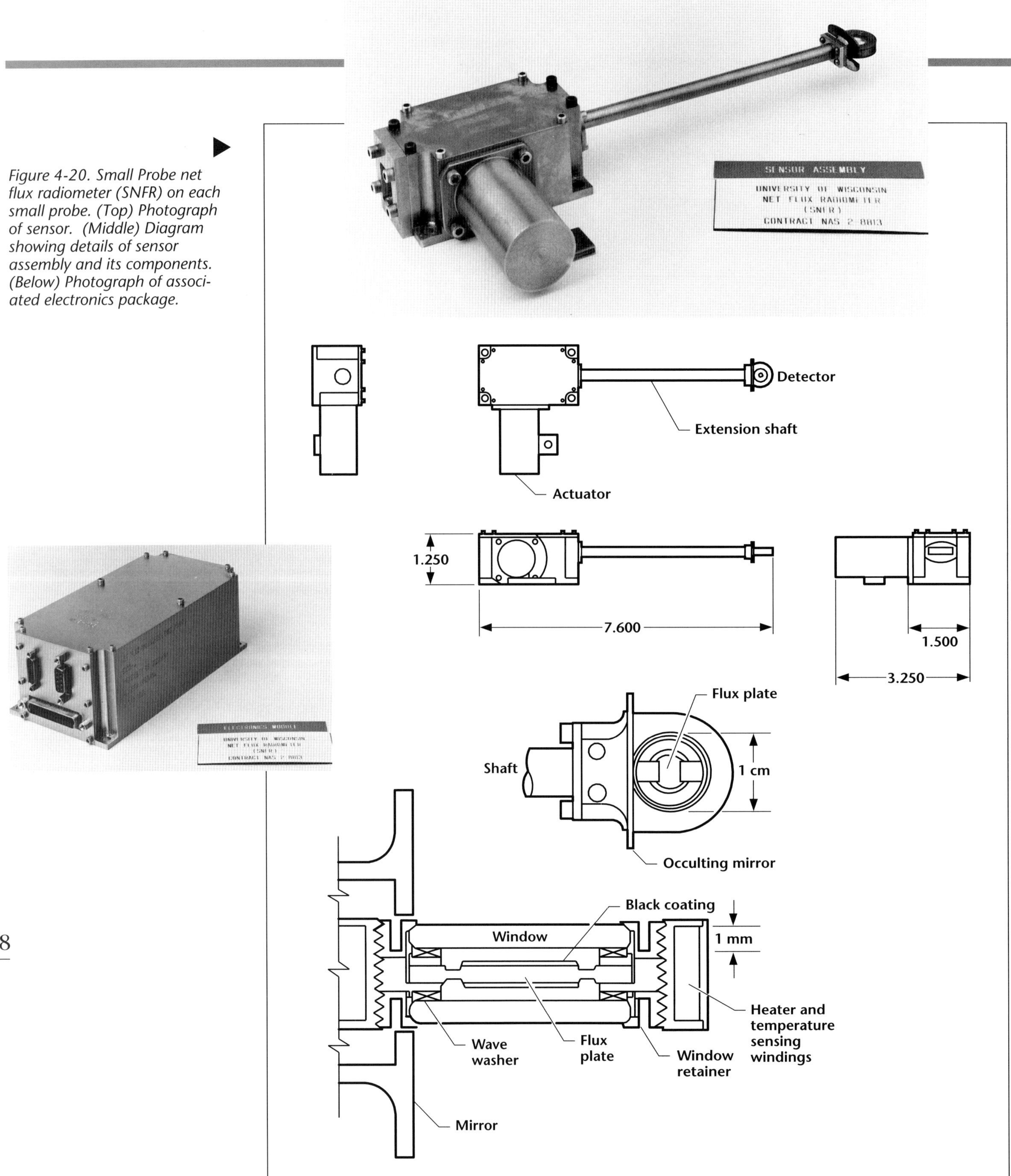

Figure 4-20. Small Probe net flux radiometer (SNFR) on each small probe. (Top) Photograph of sensor. (Middle) Diagram showing details of sensor assembly and its components. (Below) Photograph of associated electronics package.

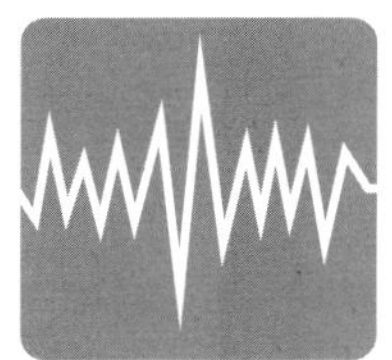

propagation experiment, and an atmospheric turbulence experiment. The principal investigators for these experiments were, respectively, C. C. Counselman, Massachusetts Institute of Technology, T. A. Croft, SRI International, and R. Woo, Jet Propulsion Laboratory.

Differential Long-Baseline Interferometry
The differential long-baseline interferometry experiment measured wind velocity and direction in Venus' atmosphere. This measurement occurred as the four probes descended through the atmosphere. Experimenters compared the probes' descent paths with simultaneous measurements of atmospheric temperature and pressure from probe sensors. This information was to help develop an improved model for atmospheric circulation.

While the four probes descended to the surface, the Multiprobe Bus remained above the atmosphere. It followed a ballistic trajectory that scientists could determine accurately relative to the planet. Probe velocities were measured differentially with respect to the Bus, while velocities relative to the planet were determined by reference to the known Bus trajectory. Probe trajectory deviations from the mathematical model in a still atmosphere were attributed to winds.

Two DSN stations, Goldstone and Canberra, and two Spaceflight Tracking and Data Network stations, Santiago and Guam, tracked all spacecraft at the same time. Experimenters inferred the component of the velocity vector along the Earth-Venus line of sight from the received signals' Doppler frequency shifts. To find the other two components of each probe's velocity vector, they used differential long-baseline interferometry.

Atmospheric Propagation Experiment
The atmospheric propagation experiment attempted to obtain information about the surface and the atmosphere. It did this by studying the effects of the atmosphere on the probes' radio signals. As the probes descended, some of the transmitted power from the relatively broad antenna beam reflected from the planet's surface. Doppler effects shifted this signal away from the probe signal by up to 200 Hz. Since they provided a second component of the Doppler shift from a different angle, these reflections provided information about atmospheric winds. Data also came from atmospheric refraction and attenuation due to clouds.

Atmospheric Turbulence Experiment
The atmospheric turbulence experiment, which R. Woo, Jet Propulsion Laboratory, directed, studied turbulence in Venus' atmosphere. It achieved this by observing scintillations of the probes' radio signals as each probe penetrated deep into the atmosphere. These data complemented the radio scintillation measurements made above 35 km (22 miles) during Orbiter occultations.

Interdisciplinary Scientists

For the Pioneer Venus program, mission officials selected several interdisciplinary scientists for both the Multiprobe and Orbiter missions. These scientists helped analyze Venus' environment and generate a broader picture of the results from individual experiments.

Pioneer Venus was the first NASA program to formally select interdisciplinary scientists for participation from a program's beginning. The objective was to include senior scientists with a broad perspective cutting across disciplines represented by individual experiments. Mission personnel viewed the science payload of the Pioneer Venus Orbiter as an integrated set of instruments. This set was to address more global scientific questions than any single experiment could handle. The interdisciplinary scientists played major roles in producing the mission's scientific results. They also assumed key management and advisory roles in the project and program offices, and in the Science Steering Group.

The tasks of these scientists included serving as members of a continuing Science Steering Group throughout the nominal and extended missions. Tasks also included analyzing data from different scientific disciplines to provide overviews of the scientific results. Several scientists served as chairmen of working groups. Scientific investigations included developing models for the transport and chemistry of hydrogen, oxygen, and carbon monoxide. These investigations helped resolve questions concerning stability of the carbon dioxide atmosphere, theory of the atmosphere's evolution, and formation of some of its components and clouds. T. M. Donahue was the scientist undertaking these tasks. Another interdisciplinary scientist, D. M. Hunten, coordinated preparation of a monograph on Venus. He based his monograph on two scientific conferences. He also analyzed the voluminous data the Orbiter gathered on the neutral thermosphere. He then examined these data to plan further measurements with the Orbiter's aeronomy instruments.

Siegfried Bauer studied, analyzed, and interpreted data from Bus and Orbiter experiments. His goal: to determine the detailed properties of Venus' ionosphere and its interactions with solar wind. He accomplished this by investigating neutral gas composition, thermal structure of both neutrals and plasma, and mass transport. He also studied the role of solar wind and the magnetic field in physical processes responsible for the origin, maintenance, and variability of the planet's atmosphere.

Nelson Spencer concentrated on atmospheric motions. His research goals were many. Among them were assessing probable wind-vector parameters and calculating atmospheric motions. He also wanted to find out how these events correlated with other data and how they related to basic questions about Venus' atmosphere. To achieve his goals, Spencer analyzed data from the Orbiter's neutral mass spectrometer.

In a broad study of radar data, G. H. Pettengill first analyzed data from the Orbiter's onboard radar instrument. He then submitted his abstract data for other scientists to use. Harold Masursky processed radar data and correlated radar altimetry. He used image data to produce maps and Venus globes. He also used radar data to create topical studies of particular Venusian regions and geologic maps of the planet's surface. By plotting radar altimeter data of selected small regions, George E. McGill interpreted Venus' topography. His efforts resulted in a detailed analysis of topography and surface properties. He also studied Venus' tectonics and supported other scientists working with radar data.

A. F. Nagy developed theoretical models of the ionosphere and performed comparative studies with parameterized models of the planet's atmosphere. He also chaired one of the working groups of scientists.

Guest Investigators

The guest investigator program began in 1981. Again, the purpose was to involve new scientists in the program. These scientists would bring a fresh perspective to data analysis and interpretation. The guest investigators fulfilled this expectation admirably.

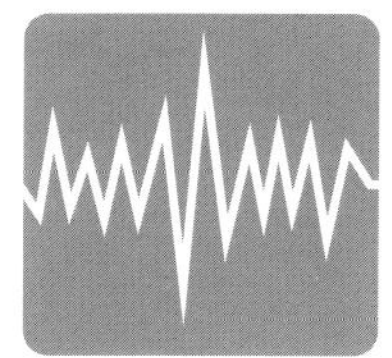

Since the guest investigators were not necessarily associated with a specific instrument, the work of only a few appears here. Results of their work are in the general science results in the next chapter.

S. Kumar, as an example, investigated escape of hydrogen from Venus. R. S. Wolff investigated the dayside ionosphere's properties and variability as a function of solar-wind conditions. He correlated a morphological classification of ionospheric density and temperature profiles with several events. These events included solar-wind dynamic pressure, interplanetary magnetic field direction, Sun zenith angle, and planetary latitude. From this classification, he constructed a model to show ionospheric dynamics.

Paul Rodriguez analyzed measurements of plasma waves in the ionosheath. He was able to derive the characteristic spectrum of these waves. From this, he determined the important wave-particle interactions between solar wind and the ionosphere. He compared these with conditions in Earth's atmosphere to gain a new understanding of how solar wind interacts with nonmagnetized planets.

Other guest investigators looked at many more aspects of the Pioneer Venus data: M. Dryer studied the viscous interaction of shocked solar wind with Venus' ionosphere; J. C. Gerard examined chemistry and transport of thermospheric odd nitrogen; A. T. Young analyzed Venus' clouds and atmosphere; J. L. Fox observed the role of metastable and doubly ionized species in the chemical and thermal structure of Venus' atmosphere compared with Mars; S. S. Limaye studied morphology and movements of polarization features; and C. O. Bowin investigated Venus' gravity, topography, and crustal evolution.

CHAPTER 5

MISSION TO EXPLORE VENUS

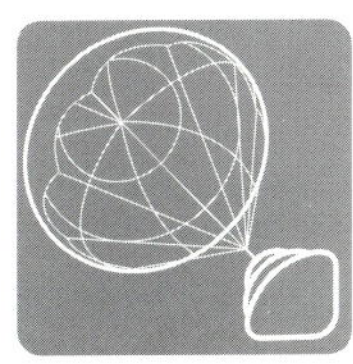

In mid-November 1978, both the Pioneer Venus Orbiter and Pioneer Venus Multiprobe converged on their target. Venus had passed a closest approach to Earth and emerged from the Sun's glare, rising as a morning star just before the Sun. Although launched 2-1/2 months after Orbiter, the Multiprobe was catching up with the Orbiter. By November, it was following closely behind it. The Orbiter would go into orbit around Venus on December 4. Five days later, the probes from the Multiprobe would make their entry and hour-long descent through Venus' atmosphere. Mission controllers prepared the Multiprobe for the first of its four probes to separate.

The Interplanetary Voyage

There had been dramatic incidents during the long flight of the two Pioneer spacecraft through interplanetary space (Figure 5-1). One incident occurred at the time of the Orbiter's first significant ground-commanded maneuver after it left Earth. Soon after the spacecraft was launched on May 20, 1978, its long magnetometer boom deployed. The dish antenna despun to face Earth from the spinning spacecraft. Mission controllers commanded checks of the Orbiter and several of its scientific instruments. Telemetry indicated all operated according to plans. Next, they tested the spin-scan imaging system by obtaining several pictures of Earth illuminated as a thin crescent.

To change the velocity of the Orbiter by 3.33 m/sec (7.5 mph), controllers commanded a first in-course correction on June 1, 1978. This maneuver was to aim the Orbiter more accurately at the point near Venus where the spacecraft had to fire its rocket motor to orbit the planet.

The maneuver did not work out as mission controllers planned. The cause turned out to be trivial. It was the first of many operational lessons that the project engineers controlling the mission learned during the interplanetary voyage. Engineers had designed the roll reference system with the safety feature of an automatic shut-off. A servomechanism followed changes about the roll axis at a restricted rate. Should the spacecraft change orientation too quickly, the servomechanism would lose synchronization. If this occurred during a maneuver, the protective design halted the maneuver. In the problem with the first maneuver, part of the spacecraft's structure deflected the propulsive jet from the thrusters. This caused a propeller-like action that changed the roll rate sufficiently to drive the servomechanism too hard. As a result, the first maneuver automatically aborted. Once controllers had identified the cause, they successfully avoided a repeat of the problem by issuing commands to disable the automatic cutoff circuit when it was safe to do so.

The Pioneer Mission to explore Venus provided a wealth of information on Earth's sister planet. The mission far exceeded its original aims. In its 14 years, the spacecraft sent a continuous torrent of data from the planet. This information ranged from pictures of the cloud cover to detailed radar maps of Venus' surface. This chapter details the events that occurred between late 1978, when the spacecraft converged with Venus, and October 1992, when the Orbiter completed its last orbit. During those 14 years, mission personnel gained invaluable experience in space exploration. In this chapter, you learn about their challenges, the problems they solved, and anomalies that remain unexplained.

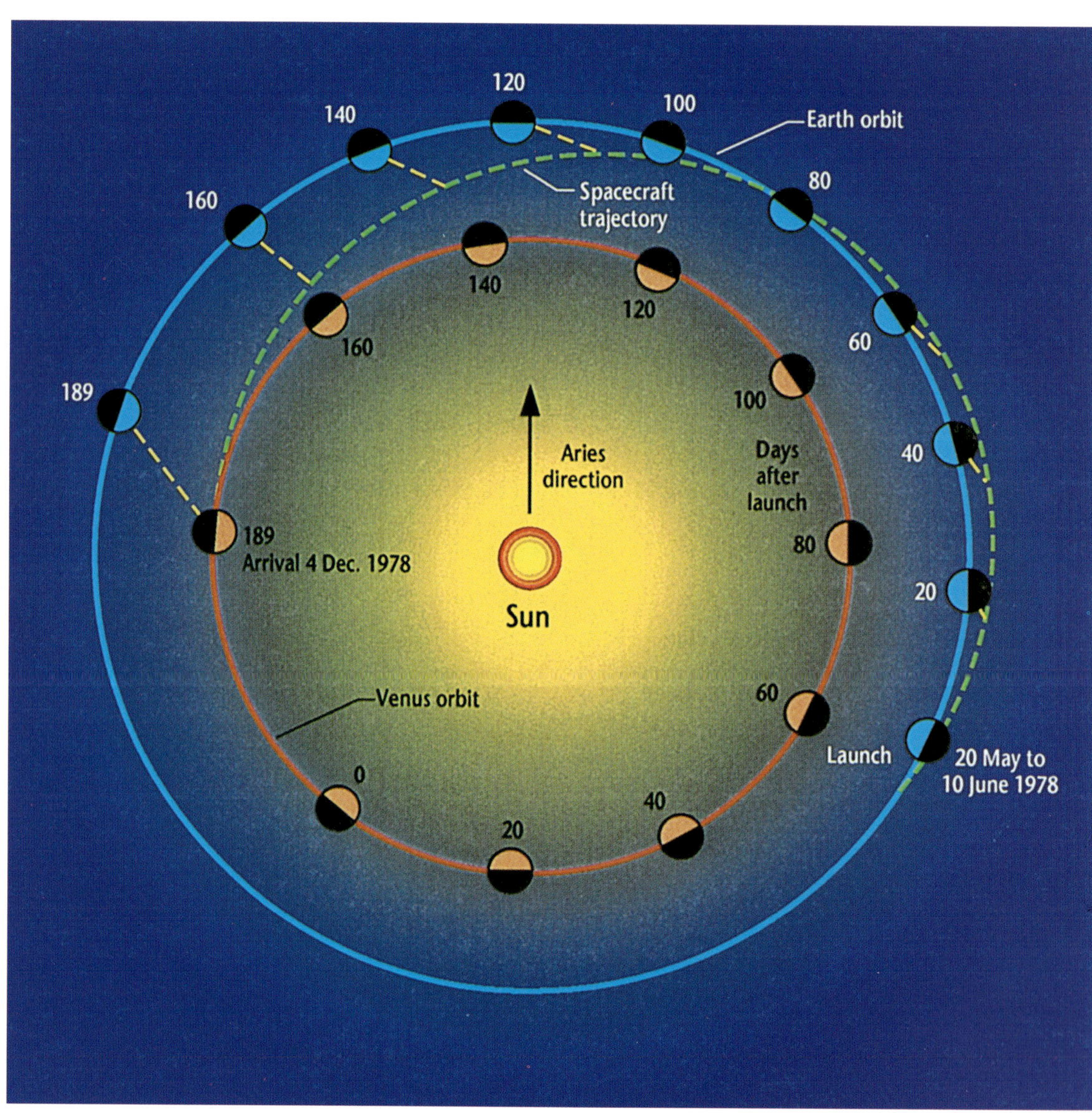

Figure 5-1. Pioneer Venus Orbiter's path from Earth to Venus carried it more than halfway around the Sun on its seven-month journey. At first, the spacecraft moved outside Earth's orbit, crossing inward approximately 90 days after launch. Then it moved toward the Sun for a rendezvous with Venus in December 1978.

The necessary maneuver was then successful, but it required 8 hours to complete with a series of rocket thrusts in two directions. The spacecraft's initial course carried it toward Venus' southern hemisphere. The maneuver corrected the spacecraft's path to the required orbital injection point some 348 km (216 miles) above the planet's northern hemisphere. The change in flightpath positioned the spacecraft so it could achieve its planned elliptical orbit on arrival at Venus. Science investigations required an elliptical orbit tilted 75° to the planet's equator. This would take the spacecraft to within 241 km (150 miles) of the planet at periapsis. At apoapsis, it would be as far away as 66,000 km (41,012 miles).

The in-course maneuver also slowed the spacecraft, allowing it to fall toward the Sun. Solar gravity accelerated the spacecraft so it would arrive at Venus at 8:00 a.m. PST on December 4, 1978.

By early June, the Orbiter detected an extremely powerful burst of gamma radiation. This was an early and important scientific result from one of its onboard experiments. Scientists discovered such gamma-ray bursts in 1973. They possess enormous energies and occur, on the average, about once per month. Astronomers thought the bursts came from random points in the Galaxy or even from beyond. Two other spacecraft also observed this gamma-ray burst. These were Vela, a Department of Energy satellite circling Earth, and Helios B, a NASA-European spacecraft orbiting the Sun. By triangulation of several such observations, scientists expected to locate the bursts' origins. From these origins, they could deduce what great physical event might produce such high-energy phenomena.

During its voyage to Venus, the Orbiter recorded a total of six gamma-ray bursts. Two of them were among the strongest so far recorded. On March 5, 1979, Orbiter's instrument recorded a burst of gamma rays that, when coupled with observations from other spacecraft, appeared to come from the direction of the Large Magellanic Cloud.

These observations were important in preliminary investigations of such strange explosions in space. They supplemented later, more detailed studies that used data from the Compton Gamma Ray Observatory (launched in April 1991). Scientists could not associate the bursts with any object visible at other wavelengths. They now believe the bursts originate from extremely distant objects. Such objects are far beyond the Magellanic Cloud, possibly 5 to 10 billion light years away. Later observations showed that the dimmer high-energy gamma ray bursts last longer than the brighter bursts. This observation supports the time dilation effects predicted by relativity theory.

The Multiprobe spacecraft successfully completed its first course change on August 16, 1978. Without a course adjustment, the Multiprobe would have passed Venus at a distance of about 14,000 km (8,700 miles) from the planet's surface. This course correction required a day-long procedure, featuring a series of timed rocket thrusts in two directions in space. It increased the spacecraft's speed by 2.25 m/sec (about 5 mph).

There was a minor incident during the Multiprobe's interplanetary voyage. Both the Orbiter and the Multiprobe carried redundant equipment to provide backup should a critical piece of equipment fail. For example, the communications system had duplicate power amplifiers. Either would work if the other failed. There were no receiver problems on the Orbiter, but the command receivers were switched for operational purposes. However, when engineers noticed a problem with the Multiprobe's operating receiver, they turned on the redundant receiver. Since the backup worked well, mission controllers did not later bring the original receiver back into operation. Moreover, the Multiprobe was fast approaching its rendezvous with Venus and needed many commands.

Separation of the Probes

Splitting the Pioneer Venus Multiprobe into its five independent spacecraft provided two of the most crucial and exciting operations of the Venus mission. Rather small errors would have made the probes miss their targets or fail on entry. The Large Probe was scheduled to be released on November 15, 1978. More critical was the scheduled release on November 19 of the three Small Probes. To reach the target areas on Venus, the Small Probes had to eject within a few hours of a preselected time. They had to do this within a fraction of a degree in roll.

Before controllers separated the probes, they placed precisely calculated numbers in timers aboard each probe. These numbers represented millions of seconds between release of a probe and the time when its various systems would start operating for its entry mission. The probes could be released over a period of three or four days. However, once engineers selected a time, they had to set the timers precisely for that time. Systems within each probe had to activate at the preestablished number of minutes before each probe entered Venus' atmosphere. "It was extremely critical," said Project Manager Charles Hall. "If the times were set short we would have started using the battery (in each probe) too early and run out of power by the time we reached the atmosphere. If we had set the times too long, we would have missed a lot of data as the probes began to enter the high atmosphere."

The probes did not accept uplink commands directly from Earth, only via the Multiprobe. As a result, controllers had to set the probes' timers before sending commands to the Multiprobe to release each probe from the Bus. They had to calculate release time from the instant each timer started counting. That counting started when uplink commands turned on an on-board clock pulse. Activating commands had to allow for the one-way travel time of signals from Earth to the Multiprobe spacecraft. That time amounted to several minutes. To minimize human error in those calculations, three people derived them independently.

The Large Probe could not automatically separate in the right direction from the Multiprobe Bus. On November 15, controllers had to orient the spin axis of the Bus so the Large Probe would separate in the right direction. On the journey from Earth to Venus, they kept the axis perpendicular to the ecliptic plane. On November 9, commands moved it through 90°. This allowed the spacecraft's medium-gain, aft horn antenna to communicate with Earth. The omnidirectional antenna was not suitable for Earth communications in checking the probes before their release.

About 13 million kilometers (8 million miles) from Venus, controllers aligned the spin axis again. This enabled the Large Probe to enter Venus' atmosphere along a special trajectory. That trajectory would allow controllers to orient the probe's heat shield correctly relative to the entry flightpath. However, when the spin axis changed for the Large Probe's release, tracking data from the Deep Space Network (DSN) were startling. Said Charles Hall: "These data did not seem to add up to what we were doing . . . there was some question as to the precise direction the Bus was pointing." Mission controllers had to decide quickly whether to command a compensating maneuver.

Navigating the Spacecraft

A big problem in determining orbits is measuring the north-south component of velocity relative to Earth. To do this, navigators compare the difference in Doppler shift from a tracking station in Earth's Northern Hemisphere with another in the Southern Hemisphere. The Pioneer Venus Multiprobe needed many maneuvers, particularly for targeting entry points. First, controllers had to reorient the antenna and spacecraft to target the Large Probe. Then they had to reorient to release the Small Probes. Finally, they had to reorient the Bus so that it entered the atmosphere in a special way. This special orientation allowed it to gather the maximum amount of data about the high atmosphere. Complicated bookkeeping kept track of changes to the spacecraft's velocity vector. It also monitored how the spacecraft was approaching the planet. The preseparation maneuvers to release the probes compromised the long trajectory tracking history during the voyage from Earth. Navigators were concerned that they had not measured the orientation precisely enough. Another possibility was that the plume of the thrusters had bounced off the structure of the spacecraft and created a sideward kick.

When controllers were tracking the spacecraft, they were not accurately measuring its current position from an angular viewpoint. Instead, they built the trajectory to a current position based on the spacecraft's previous positions. Traveling from Earth to Venus, the spacecraft obeyed the laws of celestial mechanics. It moved along a trajectory calculated from those laws. The tracking stations that observed it were on a rotating Earth. Also, the Earth itself traveled in orbit around the Sun and wobbled in concert with the Moon. To solve the problem, navigators modeled the trajectory. They then compared their observations with the model. They continued to refine the model until the two fit.

Extraneous effects that were not in the model only began to show up after they had influenced the trajectory for some time. Navigators measured frequency shifts resulting from the Doppler effect. Doppler residuals are the differences between the Doppler shift according to the model and the Doppler shift in the spacecraft's signal. Navigators continually determined, evaluated, and used these residuals to update the model trajectory. They aimed for and achieved accuracies within a fraction of a thousandth of a meter per second.

Before they made any maneuver, controllers calculated the anticipated Doppler effect. If the observed and the expected Doppler residuals differed after the maneuver, there were two possible explanations. Either the maneuver did not occur in the planned direction, or the thruster did not perform properly.

The Pioneer Venus Project Navigator, Jack Dyer, explained: "There is a lot of judgment involved in deciding on the cause. If you know the orientation, the residual must be due to the thrusters. That is especially so if the alignment of the spin axis is, say, 60° from the direction in which you are observing the Doppler effect. It is only when the direction is perpendicular to the line of sight from Earth that there is an unknown situation." So navigators tried to do all maneuvers in a spin-axis alignment turned somewhat toward or away from Earth.

The classical way to turn a spacecraft is to fire two thrusters opposite each other. "At my insistence," said Dyer, "we fired only one thruster to cause an unbalanced turn, and allowed the spacecraft to be propelled. We had a very accurate means of determining orientation of the spacecraft and had a capability of very precisely returning from one direction to another a few degrees away. These directions could be measured by the star sensors to within 0.01°. From such measurements, we could calculate very accurately how much impulse had been imparted to the spacecraft and therefore how much velocity had been applied in the maneuver." From launch, navigators applied this unbalance technique for all spacecraft maneuvers.

Pioneer project management considered one possibility for the unexpected Doppler data from DSN after the preseparation maneuver. A propellant leak could have generated an unwanted thrust. This thrust, in turn, could have pushed the spacecraft from its commanded orientation. Controllers needed an answer before they could separate the Large Probe. They scheduled it for release from the Multiprobe Bus at 6:00 p.m. PST on November 15. However, Project Manager Charles Hall decided to hold the release until they could identify the problem. "There were so many unknowns at that time that I decided we had better not separate until we had a better handle on the problem. It took us about 12 hours to see some evidence of what the problem really was. It is amazing how these small things take so long to sort out. It was an all-night session. I can recall that we had a large number of engineers and scientists in the mission control area. It was too noisy to think, so I brought a cadre of top project people into my office and we started going over all the calculations. We pieced the whole story together until it finally appeared that all the diverse facts showed we were on the right track."

Releasing the Large Probe

Because navigators could target the Large Probe to enter the atmosphere at locations over a large area of Venus, the precise aiming point was not critical. Setting the timer, however, was. Controllers decided not to attempt another correcting maneuver. Rather, they chose a timing setting that straddled the situation. By contrast, the timing problem would be serious with the Small Probes because navigators had to target them with extreme precision if they were to complete their missions.

A pyrotechnically released spring mechanism launched the Large Probe toward an entry near the equator on Venus' dayside. Separation was normal. The Large Probe became an independent spacecraft silently pursuing its path toward the cloud-shrouded planet. Its internal timer counted the seconds before its systems had to switch on. This would happen just before the probe encountered the rarefied upper regions of the Venusian atmosphere.

Targeting the Small Probes

With the Large Probe successfully on its path to Venus, controllers prepared to launch the three Small Probes. During the four days before release of the Small Probes, mission management studied the Doppler residual uncertainty problem. They recognized it was probably an effect of solar radiation. The problem occurred when they changed the Multiprobe's aspect angle during the pre-separation maneuver. The actual force of solar radiation differed from what scientists had modeled in the orbit determination program. Since the spacecraft had not previously experienced this aspect angle and solar pressure modeling had otherwise been successfully treated, the discrepancy came as a surprise.

One problem was to achieve precise dispersion of the Small Probes. Their trajectories did not allow for flexibility in targeting. This was especially so for the probe that would enter the atmosphere in the daylight hemisphere. Careful judgment could prevent incorrect interpretation of the change in Doppler data. One option was to diminish the size of the circle over which the probes would release. Navigators could achieve this by staying inward of the mission's desirable boundaries.

Alignment of the spin axis for release of the Small Probes was crucial. Improper alignment could orient the spacecraft relative to the Sun so its solar panels might produce too little power for the Bus battery. That would have limited the time the battery could stay charged at the needed confidence level. When they had reoriented the spacecraft, navigators had to measure and, if necessary, adjust both the attitude and the spin rate. They had to release the probes within a period that would not deplete the battery.

Before separation from the Bus, and still 22 days before entry, the Small Probes were checked out by radio command. All passed their tests. Two days later, navigators reoriented the Bus. They targeted the Small Probes to their entry points (see Figure 5-2). One was

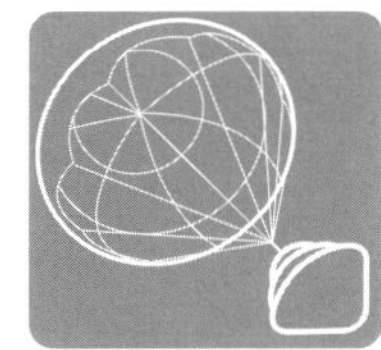

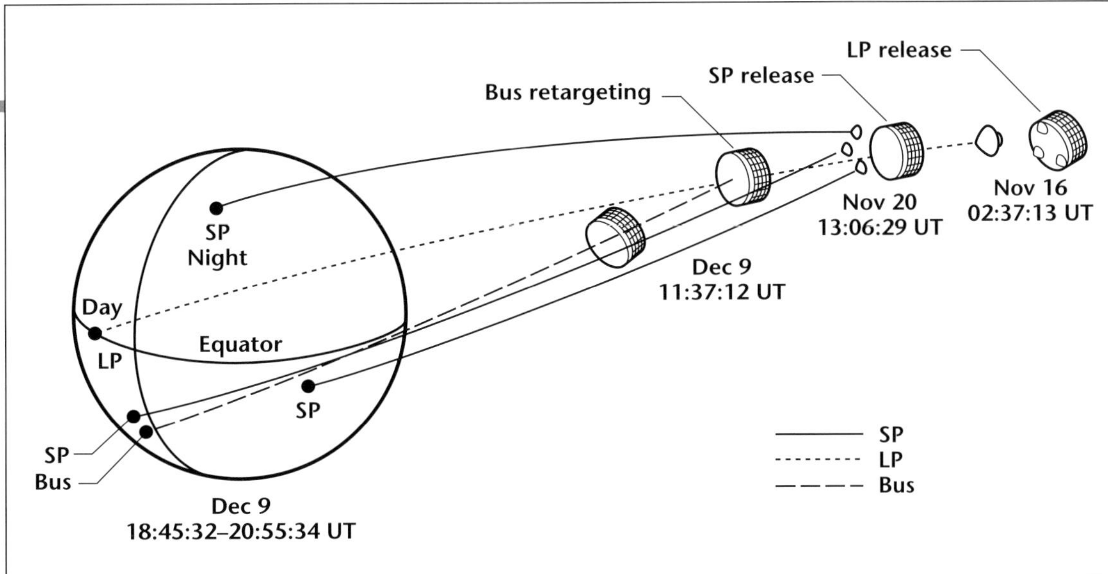

Figure 5-2. In the sequence of releasing the four probes, the Large Probe was first. Next the Small Probes separated, and finally mission navigators retargeted the Bus to enter the Venusian atmosphere.

on the dayside at midsouthern latitudes (the Day Probe). The second was on the nightside, also at midsouthern latitudes (the Night Probe). The third was on the nightside at high northern latitudes (the North Probe).

Astronomers were aware of the predicted positions for Venus before the Pioneer mission. Earlier Venus flybys by Mariner spacecraft had more precisely determined the planet's ephemeris. Navigators predicted that the error in this ephemeris could contribute about a 30-km (18.6-mile) uncertainty in the direction of the spacecraft's arrival at Venus. However, the gravity of the planet helped; it focused each probe toward Venus and halved the uncertainty.

However, gravity did not reduce errors in the downtrack. There the uncertainty was greater, amounting to hundreds of kilometers. Estimating the downtrack uncertainty and then planning the encounter to this uncertainty gave navigators and mission planners a significant problem. They had to choose targeting options for the five entry vehicles. After much discussion, scientists and mission management finally selected the entry points. If the probes entered at different latitudes and longitudes on the planet, the mission would obtain the best scientific data. The probes could gather data in day and night hemispheres and at equatorial and high north and south latitudes. There were, however, geometrical and communications constraints. The Bus spacecraft communicated to Earth from a certain angle around Venus' hemisphere from the point directly facing Earth (the sub-Earth point). Controllers had to target the probes inward from a design boundary of communications. They had to do this by enough margin to allow for the estimated downtrack uncertainty.

With the Multiprobe spacecraft oriented correctly and spinning at about 48 rpm, clamps opened to release the three Small Probes. They left within a millisecond of each other at a predetermined point in the spin cycle of the Bus. The spin of the spacecraft and the precise timing of release directed the probes onto their target trajectories. The timers in the probes began counting the seconds to atmospheric entry.

Mission of the Multiprobe Bus

After all probes had left the Bus, navigators maneuvered it for its own entry into the atmosphere. They slowed the Bus slightly so it would reach Venus a short time after the probes. Unlike the probes, the Bus did not carry a heat shield to protect it from the heating effects of high-speed entry. Mission scientists expected it to burn up within a few minutes. However, during those few minutes, its two scientific instruments—ion and neutral mass spectrometers—would gather data about the atmospheric composition. They would gather these data between the 140-km (87-mile) and 115-km (71-mile) levels.

One problem challenging navigators was how to direct the Bus for its entry into the atmosphere. It had to enter at as shallow a flight-path angle as possible. This angle would reduce the heat load and extend the period of data gathering. However, at too shallow an entry angle, the Bus could skip off the top of the atmosphere. If it did, it would not get the required low-altitude atmospheric data. The most desirable trajectory would cause the Bus to enter the atmosphere, penetrate to the 115-km (71-mile) level, and then skip out again. This would allow scientists to obtain data along incoming and outgoing paths. Commented Jack Dyer, "We could see that it was not possible to navigate so accurately. The risk would be too great that the depth of penetration needed would be missed. So we decided to go for as shallow an entry as we confidently could."

Navigators selected 9° below the local horizontal for the flightpath at 200 km (124 miles) above Venus' surface. They issued commands for the spacecraft to get as close as possible to that path. Also, they set the spin axis of the Bus so the angle of attack would be precisely 5°. They did this so atmospheric molecules would enter the scientific instruments properly. After navigators had completed these maneuvers, all the probes and the Bus were on their way to their targets.

Arrival of the Orbiter

Meanwhile, Pioneer Orbiter approached its rendezvous with Venus. Controllers would maneuver the spacecraft into orbit before the probes arrived at the planet.

December 4 was the date the mission selected for the speeding Orbiter to slow into an elliptical path around Venus (Figure 5-3). The maneuver had to take place behind Venus as viewed from Earth, and this worried controllers. The spacecraft was out of communication for almost 23 minutes at this extremely critical milestone. During this essential maneuver, a 180-kg (400-lb) solid-propellant rocket motor fired. It slowed the Orbiter sufficiently for Venus' gravity to capture the spacecraft into orbit around the planet. This event was the first time a solid-propellant rocket had been fired after being in space for seven months—the time between the launch from Earth and arrival at Venus.

On December 2, the Orbiter started maneuvers for its insertion. It began with an orientation to point the rocket nozzle in the direction of travel. Controllers lowered the communications bit rate from 1024 to 64 bits/sec. This allowed the omnidirectional low-gain antenna to maintain communications during the reorientation maneuver instead of the high-gain antenna. Next, the high-gain antenna was released and spun up to match the spacecraft's spin rate. The spin rate increased to 30 rpm. Next the high-gain antenna was despun, and the bit rate returned to 1024 bits/sec.

The Orbiter's flight from Earth had been free of major problems. However, there had been minor problems in the command memories on the way to Venus. These problems could have led to serious difficulties in obtaining a correct injection into orbit. High-energy solar cosmic rays had caused "bit-flip" errors in the spacecraft's memories. They had changed ones to zeros and vice versa. These errors occurred on an average of about once every two weeks. They could have resulted in a command sequence being interrupted or changed. Fortunately, when these bit-flips occurred,

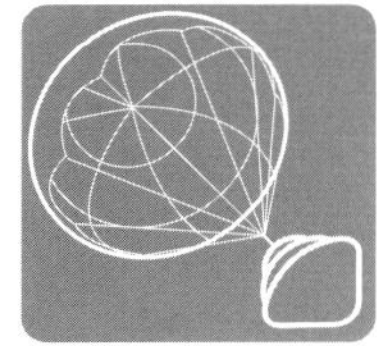

controllers could correct them or the command had already been executed. However, if such an error occurred in the command timing sequence for the rocket motor, it might have caused premature or delayed rocket firing for the orbital insertion maneuver. The results would have been disastrous.

Bit-flip errors occurred on both the Orbiter and the Multiprobe in transit to Venus. The problem surfaced so late in the Pioneer Venus program, design changes to overcome it were not practical. Although the bit-flips occurred on the Orbiter in flight before the Multiprobe was launched, it was much too late to make design changes for the Multiprobe. Fortunately, they were not as critical for the Multiprobe's operation.

Bit-flips had probably affected interplanetary spacecraft before. Scientists had to be able to compare what went into a spacecraft's memory with what came out of it. Only then could they clearly identify such events. Until Pioneer Venus, there had been no opportunity during a mission to check spacecraft memories for these bit-flip effects. Actually, bit-flips had been discovered on some Earth-orbiting satellites. Ironically, they resulted from the same high technology that can minimize energy to flip a digital circuit from one state to the other. A high-energy cosmic ray particle could provide sufficient energy.

To overcome these bit-flips on Pioneer, controllers took great care in how they stored commands in the command logic. Before execution, they always checked commands that they had stored for any period. This procedure ensured that nothing had changed in the commands.

A bit-flip could have serious consequences during the Orbiter's injection maneuver. This was particularly true if it changed the timing

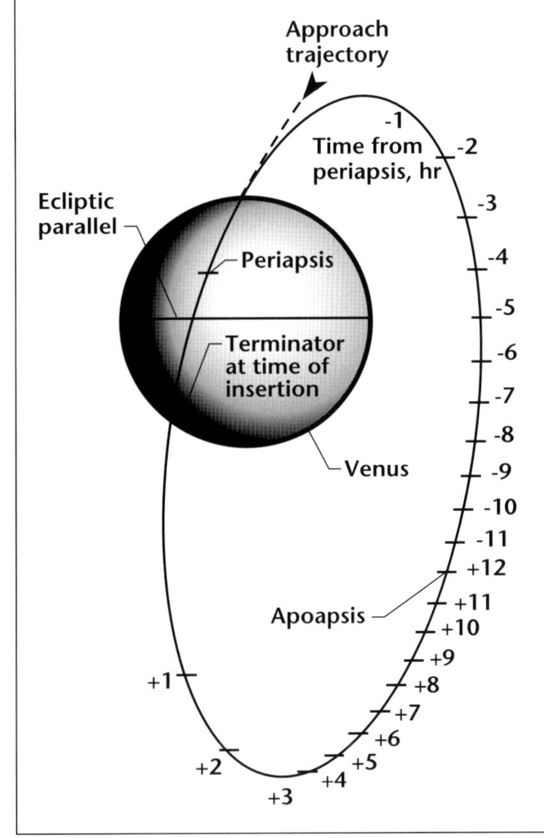

Figure 5-3. This perspective of the orbit around Venus identifies the orbital insertion point, the periapsis, and the apoapsis relative to the planet. The hour marks along the orbit show how the Orbiter traveled quickly through periapsis and more slowly around apoapsis.

sequence to ignite the motor. This sequence had to start while the spacecraft was in radio communication with Earth. Also it had to start before the spacecraft went behind Venus.

A bit-flip could change a time delay that programmers put into the spacecraft's memory to control the motor's ignition. Such a change was unacceptable. Alternatively, a sequence of small time delays, whose sum would be the total time, could command the ignition countdown. Analysis showed that greatest reliability would result from a series of time delays in two redundant command memories. Should a bit-flip affect the time delay in either parallel memory, it would have had no ill effect. By contrast, a jump to early rocket firing by either memory alone would have been disastrous.

On December 3, at 11:00 p.m. PST, controllers loaded the Orbiter's two command memories with the command sequence for firing the orbit insertion motor. The firing would occur at 7:58 a.m. PST on December 4 (Figure 5-4). Of over 40 command delays, the first few were for 1 hour, the next for 45 minutes, then 30 minutes, then delays of 1 minute, then another batch of 3 seconds each. The command memory countdown started at 1:00 a.m. PST on December 4. Each time the memory counted out one of the delays without error, the spacecraft signaled the successful timing execution.

Insertion into Orbit

At 7:51 a.m. PST on December 4, the Orbiter passed behind Venus, and communications with Earth were interrupted. If all went well, the orbit insertion commands in the spacecraft's memory would fire the rocket motor 7 minutes later. The motor's propellant would burn for almost 30 seconds and change the spacecraft's velocity by about 3780 km/hr (2349 mph).

Controllers had set the spacecraft orientation and the altitude of the closest approach of the flyby trajectory. They had timed the firing of the retrorocket precisely. It would thrust the Orbiter into an orbit as near as possible to the mission's nominal orbit. Controllers would have to correct later any errors made in timing the firing of the retrorocket. They also would have to wait to correct any error in the total impulse developed by the rocket motor. Since corrections would need propellant and would reduce the reserve for maneuvering in orbit, they would be undesirable. This would shorten the time during which navigators could control the Orbiter's periapsis altitude to obtain upper atmospheric science data.

As the spacecraft approached Venus, it had a good propellant reserve because the launch had been early in the launch opportunity. Mission scientists wanted to preserve the capability of maintaining orbit for one Venusian sidereal day. (This was the mission design capability.) As a result, they made no attempt initially to stretch the mission to ultimate design requirements. A Venusian sidereal day is different from a Venusian solar day. The sidereal day is the planet's rotation period relative to inertial space. A solar day is the rotation period relative to the Sun. The Venusian sidereal day is 243.1 Earth days. The solar day is 116.8 Earth days.

Maintaining propellant reserves was important because there were data transmission limits to the mission. Experiments could gather more data than the radio link to Earth could handle. It was a foregone conclusion that experimenters would want the spacecraft to continue in orbit after the first sidereal day. Their goal was to gather and transmit data into a second sidereal day. In such an extended mission, investigators would change emphasis on the types of data they gathered and transmitted. To preserve this capability, controllers had to budget propellant usage and conserve reserves.

Getting the spacecraft into orbit was exciting for project management, said Charles Hall. "We had never done anything like this before. Ignition of the rocket motor behind the planet meant there was always the question of whether or not the motor had ignited." To ensure that the spacecraft got into orbit, controllers sent a second ignition command. They timed it to arrive at the spacecraft after its emergence from behind Venus. This backup command would start ignition if the earlier command had not worked behind the planet. The orbit would not, of course, have been as good from such a late ignition. But it would have prevented the spacecraft from flying past Venus and going into solar orbit.

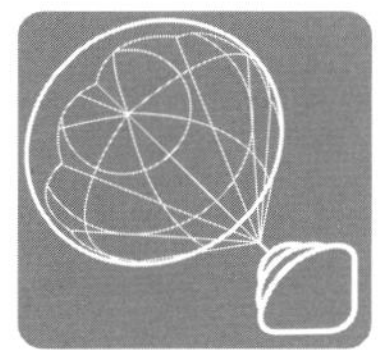

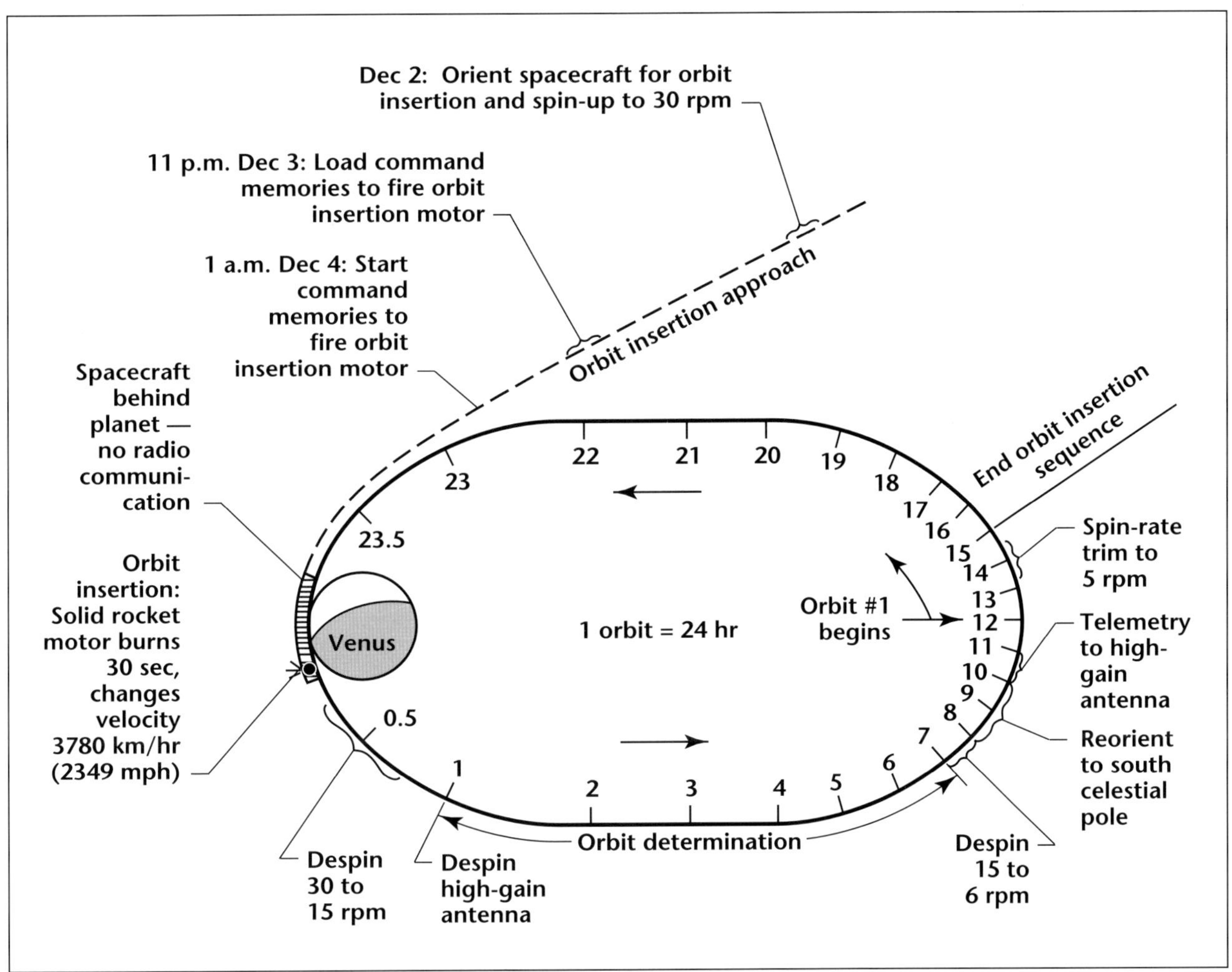

Hall explained how ignition was confirmed. "If we had ignition, then, when the spacecraft emerged, the frequency of the carrier radio wave (from the spacecraft) would be different from that if ignition had not occurred (because of Doppler effects). I recall that we had two receivers on the ground waiting to pick up signals on one or the other frequency. At 8:14 a.m. PST, the spacecraft emerged from behind Venus. It took 3 minutes for the radio signals to travel the 56 million km (35 million miles) to Earth. Everyone waited for one of the two ground receivers to lock onto the spacecraft's signal. When it was clear that the right receiver had locked onto the signal from Pioneer Orbiter, there was a big cheer because we knew then that the spacecraft had gone into orbit."

At 8:30 a.m. PST, navigators adjusted the Orbiter's spin rate to 15 rpm. The high-gain antenna despun and pointed toward Earth. Within the next few hours, navigators analyzed tracking data to determine the parameters of the orbit around Venus. The highly elliptical orbit, inclined 75° to the equator of Venus (105° retrograde) was almost, but not quite, as they expected. Table 5-1 gives the

Figure 5-4. Operations of the Orbiter spacecraft before, during, and after the period of insertion into orbit.

Table 5-1. Planned and Initial Orbit Parameters

Parameter	Planned	Actual
Periapsis altitude, km (miles)	350 (217.5)	378.7 (235.3)
Periapsis latitude, deg	18.5 N	18.64 N
Periapsis longitude, deg	203.223 E	207.990 E
Inclination, deg	105	105.021
Period, hr:min:sec	24:00:00	23:11:26

orbit parameters the injection burn achieved, compared with those the mission planned.

Navigator Jack Dyer explained the problems of entering an orbit around another planet. "We had to be very precise with navigation so that the burning of a given weight of propellant would put the spacecraft into orbit. We spent a lot of time determining how accurately we thought the manufacturer of the retrorocket could predict the amount of impulse it would deliver."

The retrorocket performed better than navigators predicted. This was as bad as underperforming. The over-performance had slowed the Orbiter too much and resulted in the apoapsis being lower than they had planned. The periapsis was higher, too. It also resulted in a shorter orbital period of 23 hours 11 minutes. As a result, navigators had to use more propellant from the attitude control subsystem to correct the orbit's period to the required 24 hours.

Mission navigators had to adjust the period at periapsis. Because the first orbits were times of great scientific activity, they had to delay the adjustment for two orbits. In the meantime, however, they began a preplanned maneuver to lower the periapsis from 378 km (234 miles) to 250 km (155 miles). This took place at apoapsis on December 5 by firing two of the spacecraft's thrusters for slightly longer than 3 minutes.

Initial orbital operations followed a carefully preplanned sequence (Figure 5-5). At 3:00 p.m. PST on December 4, commands reduced the spin rate to 6 rpm from 15 rpm. Others adjusted the spin axis to point toward the celestial poles. Then, a couple of hours later, controllers pointed the high-gain antenna toward Earth. Communications switched to it from the omni-antenna. In the following hours, scientists activated some of the scientific instruments. The first were the spectrometer and the electron temperature probe. Next, they unlocked the radar antenna and deployed the boom of the electron temperature probe. The neutral mass spectrometer and the radar went through calibration sequences. Just after the first orbit began, at the first apoapsis after the spacecraft entered orbit, controllers activated the magnetometer and the retarding potential analyzer. The first orbital data-gathering sequence had started. The remaining instruments were turned on later in that first orbit at times requested by the principal investigators.

The short orbital period caused the time of periapsis to occur earlier each Earth day than desired. As a result, it affected assignment of tracking stations. Mission controllers had to rearrange the relative geometries of the spacecraft, of Venus, and of Earth's tracking stations so key tracking stations at Goldstone, California, and Canberra, Australia, could receive signals from the spacecraft at a preselected part of its daily orbit around Venus. After two orbits of the spacecraft, navigators fired thrusters at periapsis on December 6. This increased the

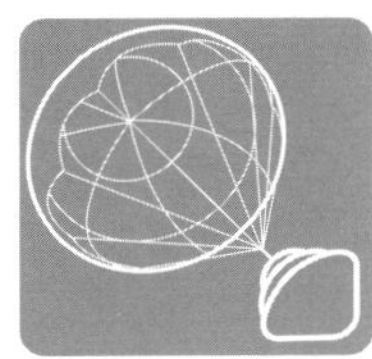

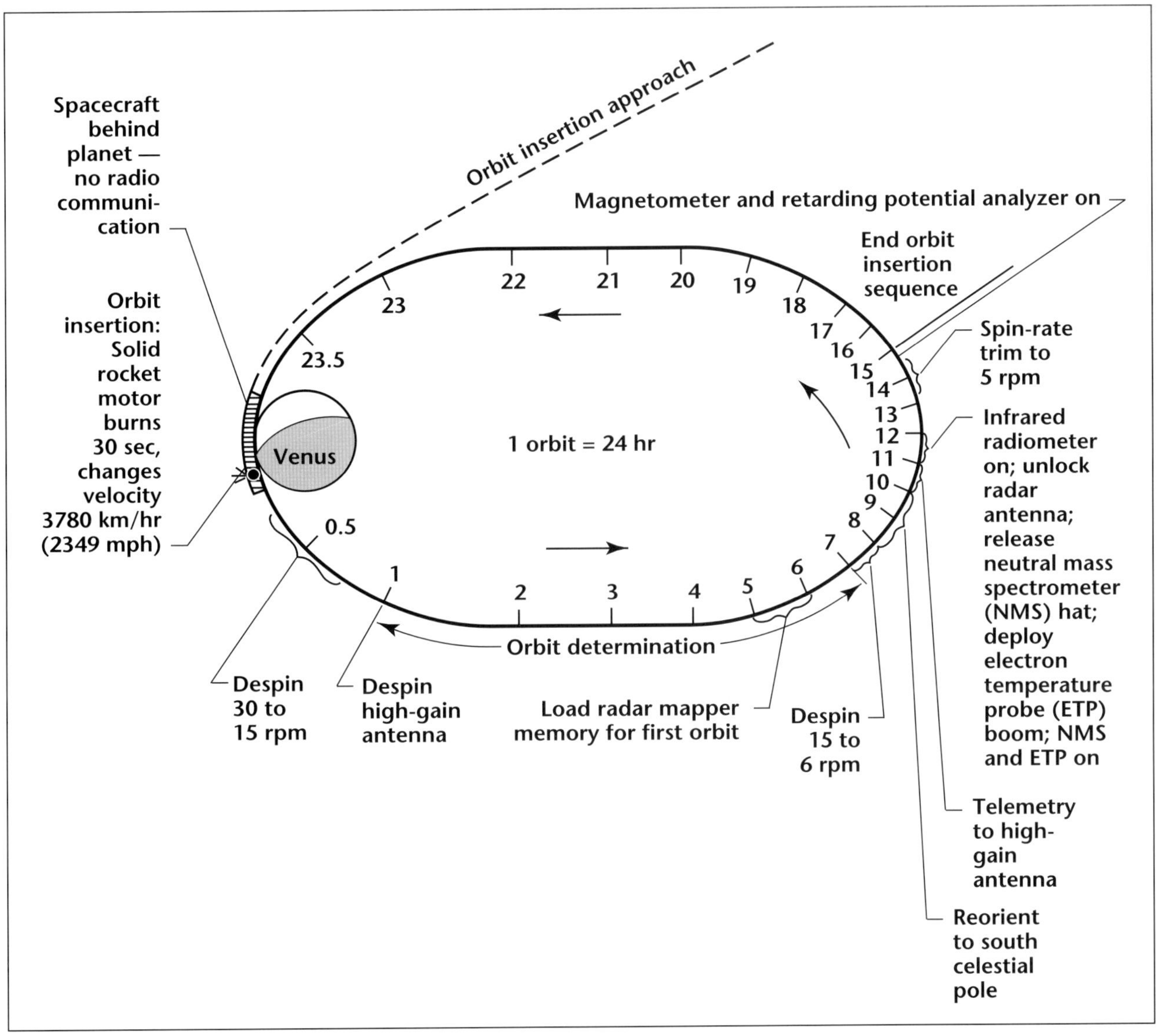

Spacecraft behind planet — no radio communication

Orbit insertion approach

Magnetometer and retarding potential analyzer on

End orbit insertion sequence

Spin-rate trim to 5 rpm

Orbit insertion: Solid rocket motor burns 30 sec, changes velocity 3780 km/hr (2349 mph)

Venus

1 orbit = 24 hr

23.5 23 22 21 20 19 18 17 16 15 14 13 12 11 10 9 8 7 6 5 4 3 2 1 0.5

Infrared radiometer on; unlock radar antenna; release neutral mass spectrometer (NMS) hat; deploy electron temperature probe (ETP) boom; NMS and ETP on

Orbit determination

Despin 30 to 15 rpm

Despin high-gain antenna

Load radar mapper memory for first orbit

Despin 15 to 6 rpm

Telemetry to high-gain antenna

Reorient to south celestial pole

Figure 5-5. The orbit insertion sequence ended 15 hours after achieving orbit. The first operational orbit began after controllers turned on several instruments just before the first apoapsis passage. These highlights are called out in this schematic drawing.

orbital period to just over 24 hours. Afterward, the time of periapsis gradually moved within an acceptable range.

Once the 24-hour orbit was achieved, mission operations divided it into two segments. Each segment reflected the kind of measurements they were taking (Figure 5-6). The periapsis segment was about 4 hours long. The apoapsis segment was 20 hours long. Mission operations used one of five data formats during each short periapsis segment. The formats made it possible to emphasize certain experiments when desirable. For example, one format was for intensive aeronomy coverage at periapsis and another was for optical coverage.

Normally, scientists used only one of two data formats in the 20-hour apoapsis segment of the daily orbit. The first was for obtaining

Figure 5-6. This orbit schematic illustrates how operations proceeded through the mission on a regular basis. Sequences were modified at times to satisfy requirements of experimenters and periods of occultations and eclipses.

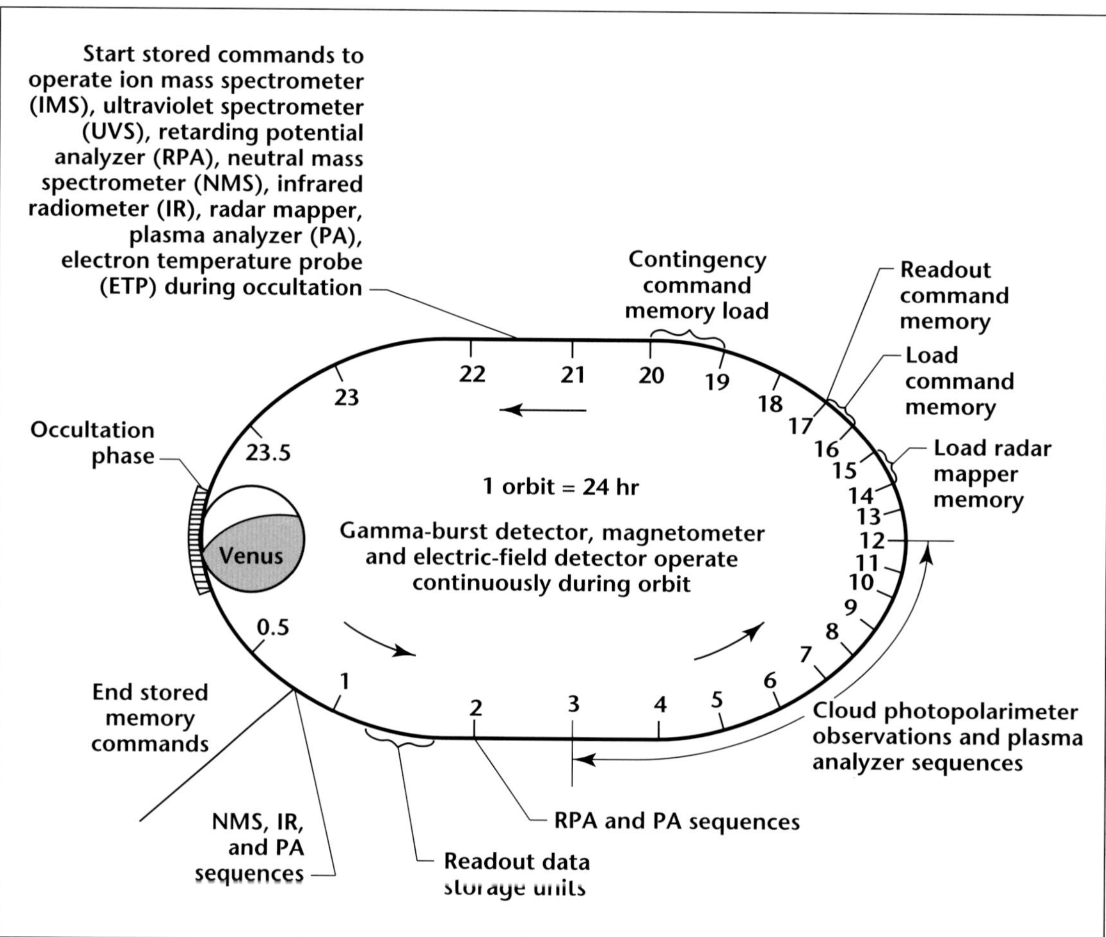

images of the planet's whole disk in ultraviolet light to record cloud features (Figure 5-7). It allocated 67% of the data stream to imaging data. It divided the balance of the data transmission among the three instruments that measured solar-wind and planet interactions and the gamma-burst detector. The other format allocated data return to all instruments except the imaging instrument and the infrared radiometer.

By December 6, NASA had successfully received the first image of Venus (Figure 5-8), and science data were flowing to Earth. All was going well with the Orbiter spacecraft.

Entry of the Probes

When the probes separated from the Multiprobe Bus, they went "off the air." This happened because they did not have sufficient on-board power or solar cells to replenish their batteries. There was no way to command the probes from Earth. Preprogrammed instructions were wired into them, and their timers had been set before they separated from the Bus. The on-board countdown timers were scheduled to bring each probe into operation again. This would occur 3 hours before they began their descent through the Venusian atmosphere. This was timed for 7:50 a.m. PST on December 9, 1978. The timers had to turn on heaters to warm the battery and the stable oscillators of the radio transmitters. This ensured that the carrier frequencies would be correct when the transmitters began sending signals to Earth shortly before entry. Later, the command unit started warmup and calibration cycles for the three instruments on each probe.

At 8:15 a.m. PST, the command timer on the Large Probe began warmup of the Probe's battery and radio receiver. The latter received a

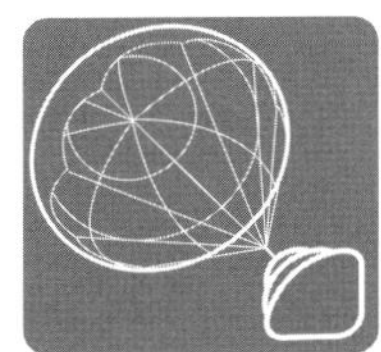

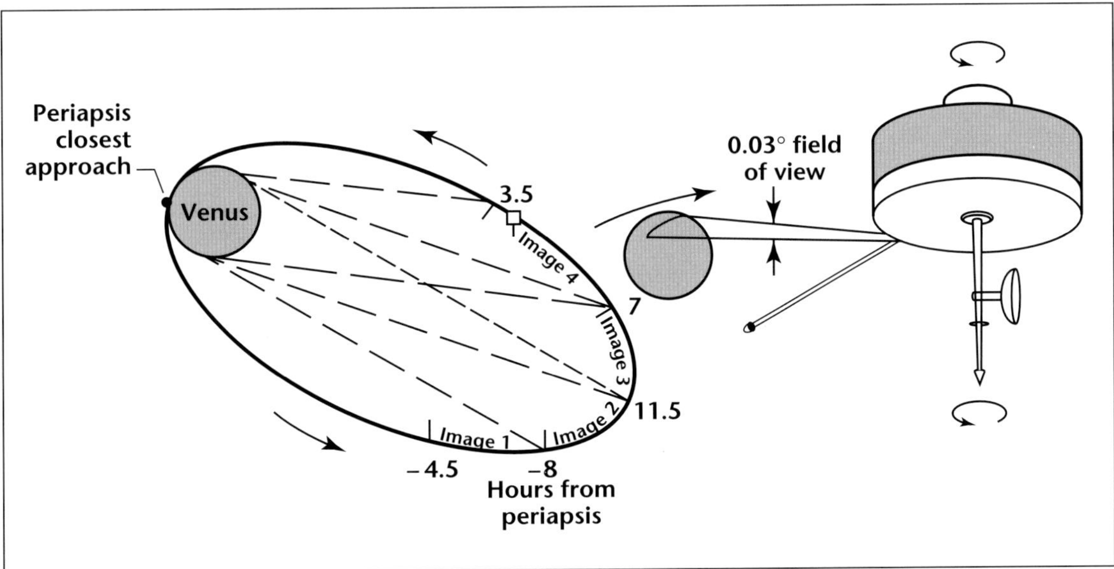

Figure 5-7. Cloud photopolarimetry used motion along the Orbiter's flightpath and rotation of the spacecraft to scan the planet in ultraviolet radiation. The instrument could make five planetary images in each orbit with a resolution of about 30 km (19 miles). The instrument determined cloud particle characteristics from polarization measurements, made images of haze layers at the planet's limb with a resolution of 15 km (9.3 miles), and observed several comets.

carrier frequency from Earth to spacecraft that provided the reference frequency for the downlink signal from spacecraft to Earth.

At 10:23 a.m. PST, the Large Probe began to transmit radio signals to Earth for two-way Doppler tracking at 256 bits/sec. This occurred just 22 minutes before entry. The 22-minute interval was a compromise between consuming precious battery power and providing DSN stations with sufficient time to lock onto the signals before the probes began to send entry data. Within the next 11 minutes after the Large Probe's transmission began, all the Small Probes started transmitting. First came the signal from the North Probe, then the Day Probe, and finally the Night Probe.

Seventeen minutes before hurtling into the Venusian atmosphere at 42,000 km/hr (26,099 mph), each Small Probe began transmitting data at a rate of 64 bits/sec. The Large Probe transmitted at 256 bits/sec.

Charles Hall related how, several months before the encounter with Venus, a group from the Pioneer project traveled into California's Mojave Desert. The purpose of the trip was to visit DSN's isolated Goldstone Tracking Station. There the group reviewed the station's equipment and operating procedures for obtaining data from the probes during their entry into Venus' atmosphere. The operators at Goldstone went through encounter simulations to demonstrate how the actual mission would occur. The aim was to identify and eliminate potential operational and ground equipment problems.

Operators simulated the five frequencies from the four probes and the Bus. This simulation represented the expected form of the frequencies when they arrived from the distant spacecraft fleet as it approached Venus. Equipment received radio signals from these spacecraft in an open-loop mode. That is, reception occurred without using the output to correct the input. If the frequency of a carrier emitted by any spacecraft were detected, a small blip would appear among radio noise on a monitor screen. "When I first saw this screen and the blip, it looked like a rowboat in the middle of the

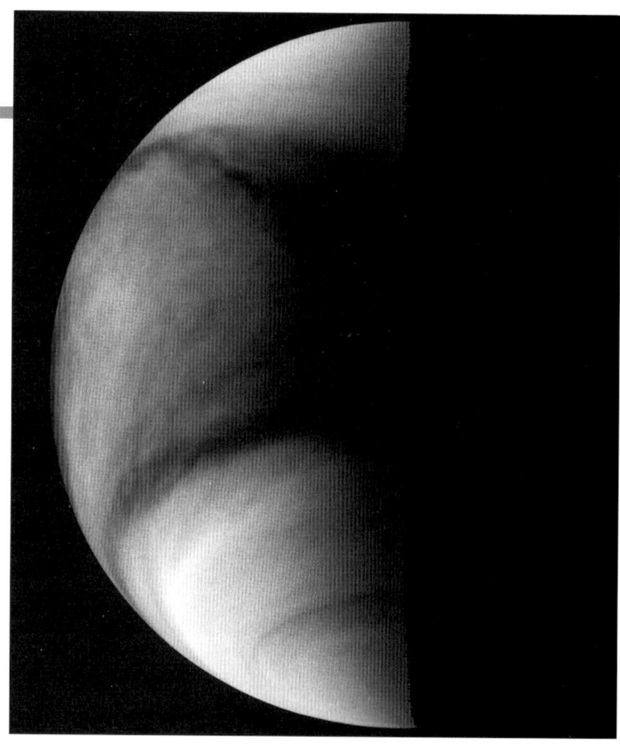

▲

Figure 5-8. (Left) The first image of Venus from the Pioneer Orbiter reached Earth on December 6, 1978. It showed the planet in a crescent phase. (Right) Subsequent images at increasing phases showed much greater detail of the Venusian cloud systems. This image was received on December 25, 1978.

Atlantic Ocean during a storm," said Hall. "We could hardly see the blip for all the noise. A crowd of dots moved up and down on the screen and only one of them was still. Highly skilled operators had to be very alert to see the stationary blip."

He recounted how the operators became very skilled in finding the blip among the noise. "They homed in on it by reducing the bandwidth so that the blip stood out clearly from the noise, bringing a pointer to the correct frequency and pressing a button. This started an automatic calculation so that the operator of the closed-loop receiver could have information to set into his control dials and get the real-time data flowing from the simulated probes. In this way, the operators were able to change to a closed-loop system and lock onto a simulated signal within seconds."

These extensive practice runs paid off when the probes reached Venus. During the encounter, friendly competition developed between the two tracking stations at Goldstone and Canberra. Which station would be first to detect the radio signals when the probes entered the atmosphere? Said Hall, "I guess the most exciting part of the mission was to hear the DSN (audio communications) as the probes were turned on and their signals were received and locked onto."

The first signal came from the Large Probe. It left the probe at 10:24 a.m. PST on December 9 and arrived at Earth 3 minutes later.

Said Hall; "When we got the message—'We've locked up on the Large Probe'—everyone cheered. Then three or four minutes later, we heard 'Forty-three (ID for the Canberra station) has locked up on a Small Probe,' and so on, right down the line. First one station and then the other announced a lockup. In retrospect, it was a tie between the stations."

One by one, and within a few minutes, each probe reestablished communications with the Pioneer Mission Operations Center (PMOC) at Ames Research Center in California. Shortly after each probe had been acquired, it was sending data to Earth. By 10:45 a.m. PST, the Operations Center reported that all instruments were operating satisfactorily.

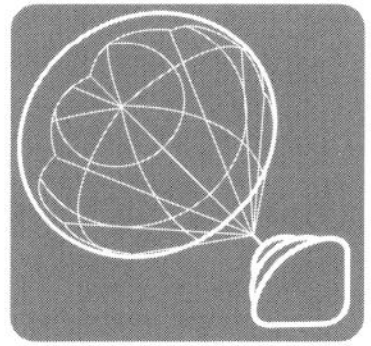

"We had been waiting for 24 days (for the Large Probe) and for 19 days (for the Small Probes). To have them come on within a split second of the times they were supposed to, and particularly to have the ground stations lockup, was quite an achievement," commented Hall. "I think that the lockup of the four probes was probably one of the most difficult tasks that the DSN has ever had to deal with."

Five minutes before each Small Probe entered the atmosphere, it deployed the two cables and weights of its yo-yo despin system. These enabled it to reduce its spin rate from 48 to 15 rpm. The Bus imparted high spin rates to disperse the probes to entry points widely spaced over the planet. However, this wide dispersion had another consequence. It meant that the smaller probes entered the Venusian atmosphere somewhat tilted off their flightpaths. The spindown of the probes allowed aerodynamic forces to line up their axes with the desired flightpaths. This had to occur quickly before heating at the edges of a probe's conical heat shield could become serious. The probes jettisoned the cables and weights immediately after spindown.

At 200 km (124 miles) above the surface of the planet, the probes plunged into the atmosphere at almost 42,000 km/hr (26,099 mph). Expected entry communications blackout occurred as the heated atmosphere flowing around the heat shield ionized. The plasma blocked the communications signal for about 10 seconds. After this blackout, the probes were moving more slowly. Now the tracking stations had to reacquire their signals at a different radio frequency. The DSN successfully locked again on all the probes' signals after each went through its individual radio blackout.

Now the most exciting part of the mission began. Enormous pressure and intense heat coupled with acid chemical corrosion in Venus' atmosphere were the great environmental challenges to engineers responsible for designing and building the probes. For example, the Large Probe had to jettison its parachute to speed its descent through the thick, lower atmosphere. In this way, the probe could telemeter data all the way down to Venus' surface. A slower descent would have heated the probe to dangerously high temperatures before it reached the lower atmosphere. This would have prevented it from obtaining information there.

An earlier chapter recounted how the probe pressure vessels were constructed from titanium. Titanium is a light but strong metal that is very difficult to machine. Deep in Venus' atmosphere, the probes would encounter enormous pressures. To withstand these pressures, designers applied experience from building bathyspheres for exploring Earth's deep oceans.

Each pressure vessel needed multiple ports so scientific instruments could access the ambient atmosphere. There were 19 such penetrations in the Large Probe's pressure vessel and 7 in each Small Probe. Protecting the vessels against the great range of outside pressures had presented many engineering difficulties. Sealing windows against pressure and heat was perhaps the most demanding task. For example, the sapphire windows often cracked when engineers tested them at high temperature. As a result, designers thickened them so they could survive the conditions on Venus. A brazed seal for use with the diamond windows had deteriorated when tested, too. Engineers replaced it with complex seals of Graphoil, Anviloy (containing 90% tungsten), and Inconel.

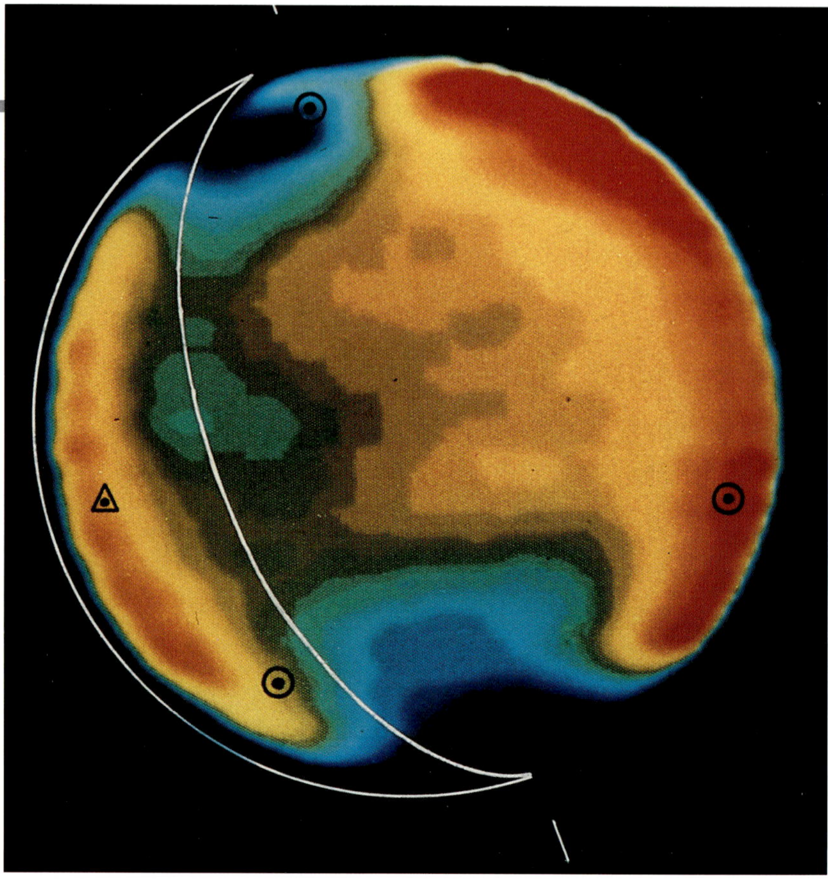

Figure 5-9. This ground-based picture of Venus was taken by Jay Apt with the 60-in. Mt. Hopkins Observatory telescope, Tucson, Arizona. He took it when the probes plunged into the Venusian atmosphere. The image was obtained at a wavelength of 11.5 micrometers. Circles show the entry points of the Small Probes. A triangle shows the Large Probe's entry point.

As the probes plunged toward Venus, engineers anxiously awaited results that would confirm the success of their designs. Although the probes had withstood rigorous tests before launch, there was always the possibility that Venus' environment could hold some surprises.

The probes were protected in several ways against heat arising from their high-speed entry into the atmosphere and from the high ambient temperature deep in that atmosphere. Heat shields, chiefly of carbon phenolic, protected the probes against excessive heating. Transfer of entry heat to the scientific instruments was controlled by mounting the instruments on heat absorbers (sinks). These consisted of beryllium shelves for the Large Probe and aluminum shelves for the Small Probes. Multilayered protective blankets of plastic sheet that were extremely heat resistant further limited heat transfer. Filling the probe's interior with the inert gas xenon reduced conduction of heat through the atmosphere inside the Small Probes. This gas conducts only about 21% the amount of heat that air does. The aim was to keep each probe's interior below 50°C (122°F) in an ambient environment with temperatures as high as 493°C (920°F).

As the time for entry approached, excitement rose dramatically. This was particularly true at the PMOC and at the many contractors' plants that helped design the Pioneer Venus vehicles. Many years of design and exhaustive ground-based simulations were about to be put to their ultimate test. Everyone waited as the four probes plowed through the global haze and sulfuric acid clouds, through the violent winds, and the hot carbon dioxide of Venus. Entry points are on Figure 5-9.

Table 5-2 summarizes the sequence of some important events that occurred during the entry of the Pioneer Venus probes. On entry (Figure 5-10), the Large Probe decelerated from 41,800 to 727 km/hr (25,975 to 452 mph) within 38 seconds. During this period, its onboard memory stored data for later transmission after radio blackout. Its parachute opened at 10:45 a.m. PST to further slow its speed of descent. Its forward aeroshell heat shield jettisoned to expose all apertures and windows for the operation's descent phase. Forty-three seconds after entry, instruments on the Large Probe operated normally and returned data to Earth. This was at an altitude

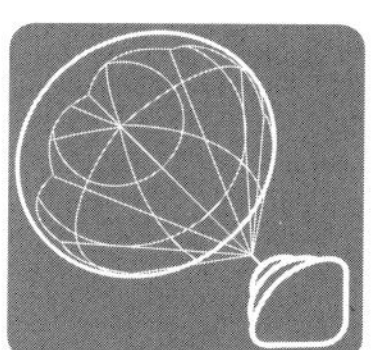

Table 5-2. Important Entry Events

Parameter	Time at spacecraft[a], hr:min:sec, PST			
	Large probe	North probe	Day probe	Night probe
End of coast timing	10:24:26	10:27:57	10:30:27	10:34:08
Initiate telemetry	10:29:27	10:32:55	10:35:27	10:39:08
200-km (124-mile) entry	10:45:32	10:49:40	10:52:18	10:56:13
Radio blackout began	10:45:53	10:49:58	10:52:40	10:56:27
Signal locked on	10:46:55	10:50:55	10:53:46	10:57:48
Jettison parachute	11:03:28	NA	NA	NA
Impact with surface	11:39:53	11:42:40	11:47:59	11:52:05
Signal ended	11:39:53	11:42:40	12:55:34	11:52:07
Bus entry (200 km; 124 miles)	12:21:52			
Bus signal ended (110 km; 68 miles)	12:22:55			
Durations				
Descent time (entry to impact)	54:21	53:00	55:41	55:52
Blackout time (signal loss to relock)	00:62	00:57	00:66	00:81
Time on parachute (large probe only)	~17:07			
Surface operations (impact – signal end)	None	None	67:37	00:02

[a]Earth-received times were approximately 3 min later than the above spacecraft times.

of about 66 km (41 miles). Seventeen minutes later, at 11:02 a.m. PST, and at an altitude of 45 km (28 miles) above Venus' hot surface, the probe jettisoned its parachute (see Figure 5-10). Rotating slowly under the influence of its spin vanes, the probe continued to plunge down. The dense atmosphere slowed its descent, just as a huge metal ball would slow if sinking into Earth's ocean.

The aerodynamically stable pressure vessel reached Venus' surface about 39 minutes after the probe had jettisoned its parachute. The probe hit the surface at only 32 km/hr (20 mph). It landed near Venus' equator on the dayside at 11:41 a.m. PST, some 55 minutes after first encountering the Venusian atmosphere. Its radio signals ended abruptly at impact.

Five minutes before the Small Probes encountered the peak deceleration pulse of atmospheric entry, each probe's command unit ordered the blackout format. This stored spacecraft data in an internal memory. It also stored heat-shield temperature and accelerometer measurements for the atmospheric structure experiment. This procedure ensured that no data were lost during the 10- to 15-second communications blackout at entry. The probes transmitted these data later during the descent.

The Small Probes, entering the atmosphere within a few minutes of each other, quickly slowed down. This occurred between 10:50 and 10:56 a.m. PST. The atmosphere retarded their fall to the surface without the use of parachutes. Because the flightpath angles of the three Small Probes varied considerably, each probe's deceleration rate and entry heating also varied widely. Peak decelerations ranged from 220 to 456 g (1 g is 32 ft/sec/sec).

At 10:51 a.m. PST, the nephelometer's window opened on the North Probe. The instrument began to gather data on locations and densities of cloud layers. The atmospheric structure and net flux radiometer housing doors opened

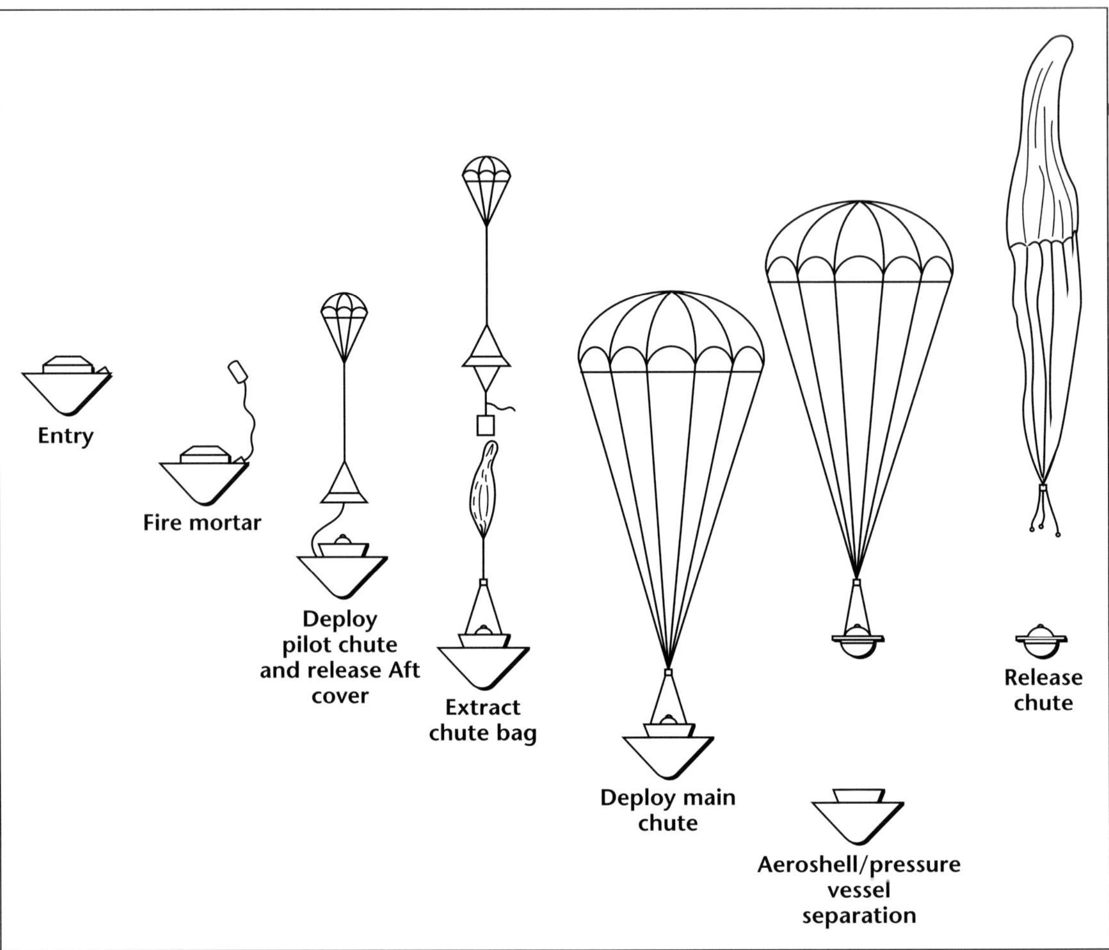

Figure 5-10. Entry sequence of the Large Probe was more complicated than that of the Small Probes. A parachute had to slow its descent. When the parachute's work was done, the spacecraft jettisoned it.

next. These instruments started telemetering to Earth data about the atmosphere's thermal structure. Instrument booms deployed. Within the next 6 minutes, similar sequences had started on the other Small Probes.

As instrument compartment doors opened on either side of each Small Probe's afterbody, their drag effects on the atmosphere further reduced each spacecraft's spin rate. A small vane on the pressure inlet prevented the despin rate from falling to zero. This would have prevented instruments from making observations over a full rotation of the probe. Now the upper descent phase began, with the three probes in the altitude range of 72 to 65 km (44 to 40 miles) and all instruments operating.

As the probes penetrated deeper into thicker atmosphere, it interfered with radio communication. Signals received at Earth were weakened. At entry plus 16.4 minutes and at an altitude of about 30 km (18 miles), the bit rate of data transmission from probes to Earth automatically reduced to 16 bits/sec. This ensured that Earth stations would receive data from the lower atmospheric regions. The DSN now had to achieve a third lockup on each probe's transmission. Again, it was highly successful, and no data were lost in the process.

From that point on, the three probes descended into Venus' increasingly dense atmosphere. They impacted the surface at 36 km/hr (22 mph) 57 minutes after their entries. Unlike the Large Probe, the Small Probes retained their heat shields to the surface. The atmosphere's density is so great that the drag of these aerodynamic surfaces slowed the probes to their desired descent speed.

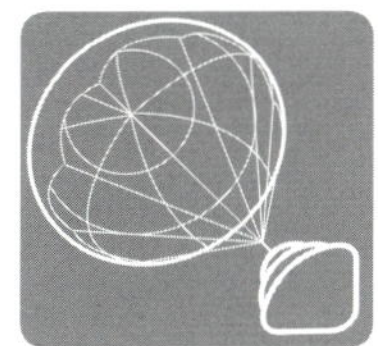

The North Probe landed at 11:47 a.m. PST in darkness near northern polar regions. The Day Probe went into the southern hemisphere on the dayside and landed at 11:50 a.m. It kicked up a dust cloud that took several minutes to settle. The Night Probe went down in darkness onto the surface in the southern hemisphere at 11:53 a.m. PST. Signals from the North Probe and the Night Probe ended at impact. However, transmissions continued from the Day Probe for another 68 minutes (Figure 5-11) before it, too, became silent. Engineering data radioed back from the Day Probe showed that its internal temperature climbed steadily to a high of 126°C (260°F). Then its batteries were depleted, and its radio became silent. The internal pressure monitors showed that the pressure within the probe rose as expected for a sealed bottle on the surface of Venus. The temperature increase gradually caused an expected increase in internal pressure. There was no evidence of any leaks into the probe from the atmosphere following the impact. It was clear that the seals had withstood the real-life test of impact with the hot surface of Venus.

Table 5-3 shows the locations on Venus where the probes impacted and the conditions at the impact points. These locations were very close to the points targeted before the probes separated from the Bus.

Meanwhile, the Multiprobe Bus hurtled toward Venus close behind the probes. On December 8, controllers reoriented the Bus to its final entry angle. They calibrated its instruments and released the cap covering the inlet to the neutral mass spectrometer. Entry was scheduled for 12:21 p.m. PST on December 9. This was about 96 minutes after the first probe entered and 88 minutes after the last probe had entered.

Figure 5-11. This is how a Pioneer Venus probe might have looked on the hot surface of Venus. Although engineers did not design the probes to withstand impact, there was a chance that one might survive and transmit some data back to Earth. One Small Probe did survive and sent data from the surface for 67 minutes.

The Bus plunged into the atmosphere on the planet's dayside at a high latitude in the southern hemisphere. Table 5-4 gives the Bus' entry position at an altitude of 200 km (124 miles) and the locations of the subsolar and sub-Earth points. These are the points on Venus' surface where the Sun and Earth would appear directly overhead to an observer.

Since the Bus had no heat shield to protect it from high-speed entry, scientists expected to gather data for only 2 minutes before it burned up. Radio transmissions from the Bus poured back to Earth carrying scientific data at a rate of 1024 bits/sec. These data carried information about the composition of Venus' very high atmosphere, including the region where the ionosphere is most dense. The other probes could not explore this region. They could gather no data from external sensors until they had been slowed by the atmosphere and were much deeper within it.

The Bus burned up at 12:23 p.m. PST, and the uniquely exciting phase of the entry part of the mission concluded. It had lasted for only

Table 5-3. Pioneer Venus Multiprobe Impacts

Probe	Latitude, deg	Longitude, E deg	Solar zenith angle, deg	Local Venus time, hr:min
Large	4.4 N	304.0	65.7	7:38
North	59.3 N	4.8	108.0	3:35
Day	31.3 S	317.0	79.9	6:46
Night	28.7 S	56.7	150.7	0:07

Table 5-4. Pioneer Venus Bus Entry and Location of Sun and Earth Subpoints

Probe	Latitude, deg	Longitude, E deg	Solar zenith angle, deg	Local Venus time, hr:min
Bus entry at 200 km	37.9 S	290.9	60.7	8:30
Subsolar	0.5 S	238.5	0	12:00
Sub-Earth	1.6 S	1.7	123.1	3:47

90 minutes. Yet in that short period, the probes and the Bus had recorded data for a completely new look at the complex atmosphere of Earth's sister planet (Figure 5-12). During the following few days, scientists completed preliminary data analysis and announced some unexpected discoveries. Then the mission settled down to the equally fascinating but more lengthy process of observing Venus from the Orbiter. This lasted many Venus sidereal days.

There were major findings from the probes. The four probes measured the atmosphere's structure, temperature, pressure, density, and wave structures. They started at altitudes of 138 km (Large Probe) down to Venus' surface. The Small Probes measured structure from 133 km (Night Probe), 126 km (Day Probe), and 120 km (North Probe). Chapter 6 discusses the science results in detail. However, there were some discoveries that produced much excitement in the days immediately following the encounter.

An unexpected result was concentrations of primordial argon and neon several hundred times those on Earth. This finding conflicted with most accepted theories about the origin of the Solar System. Those theories argued that the Sun and planets formed about the same time. They claimed the planets gradually grew through planetesimals and planetary embryos from a gas cloud surrounding the Sun and composed of the same elements as the Sun. The next chapter discusses the isotopic findings in detail.

Some Puzzling Results

How did the probes and their instruments withstand the rigors of the descent into Venus' atmosphere? Scientists had been concerned that, when the probes went through the clouds, droplets might condense on the inlet to the mass spectrometer. To prevent such contamination, engineers had placed a heater coil around the inlet. Nevertheless, the inlet did become blocked, and observers noticed a change in the amount of gas entering the instrument. Later in the descent, when the temperature had risen, they observed peaks of sulfur in the data. It appeared that a large drop of sulfuric acid had blocked the inlet. When it later boiled off, its components entered the instruments and were revealed in the data.

There were some anomalies, or irregularities, with all the probes. Anomalous events appeared in the engineering data and in the science data at approximately the same altitude in all four probes.

The first signs came from the sensors of the atmospheric structure experiment at an altitude between 12 and 14 km (7.5 and 8.7 miles). Soon afterward, external sensors of the net flux radiometer on the North Probe, Day Probe, and Night Probe suddenly failed at approximately the same altitude. In the data from other scientific instruments and from engineering transducers, other anomalies occurred just before, during, and after these failures. Table 5-5 summarizes these anomalies.

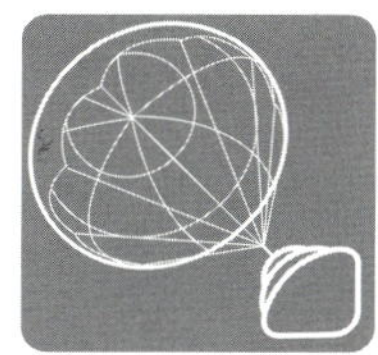

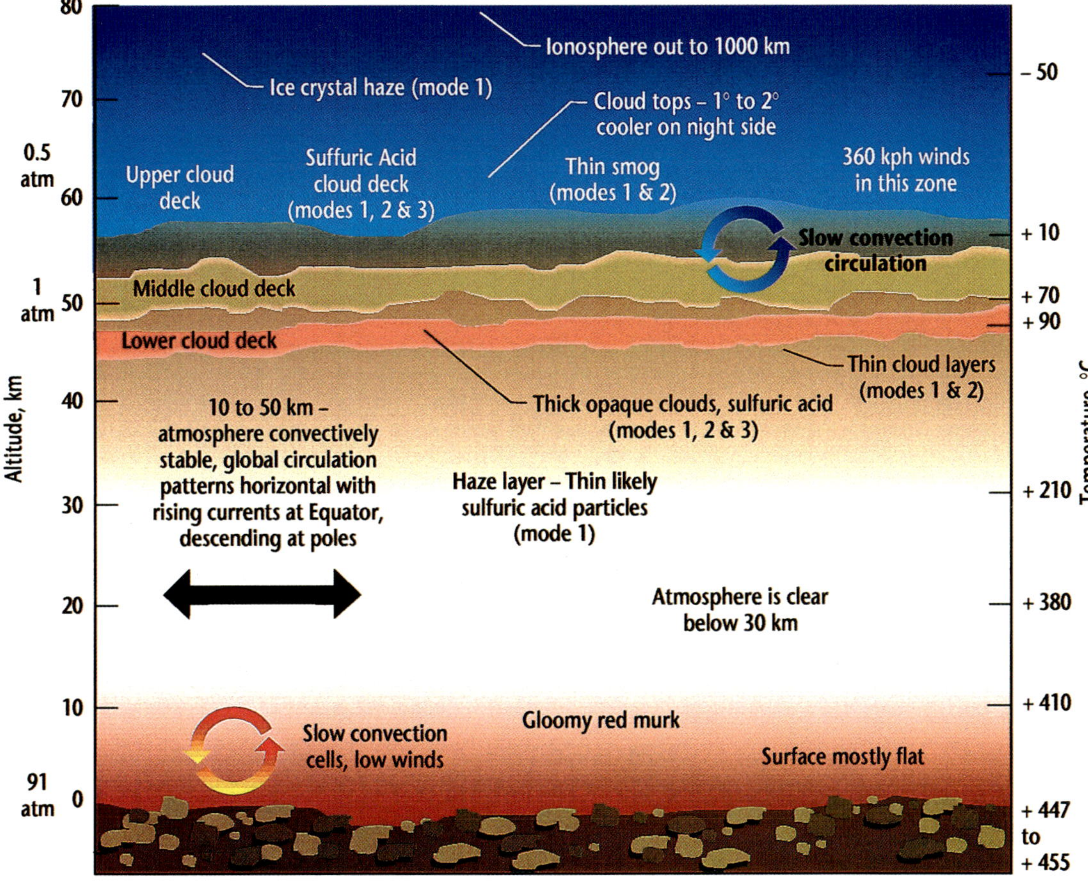

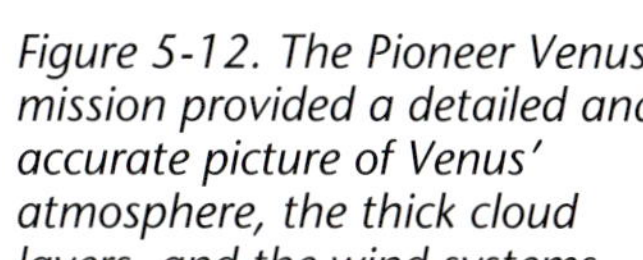

Figure 5-12. The Pioneer Venus mission provided a detailed and accurate picture of Venus' atmosphere, the thick cloud layers, and the wind systems.

It seems unreasonable to assume that all these different instruments failed together and at precisely the same condition. A cause other than simple, virtually simultaneous equipment failure seemed likely.

The temperature sensors (Figure 5-13) of the atmospheric structure experiment were exposed to Venus' atmosphere, and they showed anomalies. However, it seemed clear from the data that the temperature sensors did not physically break because an expected electrical resistance through the sensor of 25 ohms remained. Partial shorting of the insulation of the T1 fine-wire sensors while in the clouds indicated continuous acid films on the sensors. However, this cleared as the probes descended lower into higher temperatures. Also, the shorting effects within the clouds varied for the different probes, but the anomalies all occurred later at the same altitude. That is, they occurred at the same temperature and pressure levels in the atmosphere. Moreover, the T1 and T2 sensor elements exhibited anomalies almost at the same time, despite their different physical configurations. The T1 sensors each consisted of a coil of fine platinum wire wound on a frame. The T2 sensors were more robust. They consisted of platinum wire bonded as a resistance thermometer on top of a thin glass insulating layer. It is important to note that the sensors that failed at almost the same time were made of different materials and that their electronics were isolated from each other.

Another anomaly involved the sensor boom for the atmospheric structure and net flux radiometer experiments. Its telemetered change from deployed to stowed position was a mechanical impossibility. Investigators made a post-flight analysis of identical boom status

Table 5-5. Anomalies Experienced by Probes

Anomaly	Large probe	North probe	Day probe	Night probe
Apparent failure of temperature sensors	X	X	X	X
Apparent failure of net flux radiometer fluxplate temperature sensors		X	X	X
Abrupt changes and spikes in data from net flux radiometer		X	X	X
Change in the indicated deployment status of the atmosphere structure temperature sensor and net flux radiometer booms		X	X	X
Erratic data from two thermocouples embedded in the heat shield		X	X	X
Erratic data from a thermistor measuring junction temperature of the heat-shield thermocouples		X	X	X
Slight variation of current and voltage levels in the power bus		X	X	X
Abrupt changes in cloud particle size laser alignment monitor	X	NA	NA	NA
Decrease in the intensity of the beam returned to the cloud-particle-size spectrometer	X	NA	NA	NA
Noise in the data from the infrared radiometer	X	NA	NA	NA
Spikes in the data monitoring the ion pump current of the mass spectrometer analyzer	X	NA	NA	NA
Spurious reading from the thermocouples when the heat shield was dropped from the probe	X	NA	NA	NA

switches. They concluded that failure of these switches under conditions of high temperature and pressure was a likely cause.

Investigators also had an explanation for anomalies in the Large Probe's housekeeping data, particularly the strange readings from the heat-shield thermocouple and thermistor. They reasoned the probe became covered with a plasma of charged particles, but scientists no longer consider this likely. An apparent reading from a thermocouple in the Large Probe's heat shield occurred after the leads had severed and the heat shield became detached. Somehow an electrical potential of 0.2 mV had been created between the ends of the severed leads. This potential exhibited slight changes during the rest of the descent to the surface. One suggestion was that the severed leads acted as a Langmuir probe in a plasma. However, there was no known source for such a plasma.

If they could have occurred, static discharges within or outside the probe might explain several anomalies. These included anomalies of changes in the Large Probe's transponder static phase error and receiver automatic gain control. They also could explain jumps in internal pressure and temperature readings.

Investigators considered charge buildup on the probes, but the nephelometer showed a clear atmosphere below 40 km (25 miles). So a major question was how such a charge might build up in a particle-free atmosphere. Although the atmosphere was optically clear, it might be ionized. It could literally be swarming with submicroscopic ions created by cosmic-ray reactions at the molecular levels as opposed to the particle level. Such chemical reactions could build up a charge in a clear atmosphere. However, the existence of these anomalies is now in question. In-depth review suggests the measurements were within normal operating limits.

The diamond window heater for the infrared flux radiometer burned out. The instrument measured data from the window frame, and this gave a spurious input. Investigators first thought it was an anomaly.

One possible cause for the window heater failure is related to the tantalum heater sheath. At high temperatures, there is a reaction between tantalum, carbon dioxide, and acid. Both of the latter are present in quantity in the Venusian atmosphere. Engineers speculate that holes developed in the tantalum heater sheath from such a reaction. The insulation could have then become contaminated enough to provide conductive paths. Such paths could have allowed an electrical short between the heater and the spacecraft ground. This would have shorted the heater circuit and blown its fuse.

Several factors can explain most of the probe anomalies. These include effects arising from an unexpected electrical interaction between the probes and the atmosphere, or from chemical reactions between the atmospheric gases and probe materials. The source for a reaction of such widespread effect is, however, still uncertain.

The performance of these probes in the extremely inhospitable atmosphere of Venus was remarkable. They gathered a wealth of important new data just as project scientists planned. Also, technology had been proved for penetrating planetary atmospheres and gathering data under conditions of extremely high temperatures and pressures. This new technology held the potential for exploring

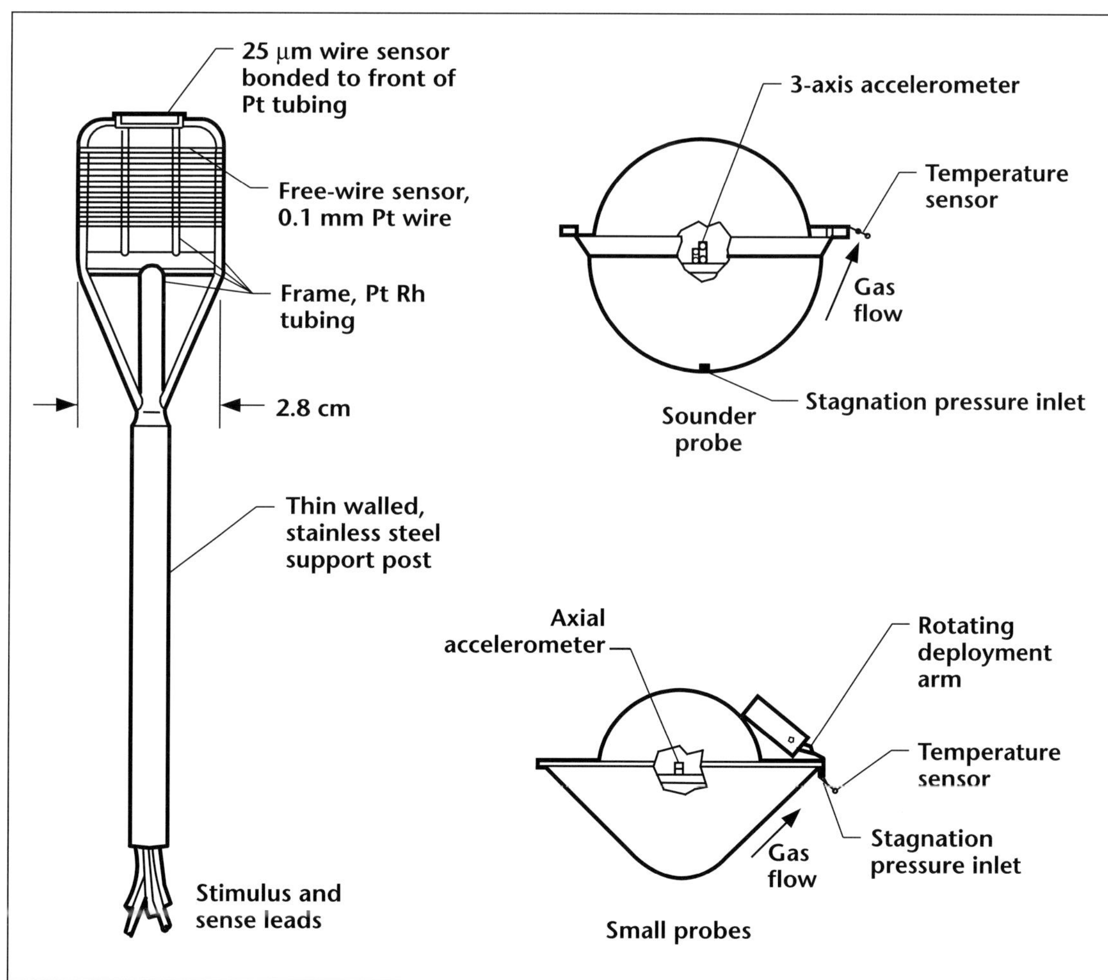

Figure 5-13. As the probes reached deep into the atmosphere, several instruments produced unexpected readings. These included the atmospheric structure temperature sensors in this figure. Sensors of entirely different design produced bizarre results at the same altitude.

the many bizarre atmospheres of the planets in the outer Solar System.

Further Analysis of the Anomalies

A workshop meeting held at NASA Ames Research Center on September 28 and 29, 1993, reviewed these probe anomalies again. This was done in connection with planning for the design of a Discovery Venus Probe. Participants included probe system engineers, project office personnel (retired and active), probe scientists, instrument designers, and atmospheric scientists. These latter included chemists, dynamicists, and electrodynamicists.

Workshop attendees reviewed anomalies that occurred at or below 12.5 km (7.75 miles) in detail. As this chapter described earlier, instruments outside the sealed pressure vessels had exhibited problems on all four probes. Temperature sensors had continued to report data, but not valid data. Net flux radiometers had shown a sudden decrease in net flux toward zero. Also, box cover status signals, on boxes from which the temperature and net flux radiometer had been deployed on the Small Probes, had indicated the sensors had been restored. This was an impossibility. The Large Probe's thermocouple wire, cut before parachute deployment, had indicated signals of a few millivolts. By contrast, the scientific data from all internal sensors continued without anomaly throughout the descent.

The workshop clarified the anomalies. It also corrected some mistaken impressions circulated mainly by word of mouth earlier during the mission. Participants credibly accounted for a few of the anomalies during Phase I. Yet, despite many speculations and suggestions, there was no clear-cut explanation for the remaining array of nearly simultaneous events.

The workshop participants considered several atmospheric phenomena to explain these anomalies. All appeared possible but needed further investigation. They were:

1) Chemical interactions such as clouds acting on the harness and sensors to produce sulfuric acid and carbon dioxide oxidation of titanium parts and harness materials.

2) Conductive vapors condensing on the external sensors in the deep atmosphere, leading to electrical shorts.

3) Probe charging with subsequent electrical breakdown of the atmosphere, possibly leading to sparks that could ignite fires in external materials such as the Kapton insulation. Also, many metals burn in carbon dioxide, and the flammability increases with increasing pressure. For example, zirconium, magnesium, and titanium ignite easily in pure carbon dioxide.

Soviet Venera and Vega probes and landers also carried many external instruments. Titanium was an important element in their construction, too. Soviet scientists have stated, however, these spacecraft did not experience anomalous behavior. It seems that particular probe or instrument design features must explain the Pioneer Venus anomalies. Investigators needed to identify these features.

The workshop concluded that although the data are not now sufficient for conclusive proof, investigators have identified the most probable causes of the anomalies. The most likely hardware event is insulation breakdown of the external harness. This resulted from chemical interaction with the high temperature and pressure of the carbon-dioxide atmosphere after exposure to the clouds of sulfuric acid. Laboratory testing before the workshop had not ruled out that possibility. The probable interaction between the probe and the

atmosphere resulted in a charge buildup during transit through the clouds (with charge retention to breakdown occurring at the anomaly altitude).

The workshop recommended that investigators continue to try to reproduce these anomalous effects and attribute their cause to a few credible explanations. This testing also would have the potential of identifying other possible atmospheric interactions. Engineers then could design deep atmosphere probes that would not succumb to these anomalies.

Nominal Mission of the Orbiter

Preliminary science discoveries came from the Orbiter experiments. Data from the Orbiter's first radar map (Figure 5-14) suggested that Venus' topography might be similar to Earth's. The data revealed high features similar to mountains and extensive, relatively flat areas. Some of the radar mapper's first preliminary scans were in a region of Venus previously unexplored by radar. This was a strip that extends for about 1900 km (1180 miles). In this region, much of the surface appeared relatively flat. It was similar to Earth's surface and quite different from the rough, cratered surfaces of Mars, Mercury, and the Moon.

After the first two dozen orbits, a serious setback occurred. The radar instrument stopped working. Teams of scientists and engineers tried several remedies, but to no avail. This failure greatly disappointed everyone because the radar had started to reveal tantalizing details of the planet's surface. When all corrective measures failed, controllers turned off the radar mapper. During the down time, mission scientists analyzed the instrument's design.

However, they came up with no additional corrective ideas. Yet, when controllers turned on the radar again a month later, it worked (although not quite normally). The problem seemed transient, associated with operating the instrument for periods longer than 10 hours. Controllers had operated the instrument for the first orbits and not turned it off. Analysis led to the conclusion that an electrical charge may have accumulated in its sensitive logic circuitry. So the experiment team leader, Gordon Pettengill, and project personnel decided to use new operating modes for the instrument. During each orbit, they operated it for a while and then turned it off. This periodic use resulted in normal operation of the radar mapper within about 10 days. Afterward, it operated satisfactorily. Although this failure caused a month of radar data loss, the extended mission later covered the missed areas.

Another disappointment with Pioneer Orbiter was not as happily resolved. The infrared

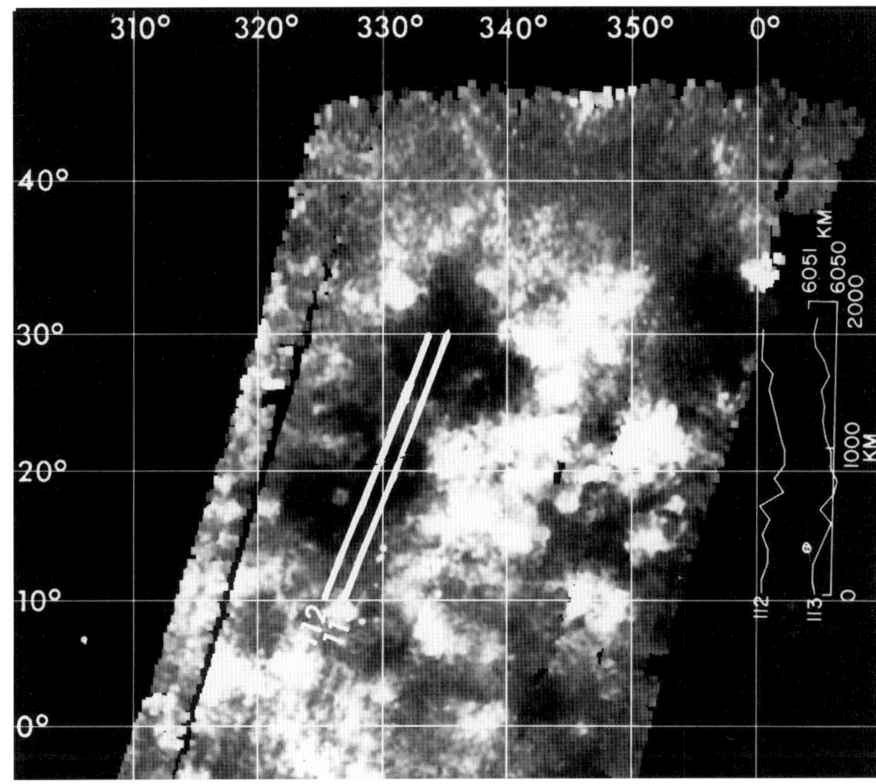

Figure 5-14. Pioneer Orbiter's first radar scans of Venus' surface produced intriguing new maps of the cloud-hidden surface. The instrument also measured elevations and revealed enormous mountains, continental masses, and deep valleys.

Table 5-6. Orbital Parameters for Nominal Mission

Parameter	Value
Periapsis, km (miles)	150–200 (93–124)
Apoapsis, km (miles)	66,900 (41,572)
Eccentricity	0.842
Average period, hr	24.03
Inclination to equator, deg	105.6
Periapsis latitude, deg	17.0 N
Periapsis longitude, deg (for orbit 5)	170.2 E

radiometer failed when the spacecraft was on about its seventieth orbit. Despite many attempts to correct the failure, personnel could not bring the instrument back into operation. Investigators believed that the problem arose in the instrument's power supply.

Other instruments experienced minor problems from time to time, but all were resolved. The instruments recovered quickly, and they gathered data throughout the mission.

Mission planners selected the initial altitude of periapsis high enough for negligible atmospheric drag on the spacecraft during the first orbit. They had to choose a very conservative altitude because information about Venus' upper atmosphere was sparse. As they received information from the spacecraft, controllers commanded seven periapsis correction maneuvers during the first 10 orbits. These reduced the periapsis to the scientifically desired 150 km (93 miles) above the mean surface of Venus. In this way, they achieved the orbital parameters for the nominal mission (Table 5-6).

Perturbations from the Sun's gravity field affected the periapsis position of the orbit. This required control by thrusters to maintain the variations in altitude within predetermined limits. Without corrections to the orbit by use of these thrusters, the Sun's gravity would have pushed the periapsis out from the planet. That is, it would have raised its altitude. To keep the periapsis within the range of altitudes desired by the scientists, periodic corrections were required throughout the entire nominal mission.

Figure 5-15 shows a plot of periapsis altitude for the early part of the mission. It illustrates how the altitude of periapsis changed through the nominal and into the extended mission. During the first few weeks of the spacecraft's operation in orbit, controllers' commands to lower the periapsis to 150 km (93 miles) were issued before it passed from the dayside to the nightside of Venus. The atmosphere is less dense on the nightside of the planet than on its dayside. Because of this, they commanded the periapsis several times to 142 km (88 miles) while it was on the nightside to allow the spacecraft to sample deeper into the atmosphere.

The Orbiter was oriented with its spin axis perpendicular to the ecliptic plane. The despun antenna was to the south end of the spacecraft. This orientation continued through the mission, except for several short periods to observe comets. Initially, the view of the north polar region was better than that of the south polar region. Two factors accounted for this. First, the scientific instruments were on an equipment shelf near the antenna's base and, second, periapsis occurred at a northern latitude.

Figure 5-16 shows how some orbit relationships varied during the nominal 243-day mission. The Sun-Venus-Pioneer orbit system

appears at four positions in the sidereal year from December 9, 1978, to July 22, 1979. Since the orbit was fixed in an inertial reference frame, the lines of apsides remained "parallel" to one another at each of these four positions. The local time of periapsis increased by 1.6° each Earth day. At periapsis, the Orbiter first sampled the dayside upper atmosphere of Venus. Then, after several weeks of moving at 1.6° per day, the periapsis crossed the evening terminator. Now the spacecraft sampled the nightside atmosphere and ionosphere at each periapsis. Later still, the periapsis crossed the morning terminator, and the spacecraft sampled the dayside again. The spacecraft crossed the evening terminator again at the end of the nominal mission. Instruments thus obtained data at periapsis and along the orbit for all Venus local times in a period of 224.7 Earth days.

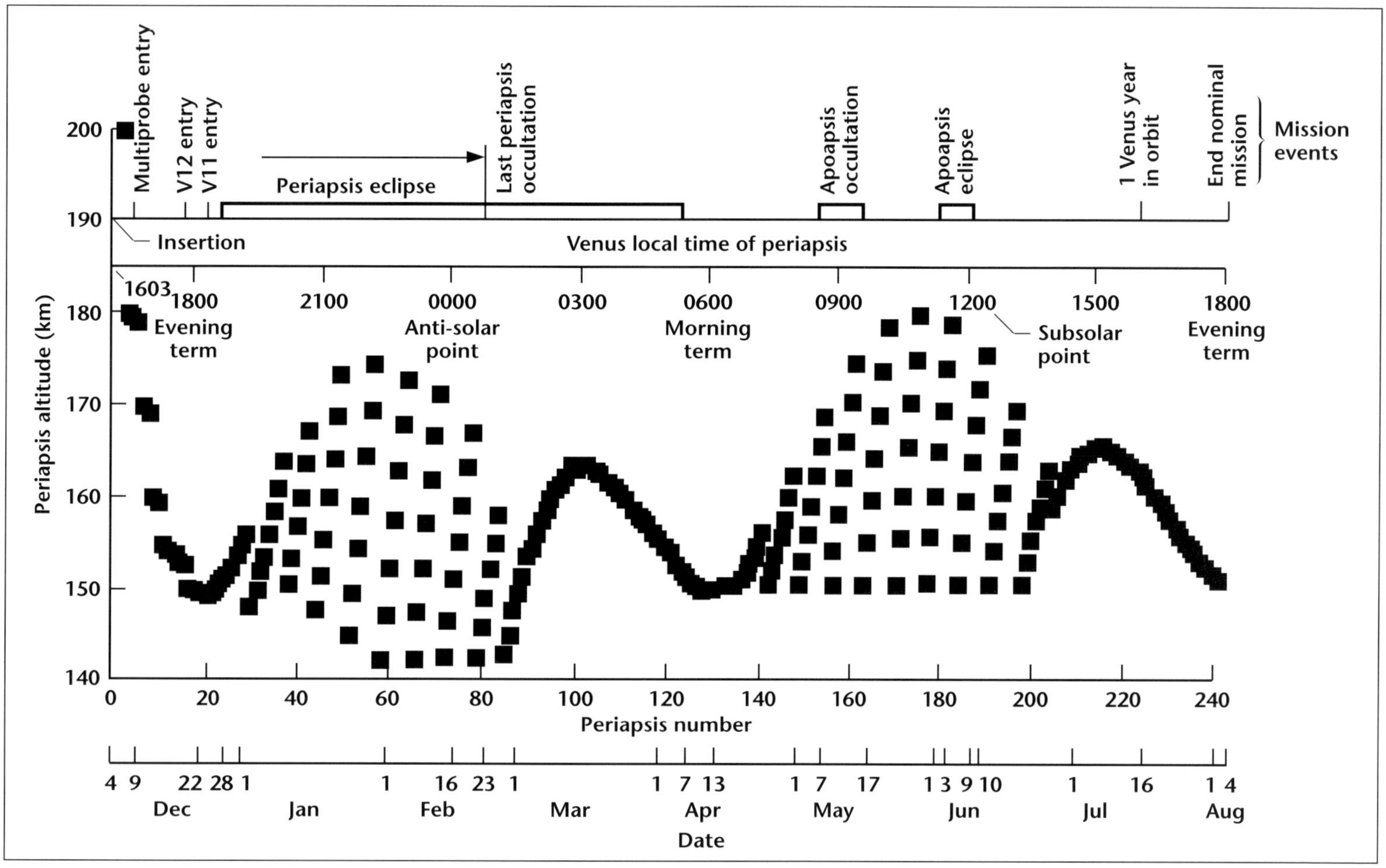

However, because of Venus' retrograde axial rotation, the longitude of periapsis moved relative to the solid body of the planet at 1.48° per day, that is, per orbit. So, the spacecraft needed 243 Earth days to observe all longitudes on the solid planet. The Orbiter completed its nominal mission on August 4, 1979. It had conserved enough propellant to stay in orbit for at least another two sidereal periods. That amounted to another 486 days. In fact, it operated for many sidereal days. This provided

Figure 5-15. This plot shows the altitude of the periapsis during the nominal mission of the Orbiter and partway into the extended mission. The periods of eclipses and occultations are identified.

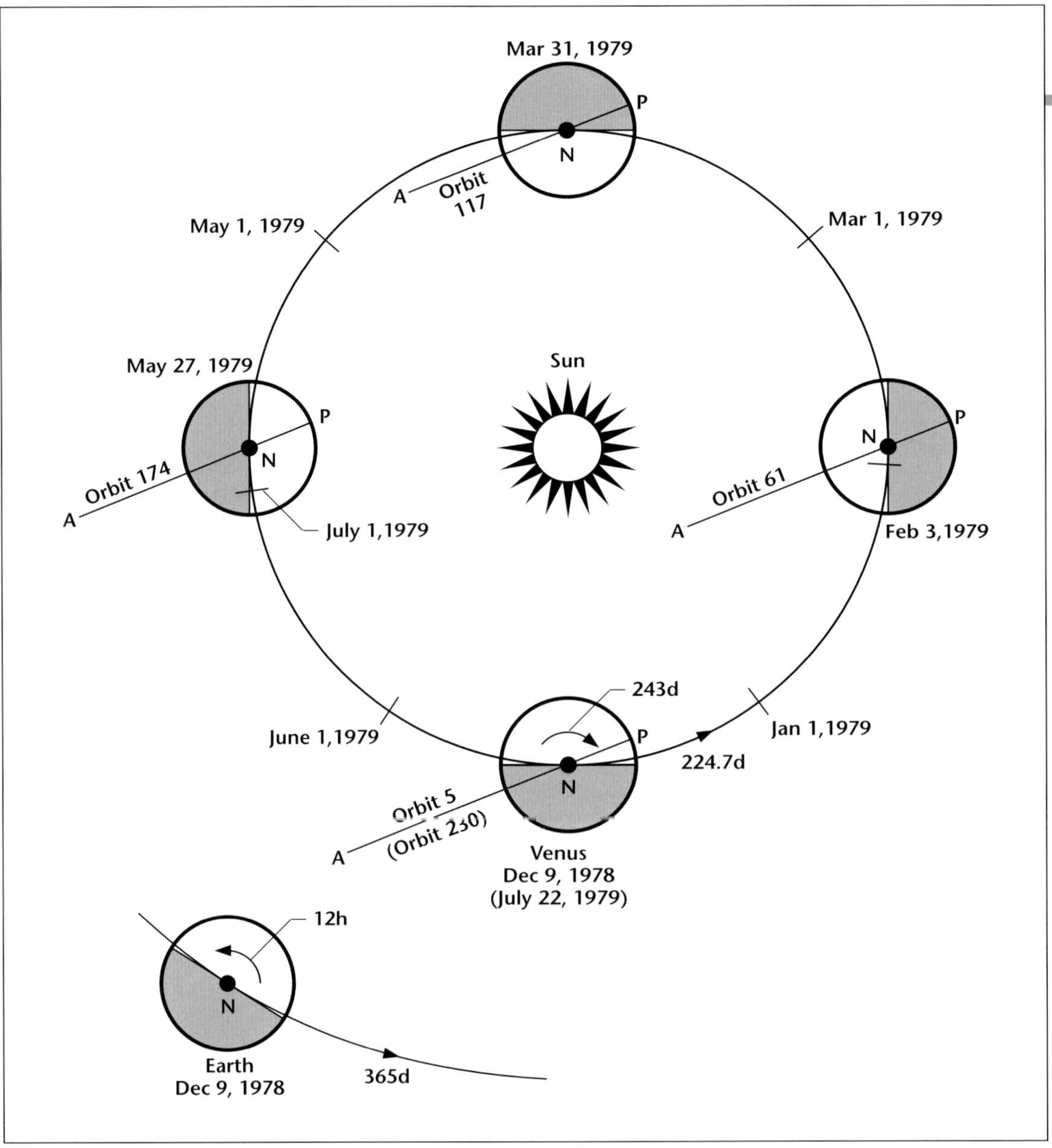

Figure 5-16. This drawing of the Sun-Venus-Orbiter geometry illustrates how the periapsis moved around the planet during the Venusian sidereal year to sample day and night hemispheres. Because the planet rotates in a retrograde direction, more than one Venusian sidereal year was required for periapsis to move over all longitudes of the planet.

a tremendous scientific bonus from a relatively inexpensive planetary mission.

The Science Steering Group (SSG) and mission management decided to continue the basic periodic control of the orbit until about orbit 600 on July 27, 1980. Then they allowed the periapsis altitude to rise slowly. Initially it rose at a rate of 400 km (249 miles) each 243 days. By 1984, it was rising at only 225 km (140 miles) each 243 days. The apoapsis descended at an identical rate, and the period of the orbit remained constant.

The Extended Mission of Orbiter

The Pioneer Venus Orbiter reached Venus on December 4, 1978, and controllers placed it into a highly eccentric orbit. Then they used changes in the altitude of periapsis as the basis for dividing the mission into three separate phases. Phase I was the initial 19 months when controllers maintained periapsis at low altitudes of about 150 km (93 miles). Phase II began when propellant began to run low. Solar gravitational perturbations were then allowed to cause periapsis to rise out of the thermosphere and the main ionosphere. Eventually,

in 1986, periapsis reached an altitude of about 2300 km (1430 miles). It then started to descend. Phase III, or the Entry Phase, began in April 1991. This was when periapsis was below 1000 km (620 miles) and instruments made direct measurements within the main ionosphere.

Project management changed for Phase II. Richard O. Fimmel became Pioneer Project Manager when Charles Hall retired from NASA.

A solar gravitational effect caused periapsis to rise during the first half of Phase II and then to fall during the second half of Phase II and into Phase III. It acted in a distinct cyclic fashion. There were periods of decline interrupted twice each Venus year by increases in altitude. These took the shape of S-curves (Figure 5-17). They occurred as the orbit plane of the spacecraft passed nearly perpendicular to the Sun. Each cycle was associated with a 12-hour sweep of local solar time. These altitude and local time changes generated opportunities to observe various phenomena of scientific importance occurring in Venus' environment.

By the middle of 1992, periapsis was again low enough for instruments to resume making measurements within the ionosphere and thermosphere. They also were able to scan the limb and observe the thermosphere in ultraviolet light. Scientists had an equally important Phase III goal. They wanted to extend the observations into much denser atmospheric regions than was acceptable during Phase I. Also, during the final phase of Orbiter's mission, instruments made measurements at a different part of the solar cycle. These occurred within those higher altitude regions that the spacecraft examined earlier in the mission.

The most critical part of Phase III was the period of final encounter, which began early in September, 1992. During that month, navigators used most of the remaining hydrazine propellant in a series of maneuvers to lift periapsis. Mission controllers and scientists hoped that sufficient propellant remained to delay entry long enough to reach the next S-curve. Then they would not need further maneuvers to maintain the altitude of periapsis. At that time, periapsis would move to the planet's dayside. This would correspond to about 100 orbits (100 days) in the range of orbit numbers 5020 to 5120. If additional propellant remained after the last periapsis maneuver, mission planners intended to use it at the end of the S-curve. That would probably happen in the middle of December, 1992. The extra propellant would extend measurements further into the midday thermosphere and ionosphere.

In 1989, NASA established a task force to identify the most important scientific goals for the Entry Phase. It also provided the Pioneer Project Office (PPO) with operational guidelines on how the mission might best use the Orbiter and its instruments to achieve these goals. The guidelines and goals had to stay within the limitations of the orbit, the spacecraft, and the DSN. Detailed planning was important to take full advantage of the measurement opportunities at that time.

The local time and altitude of periapsis largely determined the kinds of phenomena the spacecraft could encounter. At the same time, orbital mechanics placed important constraints on the quantity and quality of the data. For example, the telemetry bit rate, which controlled the temporal resolution of the measurements, varied widely with the distance from Earth to Venus. Also, instruments could not retrieve data near solar conjunction. This was because DSN antennas picked up solar radio noise as Venus moved close to the Sun (as viewed from Earth).

Figure 5-17. *The three stages of the Pioneer Venus Orbiter mission appear plotted below the graph of solar activity for the same period. Phase I provided* in situ *measurements deep into the thermosphere and the main body of the ionosphere at solar maximum. The changing altitude of the periapsis during Phase II permitted the spacecraft to explore the upper ionosphere, magnetosheath, and bow shock. Phase III allowed instruments to reexamine the ionosphere and thermosphere at moderate levels of solar activity and down to lower altitudes than during Phase I.*

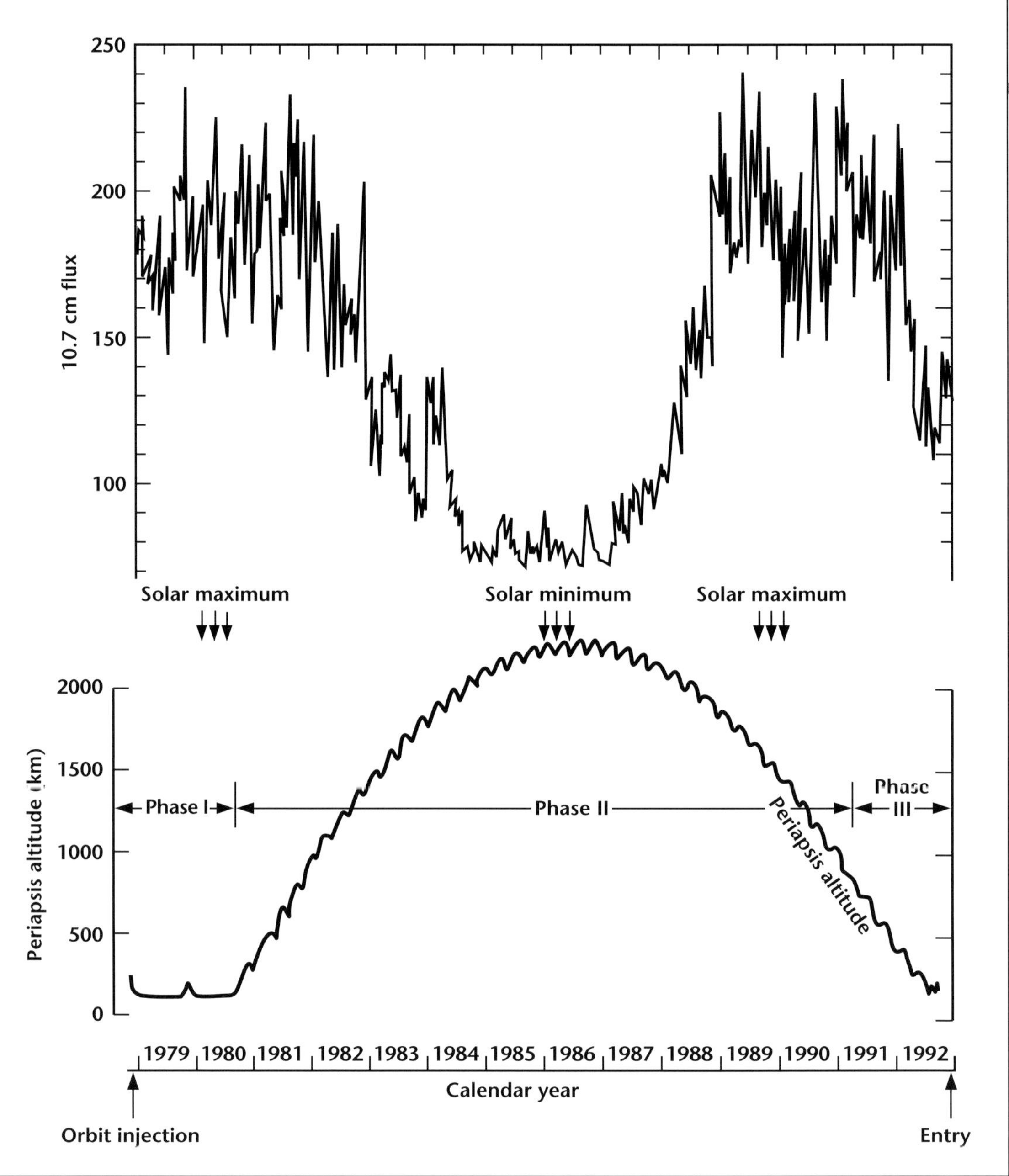

Occultations of the space telemetry signal by Venus also limited the measurement resolution near periapsis. Science data obtained at such times remained in the spacecraft's Data Storage Unit (DSU). Communications equipment transmitted the information to Earth later. The DSU's limited capacity required a compromise between full coverage of the occultation period (up to 24 minutes) and higher spatial resolution during only part of the occultation period.

Solar eclipse periods reduced the solar array's ability to recharge the spacecraft's batteries. As a result, those periods affected the time that controllers could turn on instruments, or the number of instruments that could be used during a specific orbit. These factors affected the planning of Orbiter's operations during Phase III, in addition to the scientific opportunities provided by the changing local time and altitude of periapsis.

Several other factors made the task of spacecraft operations more difficult during Phase III. The declining solar cell capability required the electrical energy budget to be more carefully balanced against desired scientific goals. Most science goals required measurements from many instruments at the same time. So, time-sharing did not offer significant power reductions. Controllers conserved energy within the spacecraft with several methods. They scheduled briefer intervals of operation about periapsis, and they reduced spacecraft operations at higher altitudes in the orbit.

Another complicating factor was the scientists' desire to obtain as many measurements as possible at very low altitudes. Maintaining the orbit for this purpose required that spacecraft maneuvers should occur every few days. This was in contrast to the weekly maneuvers navigators practiced during Phase I.

Mission personnel adopted a power-sharing plan during Phase II. The plan ensured that controllers turned on the right instruments at the right places in the orbit and in the correct local time sectors. This plan changed periodically to reflect reductions in the available electrical power and changes in the altitude of periapsis. An Entry Science Plan for Phase III served the same purpose as the Phase II power-sharing plan. However, it represented a more careful attempt to focus spacecraft operations on the unique scientific goals of the final phase.

In September 1990, a group of eight authors completed a report on the plan, entitled *The Pioneer Venus Orbiter Entry Science Plan*. The authors were L. H. Brace, University of Michigan (Chairman), R. W. Jackson, NASA Ames Research Center (Co-Chairman), G. M. Keating, NASA Langley Research Center, L. E. Lasher, NASA Ames Research Center, D. W. Lozier, NASA Ames Research Center, H. B. Niemann, NASA Goddard Space Flight Center, A. I. F. Stewart, University of Colorado, and R. J. Strangeway, University of California, Los Angeles.

This Entry Science Plan reflected the consensus of the SSG on other operational matters as well. Among them was how the project should use the remaining propellant to control the altitude of periapsis. Approaches to set the relative priority among investigations that may have conflicting operations requirements were also included. The need was identified for an Entry Encounter Activity at the end of 1992, with participation by appropriate investigator groups. The Plan also identified the need for a post-entry data analysis period. During this time, the various investigators would share the measurements from Phase III. Also, the Plan detailed the data format for submission to the National Space Science Data Center for archiving.

The project established several operational guidelines for the final entry phase. The primary scientific interest was to gather data within the atmosphere when the spacecraft was near periapsis. The relatively brief periapsis operations did not require much orbit-averaged power. Solar-cell charging current and battery capacity were devoted to operating the spacecraft. This included powering all desired instruments for an hour or so near every periapsis passage. When the DSN was not available to the Orbiter, the DSU on the spacecraft stored the periapsis data for playback later.

However, other constraints made it necessary to settle for even briefer data gathering periods during each orbit. Periapsis passages through Venus' shadow caused deep discharges of the battery. This worsened if all applicable instruments were turned on. Care in husbanding all resources aboard the Orbiter was necessary. Scientists and mission managers wanted the spacecraft to survive the first periapsis lifting interval in September-October 1992 (orbits 5020 to 5070). This would provide a chance to obtain dayside measurements in the December 1982 to January 1993 interval (orbit 5100 to entry).

A question arose whether measurements were needed on every orbit periapsis in the intervals that the scientific goal statements called for. Nearly all regions of the Venusian thermosphere and ionosphere are highly dynamic as they respond to changing solar radiation and solar-wind conditions. Each periapsis passage provided a snapshot of the conditions existing along that orbit at that time. Only through comparisons of profiles taken under diverse solar conditions could researchers hope to sort out the sources of the observed variations, and periapsis passages occurred only at 24-hour intervals. From these arguments, managers concluded that they should avoid the unnecessary omission of even one orbit, if possible. The experimenters decided that all the aeronomy instruments, which contributed such useful data, should operate during every periapsis passage. This would be the plan unless unforeseen limitations in spacecraft power or telemetry made this operation impossible or unwise.

The above reasoning had an important corollary that guided the selection of instrument operation modes for each particular passage. Three instruments had several modes of operation for acquiring specific kinds of measurements at the expense of others. These instruments were the retarding potential analyzer, ultraviolet spectrometer, and neutral mass spectrometer. Because several investigations shared the same orbit intervals, experimenters had to compromise. Instrument investigators were made responsible for coordinating their plans to select instrument modes for each orbit. Each investigator had to justify use of any special instrument modes. Often, factors restricting the science goals were limitations in the available electrical power, the telemetry rate at that time in the mission, and the occurrence of occultations that limited the quantity of data within the DSU. Limitations in the DSU's capacity forced a choice between receipt of a low data rate and an intermediate or high data rate in a slow burst.

Some general guidelines came from a variety of sources. They emerged from scientific goal statements or evolved from discussions within the Entry Operations Task Force. Others were generalized from many discussions at earlier SSG meetings. Periapsis science had the highest priority during Phase III. As a result, brief uses of instruments to obtain *in situ* and remote measurements of the atmosphere and ionosphere had higher priority than apoapsis scientific goals. The mission accepted electron temperature probe and plasma analyzer measurements an hour or two before periapsis. Experimenters needed the data to evaluate solar wind and extreme solar ultraviolet radiation conditions in the planet's atmosphere and ionosphere. However, these measurements had lower priority than periapsis science.

Orbital Geometry and Spacecraft Orientation

Because Venus is so close to the Sun, the Sun's gravitational pull noticeably perturbed the spacecraft's inclined orbit. The significant changes were in the altitude and latitude of the periapsis. For the first 20 months of the mission, the spacecraft used its thrusters to counteract these effects. This allowed it to maintain the periapsis at a low altitude. Afterward, controllers allowed the periapsis to rise by solar perturbation.

Project management logically divided the mission into three phases. (See previous sections for more details on these three phases). Phase I

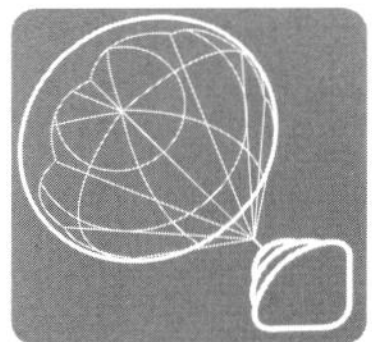

covered the period when the thrusters maintained periapsis at an altitude of 150 to 250 km (93 to 155 miles). Phase II started when project managers decided that they should no longer control periapsis. Rather, they would allow it to rise and later to fall under the influence of solar perturbations. This conserved hydrazine propellant for use in extending an entry phase (Phase III). During the second phase of Orbiter operations, the periapsis did more than just rise and fall. The latitude of the periapsis also moved from 17° north to the equator of Venus. During this period, measurements at periapsis were continually exploring new regions of the planet's environment.

Late in 1991, the periapsis began to penetrate the lower thermosphere and ionosphere. When it had fallen to about 1000 km (621 miles), Phase III of the mission began. As the periapsis continued to fall, controllers again used the thrusters to maintain periapsis. This time, they kept it within an altitude ranging from 140 to 160 km (87 to 100 miles). Also, the latitude of the periapsis continued moving southward to about 10° below the planet's equator. During Phase III, the spacecraft sampled the atmosphere to deeper levels than were prudent in Phase I.

The spacecraft's orbit was fixed in inertial space as Venus revolved around the Sun. So its orientation with respect to Earth and Sun changed as the planets moved around the Sun. The combined motions resulted in seasons of eclipses and occultations. During the eclipse period, the spacecraft was repeatedly shadowed by Venus when near periapsis. These were called periapsis eclipses. Apoapsis eclipses occurred when the spacecraft passed through Venus' shadow close to the spacecraft's apoapsis. During an occultation, the Orbiter passed behind Venus as observed from Earth. Occultation studies allowed the spacecraft to probe the atmosphere and ionosphere. Scientists observed the effect on radio waves passing through those regions on the way from the spacecraft to Earth. When periapsis occurred during occultations, data remained in the DSU. Later, communications equipment transmitted the information to Earth.

As Venus traveled around the Sun, the planet rotated slowly under the orbit of Pioneer. This permitted the sub-spacecraft point to pass over the whole planet in a period of 243 days. This amounted to the period of one rotation of Venus on its axis. However, the spacecraft sampled all local times on Venus in its year of 224 days. That is, it sampled longitudes relative to the Sun as contrasted with longitudes on the planet's surface.

Flight Operations

During the Orbiter's long mission at Venus, support services contractor personnel continued routinely to conduct flight operations. These included other Pioneer missions, too. They performed their work in the PMOC at Ames Research Center. The overall Pioneer program had begun in the summer of 1965. Since that time, computers had remained on and the facility operated 24 hours per day, 365 days per year. The only exception was for one or more shifts on major national holidays. Console operators maintained constant voice communications with the operations center of DSN at the Jet Propulsion Laboratory, Pasadena, California. These operators handled all immediate detailed coordination of tracking operations, command transmissions, and telemetry data flow. At least one flight operator and a computer operator were nominally on duty at all times at Ames Research Center for the extended Pioneer missions.

Daily operations usually included one or two passes for Pioneer 10 and Pioneer 11. These were the first spacecraft to explore the outer

Solar System and beyond. Daily operations also included an occasional (two or three per year) pass for one of the Pioneer 6-8 series. These spacecraft had first explored the interplanetary environment and the effects of solar activity on Earth. Each pass was typically 4 to 11 hours in duration. The operating schedules of DSN imposed exceptions. Occasional special computing circumstances also created exceptions to these durations.

The engineering staff of Ames Research Center defined and documented operations procedures. *PC-488, Pioneer Venus Orbiter In-Orbit Operations*, specified operations for the Pioneer Venus Orbiter. *PC-250, Pioneer F/G Standard Procedures for Flight Operations*, specified operations for Pioneers 10 and 11. *PC-053, Pioneer A Flight Operations—Standard Procedures*, defined the operating procedures for Pioneer 6. It also was applicable for Pioneers 7 and 8. However, the *DSN Network Operations Plan 616-55* was a more current reference for Pioneers 6-8.

Staff specialists at Ames Research Center or the contractor also attended special operations These included maneuvers or unusual spacecraft or instrument tests. A specially qualified contractor representative directed the more routine procedures. This person regularly reviewed plans and procedures with engineers at the Center.

Duty personnel had lists of home telephone numbers for engineers. They called these engineers when prescribed procedures could not resolve problems. If telephone communication was inadequate, engineers usually could be in the Center within about 30 minutes. As experience grew during the mission, the worst of these types of problems diminished. There was an average of several months between occurrences. Former Pioneer team members experienced in the spacecraft's development also offered their advice on request.

Maintenance of computers used in Pioneer operations was contracted to companies specializing in computers. If any one computer was down, the facility still had sufficient depth to continue working with any two spacecraft. This minimized off-hour premium maintenance expenses. A Pioneer Missions Office staff engineer monitored and directed the maintenance contract support.

The Navigation Team at the Jet Propulsion Laboratory analyzed and processed metric tracking data. This team provided trajectory predictions to DSN. Navigators needed the predictions for computations of pointing angles and Doppler shifts that they used in operations with all Pioneer spacecraft. The Navigation Team also provided periodic predictions of trajectory parameters for computer use at Ames Research Center.

Planning and Development

Throughout the mission's several phases, the staff of the Ames Research Center Pioneer Missions Office provided general plans and prepared procedures. They worked under the coordination of the Flight Director. The support service contractor at Ames Research Center was Bendix Field Engineering Corporation. This group translated the general plans into detailed schedules for command transmissions and real-time data communications and processing.

DSN produced both long-term and near-term weekly schedules for project support during the extended mission. It also had done this during Phase I. At Ames Research Center, the support service contractor produced a detailed weekly computer schedule for telemetry and command

activities. DSN resources needed to support the Pioneer spacecraft were scheduled through the Pioneer Missions representative on the Jet Propulsion Laboratory staff. This person maintained close liaison with the Pioneer Missions Office. The representative adjusted plans to tracking availability and considered constraints and special requirements of the Pioneer missions in the scheduling process. The representative also assisted in negotiating with other users of DSN to resolve conflicting requirements.

The Pioneer Missions Office and support contractor also maintained lists of software and hardware problems. For continuing and improving operations during later phases of the mission, they had to resolve these problems. Most of the modest developmental effort was dedicated to solving relatively short-term problems. These arose as circumstances changed or as long-standing complaints were solved. Long-term, larger projects for the spacecraft mission were also worked on to maintain compatibility with DSN's computer interfaces. These included DSN commands and bit error correction and longer data blocks for NASCOM. Maintaining very long term competence in the data processing software was an important objective in scheduling these efforts.

Operations

During all phases of the Orbiter's mission, SSG meetings occurred semiannually. Orbital Mission Operations Planning (OMOP) committee meetings generally took place concurrently with the SSG meetings. The committees used teleconferencing when it was necessary. These groups recommended allocation of the limited telemetry data link among the various scientific interests. These interests included the alignments of the Sun/Venus/orbital plane, the available time for tracking, and other constraints. The SSG meetings provided a general exchange of information about scientific progress, planning for publications, and special interdisciplinary investigations. The OMOP committee resolved problems that were more frequent and immediate than the issues SSG addressed. Individual investigators also provided regular and frequent instructions about the configuration to be commanded to their instruments.

Toward the end of the Orbiter's long mission, the SSG gave special attention to the available hydrazine propellant. They carefully planned its use to maintain the spacecraft's orbit and to optimize the return of scientific data. At the SSG meeting in Spring 1989, an Operations Plan Task Force (OPTF) formed. Its charter was to describe the scientific rationale for Phase III. To meet the science goals, it also defined requirements for experiment operations, science sequences, formats, data storage, bit rates, and quick-look and post-entry data analysis. The Task Force submitted its plan at the SSG's Spring 1990 meeting. The group did not immediately settle many of the plan's final details. They waited until after the SSG made decisions on priorities for competing scientific goals and instrument operations during the entry period. Many issues were involved. Among them were the collection of additional radar altimeter measurements, the relative priority of low-altitude *in situ* measurements, the importance of drag measurements, and questions of spin rate and spin axis orientation. These were all crucial to defining orbital sequences, instrument modes, and intervals between periapsis restoration adjustments.

Until the committees resolved these questions of priority, the OPTF followed a specific course. It based its plan on inputs that experimenters had given to the Entry Planning Committee at earlier SSG meetings. The Task

Force was particularly interested in data that experimenters had presented at the 1988 Spring meeting in Annapolis. The committees later distributed this information to the SSG membership. These inputs were the most complete at that date. They were adequate enough to define the long lead-time items the Task Force needed to support the plan. Such items included quick-look data requirements, spacecraft spin rate, and spacecraft orientation. Also included were changes to instrument data processing software that were required by proposed changes in data formats.

Status After a Decade in Orbit

The spacecraft design featured flexibility and redundancy. Controllers could select either of two electrical components for nearly all critical functions. These functions were receiving commands, storing and executing commands over an extended interval of time, processing data, transmitting telemetered data, storing and replaying telemetered data, controlling despin of the antenna and other spin-synchronous functions, and firing thrusters. Also, the design featured backups for the despin motor for the antenna, the liquid-propellant thrusters, and the electrical storage batteries.

By 1988, after 10 years in orbit, the Orbiter spacecraft continued in excellent working order. The conservative design and redundancy of critical subsystems had paid off admirably. All functions were serviceable with only modest degradations (when compared with conditions immediately after entering orbit around Venus). For example, the amount of power the solar panels produced had diminished during the extended mission. This occurred because they had not been designed for so many years exposure to solar radiation at the distance of Venus from the Sun. Originally, mission scientists had intended the spacecraft to orbit Venus gathering data for one Venusian sidereal day, or 243 Earth days. That was the approved primary and nominal mission. However, designers believed there could be an extended mission of gathering data for several Venusian sidereal days. The conservative design allowed the spacecraft to eventually operate beyond a complete solar cycle of 11 years. This feat provided a cost-effective bonus of scientific data.

Figure 5-18 shows how solar activity was high during Phase I when navigators maintained periapsis at a low altitude. When periapsis had reached its highest altitude during Phase II, solar activity was at a minimum. However, it increased as the periapsis descended again. During another period of low periapsis in Phase III, solar activity had passed through its maximum and had decreased to an intermediate level.

Current from the solar cells had decreased from an average of 13 A at orbit insertion to 4 A. As a result, power production from the solar cells limited operations to less than 24 hours per day. Fortunately, the design provided for regular battery operations supplementing the solar cells. This provision supported intermittent loads. Starting in 1988, scientific instruments that operated for long hours were used in a time-sharing mode to maintain power balance.

Data collecting and handling for all the science experiments was still normal after 10 years in space. However, the failure of one unit reduced storage and replay capacity. Telemetry continued normal. One transmitter had slightly diminished power, but worked with the same efficiency. The command receiving, decoding, storage, and executive systems continued to perform perfectly. However, the secondary receiver responded to only a narrow radio frequency band. The control

systems for spin axis orientation, spin rate control, antenna despin and elevation control, and synchronization timing signals all worked perfectly.

During the first decade in orbit, only four random failures occurred in the spacecraft's subsystems. There were no complete failures of critical components. All the critical subsystems still had serviceable backups to continue the extended mission toward Phase III. At that time, navigators would control the periapsis before the spacecraft finally plunged into the Venusian atmosphere.

Hydrazine propellant consumption had been conservative relative to pre-launch estimates. The spacecraft entered orbit around Venus with 32 kg (70 lb) of propellant. An estimated 2.3 kg (5 lb) remained following Phase I operations to keep the periapsis at a low level. Although the precise amount of remaining propellant was uncertain, project management estimated that enough remained for Phase III. There was sufficient fuel to control the entry sequence at the end of the mission sometime in 1992. There also was a possibility of controlling the spacecraft until final entry into the atmosphere on the planet's dayside.

There were, of course, other hazards. Although the spacecraft had survived far beyond its original design lifetime, there was always the possibility that some critical component might suddenly fail. If this happened during this final phase of the mission, it could bring the project to a premature end. Nevertheless, project management optimistically expected that they could control the Orbiter for 20 to 40 passes through Venus' atmosphere before atmospheric forces during entry destroyed it. This was very important. It would provide a unique opportunity to make measurements in much lower regions of Venus' atmosphere. These were regions where no other spacecraft had made measurements and where no planned spacecraft, such as Magellan, would be designed to do so.

Hardware Status as the Final Phase Approached

The spacecraft's attitude control system operated successfully throughout the mission. The Despin Control Electronics (DCE) consisted of a primary and a backup. Controllers switched off the primary in 1984 and used the backup from that time. Although there were no problems with the primary system, the complexity of switching back to it encouraged continued use of the secondary system. The star sensor had duplicate slits and electronics. Both were used successfully without any problems. However, solar protons affected the star sensors. As a result, the mission never used them during solar proton events that lasted more than several hours. Starting in December 1990, controllers powered off both star sensors except when they had to check attitude. This reduced the load on the spacecraft's batteries.

The spacecraft's propulsion system performed well throughout the mission. All seven thrusters continued to operate normally. However, one did show a decrease in performance, so the mission did not use it after October 1984.

One DSU failed in March 1986, and the mission did not use it after that date. The second unit continued to operate throughout the mission. It had an operational restriction: the maximum bit rate the instrument could store was 2048 bits/sec.

The command subsystem worked perfectly throughout the mission. However, there were minor problems in the communications subsystem. The spacecraft carried two receivers. Of the two, the backup, connected to the aft omni

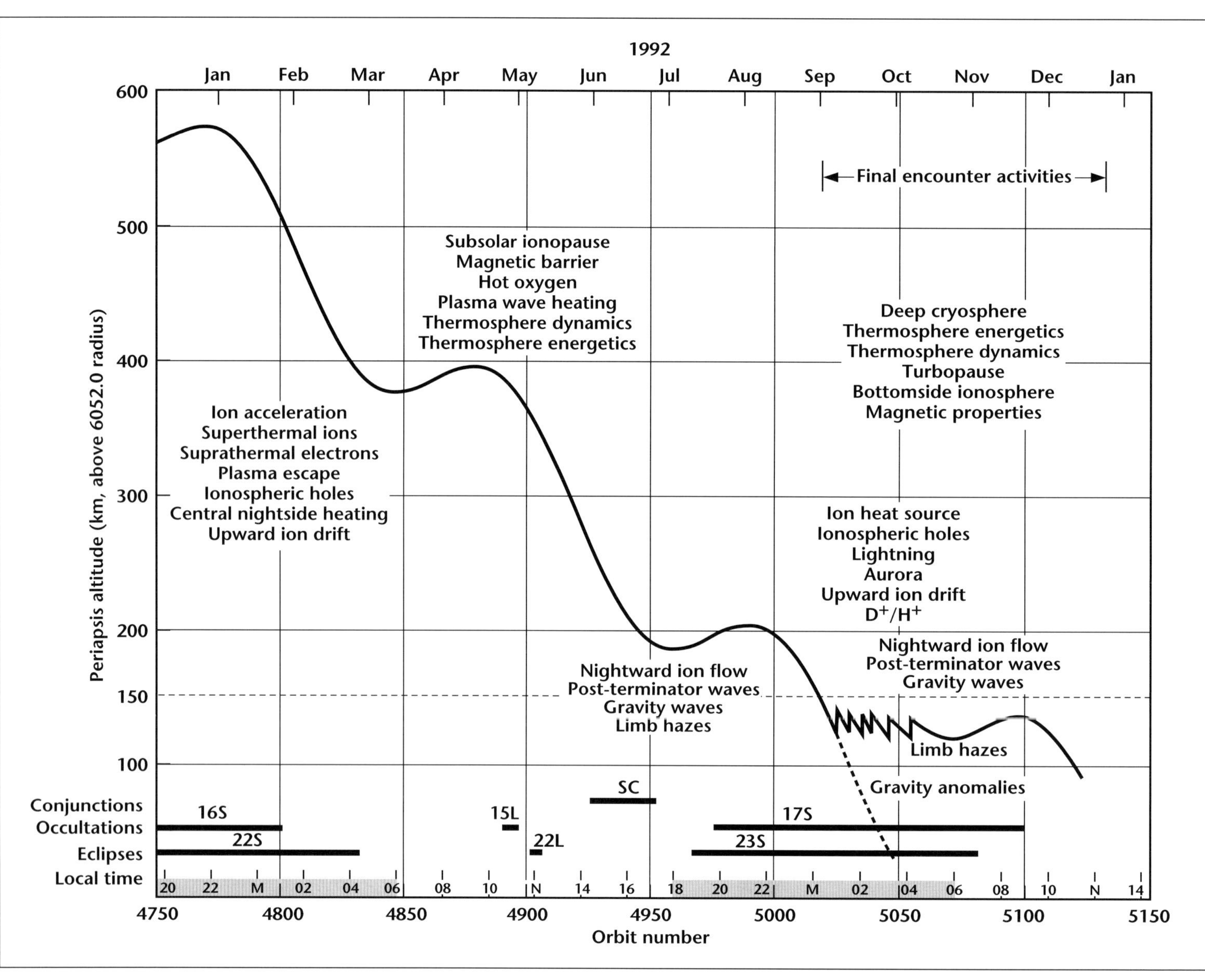

Figure 5-18. This calendar plots the altitude of the periapsis during 1992 against orbit number (bottom scale) and date (top scale). Captions identify the periods when the spacecraft investigated various lower altitude phenomena.

antenna, developed a minor problem in April 1983. The uplink bandwidth became restricted to about 250 Hz for signal detection and about 50 Hz for command processing. However, DSN developed a special procedure to allow that receiver to process at least one command if needed. The primary receiver, connected to the high-gain antenna, performed well throughout the mission.

There were four amplifiers available to power the downlink transmission to Earth. Since February 1984, controllers had turned off one amplifier because of a limitation on power. In May 1984, the output of another amplifier dropped about one decibel and became noisy. Mission controllers switched it off. The remaining two amplifiers did not develop any problems.

There were no failures in the spacecraft's power subsystem. However, both the solar arrays and the nickel-cadmium batteries degraded during the mission, as expected. Since the three solar arrays were electrically independent, solar radiation darkening the glass covering the solar cells probably caused their degradation. To compensate for the

reduced power from the solar arrays, the spacecraft had to use its batteries more frequently. However, their recharging took longer because of the reduced output from the solar arrays. The two batteries provided power for the spacecraft when it was in the shadow of Venus (eclipse periods). The spacecraft also used them to supplement power from the solar arrays when the electrical load exceeded the output from the arrays. As the solar array output declined, controllers had to use the batteries more and more. Also, use of the batteries once for each spin of the spacecraft slowly degraded their capacity. When the output voltage dropped to 27.5 V, an undervoltage switch turned off all nonessential systems to protect the batteries.

Toward the end of the mission, both the solar arrays' output and the batteries' capacity sometimes varied without warning. So, mission controllers had to balance energy usage daily. This was necessary for each battery because each had its own solar array. The procedure was to recharge each battery completely on each orbit of the spacecraft. This was done just before periapsis tracking or before an eclipse, and when the spacecraft was not using the transmitter amplifier. Controllers had to switch off other equipment during the charging period. However, mission controllers had to ensure that the batteries did not overheat by overcharging. They carefully monitored and evaluated battery loads. They calculated output of the solar arrays and organized everything to maintain battery voltage at more than 28.5 V. This was one volt above the level at which the undervoltage switch would switch off all nonessential systems.

An anomaly with the magnetometer boom did not affect operations. Telemetry signals indicated that the boom had not deployed. However, performance of instruments on the boom indicated that it was fully deployed and properly locked into place.

An Opportunity to Look at Comets

During Phase II of Orbiter's mission, opportunities arose to make systematic observations of several comets. To do this, controllers used the Orbiter Ultraviolet Spectrometer (OUVS). These observations took place between April 1984 and May 1987. The comets and their dates of observation were: Encke, April 13 through 16, 1984; Giacobini-Zinner, September 8 through 15, 1985; Halley, December 27, 1985 to March 9, 1986; Wilson, March 13 to May 2, 1987; NTT, April 8, 1987; and McNaught, November 19 through 24, 1987.

These observations had several scientific objectives. They included determining the evolution rate of water from the cometary nucleus and how it varied. Identifying the carbon/oxygen/hydrogen ratios and how they varied was another objective. Also, the observations looked for evidence of rotation of the comet's nucleus. Experimenters used the spectrometer to obtain images of the coma in Lyman-alpha radiation.

Jim Phillips was Pioneer Venus Project Trajectory Analyst. He explained that the comet missions were accomplished by sending commands to the spinning spacecraft to change the tilt of its spin axis. This allowed the spectrometer to scan across the comet instead of across the surface of Venus. The ultraviolet spectrometer had a small field of view. As a result, it gathered general pictures of the comets as a series of strips, one for each spin of the spacecraft. The comet moved slightly between each spin to expose a sequence of strips along its length. The ability to scan at high resolution across the comet's coma, tail, and hydrogen halo was important. It allowed scientists to determine the distribution of gas and particles in each region.

Unique observations of several comets by one spacecraft benefited from using the same instrument. This was an important capability for making comparisons among several comets of different ages. For example, Comet Wilson is a new comet discovered in 1986. Comet Encke is an evolved comet, one of the most evolved comets known. Comet Halley is a "middle-aged" comet. In the case of Comet Halley, the Pioneer Venus observations were special. They gathered data about the comet close to the comet's perihelion passage (Figure 5-19). Other spacecraft were not in positions where they could accomplish this.

All the comet encounters were successful, and the results appear in a later chapter.

Final Encounter

Earlier sections described how the encounter-phase science plan gave periapsis data gathering the highest priority. They also outlined how instruments would cover every periapsis passage. Data gathering elsewhere in the orbit had a lower priority. Mission controllers would not allow it to compromise periapsis data gathering. During the period from 1 September 1992 to the end of the mission in October, low-altitude periapsis passages through Venus' atmosphere dominated spacecraft activities. Velocity maneuvers using the remaining thruster propellant were important. They maintained the altitude of the periapsis above a point where drag would be unacceptable and the orbit endangered. Controllers also made precession maneuvers. These corrected the attitude shift that atmospheric drag produced and the required attitude changes that the velocity maneuvers produced to maintain periapsis alti- tude. Spin rate adjustments were also required to correct the despin caused by altitude control maneuvers. During this period, the orbit period shortened. As a result, controllers had to adjust tracking sequences to ensure that they obtained data during periapsis passage.

Generally, DSN maintained tracking periods at periapsis and at apoapsis centered about 12 hours apart. However, during the critical orbits approaching final entry, the spacecraft was in Earth occultation during the periapsis period. So, DSN could not track the spacecraft or receive data from it at periapsis.

To maintain the altitude of periapsis, the thruster changed the spacecraft's velocity at apoapsis. There were instances during this time when the spacecraft could not receive commands. This was because the undervoltage switch had switched off the high-gain antenna or the despin control. This made the prime receiver unavailable. Without the prime receiver, controllers were concerned about the critical command capability that was needed to complete the periapsis-maintaining maneuver on time. If the maneuver was delayed, there was a chance that the spacecraft would be destroyed during the next periapsis passage because it was too deep in the atmosphere.

The inability to receive a command had caused a delay of 36 hours before controllers could switch to another receiver. During earlier phases of the mission, this was not a serious problem. However, it was not acceptable during this final encounter when, even if the spacecraft survived a lower periapsis, all periapsides, 12 hours apart, had to be covered. Moreover, the backup receiver's performance had degraded. To receive just a single command, it needed special operating procedures from DSN. DSN's high-power (250 kW) transmitter was required for supporting altitude raising maneuvers.

The minimum safe periapsis altitude was established as being just above where drag-

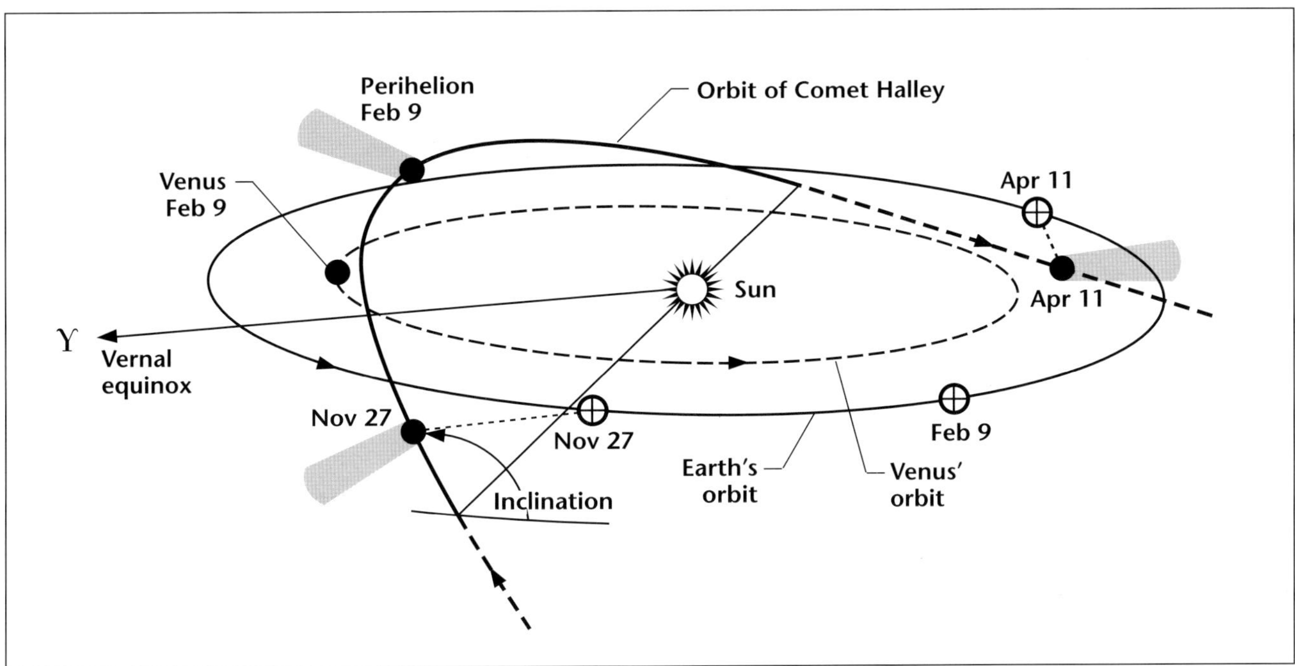

Figure 5-19. This diagram shows how the Pioneer Venus Orbiter was in an advantageous position to observe Comet Halley around the important time of its perihelion passage. European, Soviet, and Japanese spacecraft sent to fly by the comet were not able to encounter the comet and make their observations at its perihelion.

induced damage to the spacecraft would occur. It varied as a function of the hour angle of the periapsis from the sub-solar point on the planet. The exposed materials most likely to be damaged by drag heating were the Kapton thermal blankets and the small fiberglass structural elements supporting the antennas. The minimum altitude was expected to be 131 km (81 miles) in the dark and 142 km (88 miles) in daylight. This was because the density of the atmosphere at a given height varies considerably between day and night on the planet. Measurements of periapsis altitude could be made to within a few hundred meters by a single-pass Doppler observation, and to within a few tens of meters by Doppler observations over several sequential passages.

Entry Phase

By early September 1992, the altitude of the periapsis was approaching a level where it could damage the spacecraft. By this time, the remaining propellant was estimated at 1.86 kg (4.1 lb). This propellant was to be used to extend the mission as long as possible. The sequence of periapsis decay and altitude trim maneuvers during the mission's final phase appears in Figure 5-20 (relative to the local time on Venus).

Larry Lasher, Pioneer Mission Science Chief said, "My most memorable recollection of the project occurred during the final days of Pioneer Venus in early fall 1992, observing the excitement and enthusiasm of the scientific investigators as they made once-in-a-lifetime measurements as PVO [Pioneer Venus Orbiter] charted previously unexplored regions of the Venus atmosphere."

As the figure shows, the first planned periapsis-raising maneuver was performed on September 7, 1992, when periapsis was raised by 20 km (12 miles) to about 155 km (96 miles). This left sufficient propellant for five more maneuvers to maintain the altitude of periapsis. During

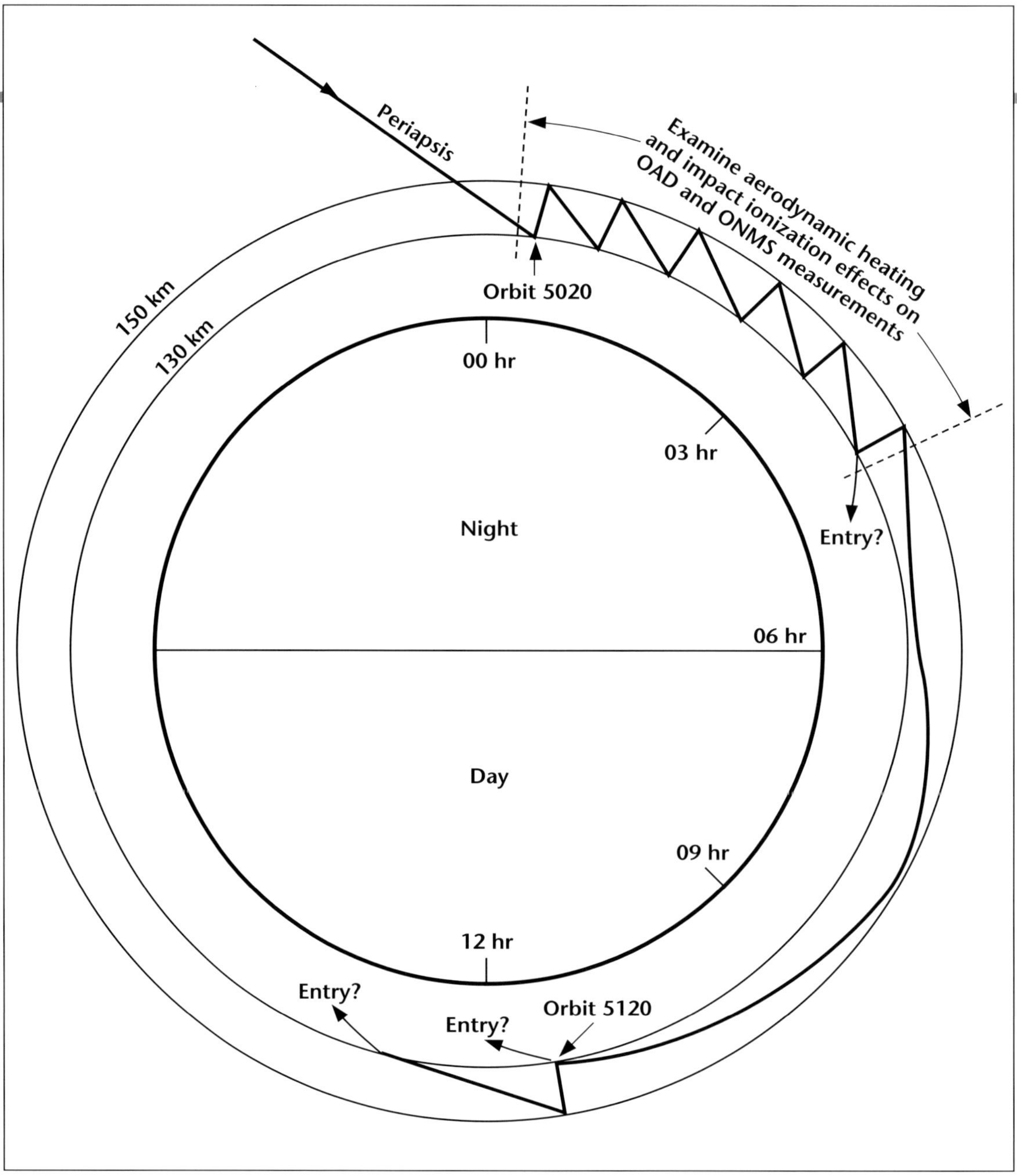

Figure 5-20. *This figure illustrates how, during the entry phase, periapsis repeatedly allowed the spacecraft to scan the very low thermosphere. This region is where differences in the neutral mass spectrometer and drag measurements occurred. The differences resulted from operation below the region of free molecular flow. Aerodynamic heating and impact ionization affected the accuracy of some measurements. In the top right quadrant, the periapsis is raised several times by use of the spacecraft's thrusters. The bottom right quadrant illustrates a hoped-for path for entry into the sunlit hemisphere without having to use thrusters. Unfortunately, there was not enough propellant left for this final maneuver and entry took place in the night hemisphere in the upper right quadrant.*

the maneuver executed on October 2, the spacecraft used the last of the propellant. As a result, the periapsis dropped lower and lower into the atmosphere. The spacecraft could not be tracked during the periapsis, nor could data be received from it. At the control center, everyone had to wait until the spacecraft's signal could be acquired again when the Orbiter moved out from Earth occultation.

Said Lasher: "There was . . . the expectation and hope of the project staff huddled around the consoles in mission control listening for the spacecraft signal, hoping for just one more orbit that fateful day . . . ."

After periapsis passage of orbit 5056 on October 8, DSN stations received no radio signals from the Orbiter when they should

have at 20:22 hours GMT. Nor did anyone hear anything further from the Orbiter.

The Pioneer Venus Orbiter had ended in a veritable blaze of glory, as a meteor flaming through the dense atmosphere of Venus (Figure 5-21). The spacecraft had orbited the planet since December 4, 1978, after its launch in May 1978. It had completed 5055 successful data gathering orbits of Venus.

The Pioneer-Venus orbital mission was declared completed on October 9, 1992, at 00:55 GMT. Said Lasher, ". . . we lost contact forever with PVO. But I recall there was a certain serenity afterward in the knowledge that Pioneer had performed so nobly for 14 years, far and above its call of duty."

The Pioneer-Venus project was a 14-year mission that far exceeded all its original objectives. It became a classic example of how management and science can design and run an advanced technology project to achieve stated engineering and scientific objectives on time and within a clearly defined budget. Commented Fred Wirth, Pioneer Deputy Project Manager, "The 14-year flood of science data from Pioneer Venus has been particularly rewarding. The radar map of the Venus surface, the characterization of the bow shock and the ionosphere, pictures of the cloud cover, and the glowing ultraviolet image of Comet Halley somehow made it all worthwhile. The Orbiter may be dead, but the legacy of scientific data it leaves behind will continue to nourish mankind in its quest for knowledge for many years to come."

▲

Figure 5-21. As the Pioneer Venus Orbiter entered the Venusian atmosphere, it produced a glowing trail like a large meteorite. This artist's rendering shows the spectacular end of the Orbiter's 14-year mission.

CHAPTER 6

Photo Credit ©Hughes Aircraft Company

SCIENTIFIC RESULTS

At the beginning of this century, E. W. Maunder of the British Royal Observatory at Greenwich commented on the Venus puzzle. He wrote: "We never see her surface; she presents but a dazzling disc, with never a marking that we can be certain is not the result of eyes tired with too much brightness. Whether her atmosphere is clear or cloudy, or what lies behind that dazzling light, we do not know."

Fifty years passed. Despite enormous increases in telescopic capabilities, Venus still remained a mystery. In 1959 and 1961, just prior to the blossoming of the space age, speakers at Lunar and Planetary Exploration colloquia summarized the problem. Commented W. E. Straly: "As opposed to the volume of material known about Mars, there is little known about Venus. Its diameter is estimated at 0.95 that of Earth, but this figure is far from exact. There is no apparent flattening of the sphere. No surface features have ever been discerned, its period of rotation is indeterminate, and little is known of its atmosphere. Its polar caps are ill-defined, and no other permanent markings have been seen. Its surface temperature has been estimated at 110°F, and surface pressure at two Earth atmospheres, but these are no more than educated guesses."

In November 1961, C. E. Anderson stated: "The rate of rotation of Venus is still a problem. On the basis of the Doppler measurements, JPL claims a period of about 225 days. However, a recent article in Izvestia stated a period of 10 or 11 days was calculated from the Russian Doppler measurements."

Not until the second decade of the space age did the veils of mystery surrounding Venus begin to lift. Scientists now were gaining insights into the planet's true nature for the first time. This new understanding resulted from experiments on the Pioneer spacecraft and measurements from earlier flyby spacecraft and from Soviet probes and orbiters.

Observing Venus at close quarters for over 14 years, Pioneer Venus revealed a bizarre world—a planet whose surface bakes under a dense atmosphere of carbon dioxide beneath clouds of sulfuric acid. Lack of an intrinsic magnetic field exposes the planet's upper atmosphere to the onslaught of the solar wind. Pioneer gathered voluminous data about the composition and dynamics of Venus' atmosphere and nearby interplanetary environment. It also provided an initial radar exploration of surface features. Later, toward the end of Pioneer's mission, NASA's Magellan spacecraft provided high resolution radar images of Venus' surface. These images enabled scientists to study in great detail the planet's geology and internal structure and probable evolution. Some of the Pioneer radar data supplemented the Magellan data or filled in parts of the surface where Magellan data were missing. Topographers also applied some Soviet data to aid the global mapping.

This chapter is the longest in this book. During its 14-year span, the Pioneer Venus mission collected a wealth of data. Analyzing these data kept scientists busy during the mission and for years after Orbiter sent its last signal to Earth. In this chapter, you learn about their many discoveries. These include details about Venus' surface, atmosphere, clouds, solar-wind interaction, magnetic fields, history, and other discoveries about Earth's sister planet.

The Planet in General

Radar data from Pioneer Venus provided the first global elevation survey of Venus' surface, from which about 90% of the planet was mapped topographically. Before the mission, Venus' surface was the least known surface of all the terrestrial planets. Optical telescopes cannot penetrate the clouds, and radar images from Earth have limited resolution and coverage. When Venus is closest to Earth, it rotates so it turns almost the same hemisphere toward us. Consequently, Earth-based radar can scan less than half of the planet's surface and only on a narrow equatorial swath. By contrast, the Orbiter spacecraft traveled in an orbit allowing coverage of most of the surface. Altimeter mapping sequences occupied about 1 hour of each orbit at altitudes below 4700 km (2921 miles). The radar data showed surface features as small as 75 km (47 miles) diameter. The smallest cell size was about 25 km (15.5 miles), but two or three were needed to define a feature other than a long narrow one such as a rift. As the orbit precessed around the planet, the radar view gradually covered nearly all the surface. However, resolution was reduced at high latitudes. Pioneer's altimetric sightings covered more than 90% of the planet's surface, from 73° north to 63° south latitude (Figure 6-1).

To map Venus, the distance from the spacecraft to the surface was measured by the radar altimeter. Since the spacecraft's orbit was accurately known from ground tracking, this allowed researchers to convert altitude measurements to radius measurements at discrete positions on the surface.

Pioneer made important discoveries about Venus' surface. At a scale at or above 100 km (62 miles), Venus is mostly smoother than the other terrestrial planets. Yet its surface topography has about as much maximum positive relief as Earth's. However, the distribution of elevations differs markedly from that on Earth. There is only one mode in surface height distribution rather than two. Both topography and gravity suggest that although Venus' interior is probably dynamic, its tectonic evolution has not been like Earth's.

Pioneer confirmed that Venus is quite round, very different from the other planets and from the Moon. Earth, for example, is flattened at the poles and bulges 21 km (13 miles) at the equator. The Moon has a bulge toward Earth. Mars bulges, too, but Venus has neither polar flattening nor an equatorial bulge. Earth has major variations between continents and ocean basins, which cover 30% and 70% of the surface, respectively. The mean levels of terrestrial continents and ocean floors are separated by 4.5 km (2.8 miles). Mars also has major variations and the colossal uplift of the Tharsis region. By contrast, Pioneer found that Venus has a very narrow distribution of surface elevations. Twenty percent of the planet lies within 125 m (400 ft) of the mean radius, and 60% lies within 500 m (1600 ft) of it. On the scale of the Pioneer radar images, the planet appeared as a monotonous world. It has only a few large continent-sized areas and smaller island areas rising above a global plain.

The highest point on the planet that Pioneer Venus measured was a summit in Maxwell Montes, 10.8 km (6.7 miles) above the mean level. The lowest point is 2.9 km (1.8 miles) below the mean level. This area is in a rift valley located at 156° east longitude and 14° south latitude. This depth is similar to that of the Valles Marineris on Mars (Figure 6-2). It is, however, only one-fifth the greatest depth on Earth (in the Marianas Trench).

Before the Pioneer Venus mission, knowledge of Venus' gravity field was scarce. Scientists

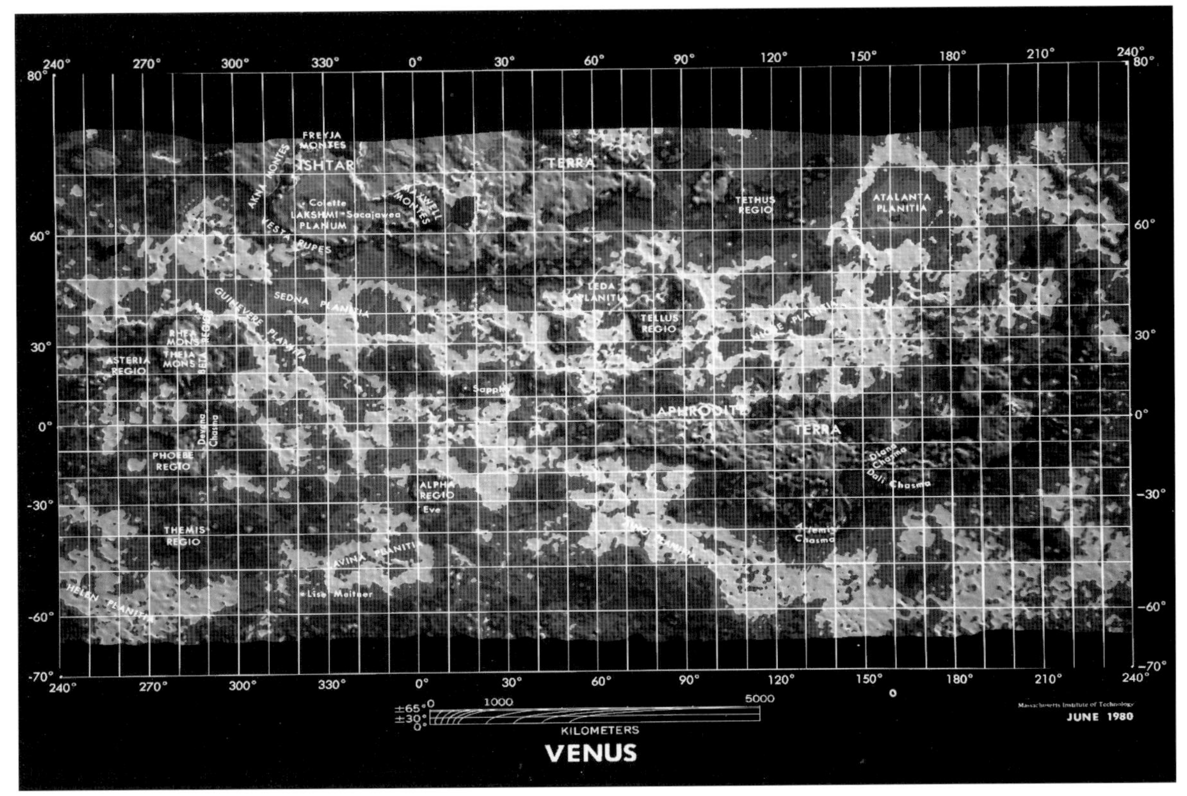

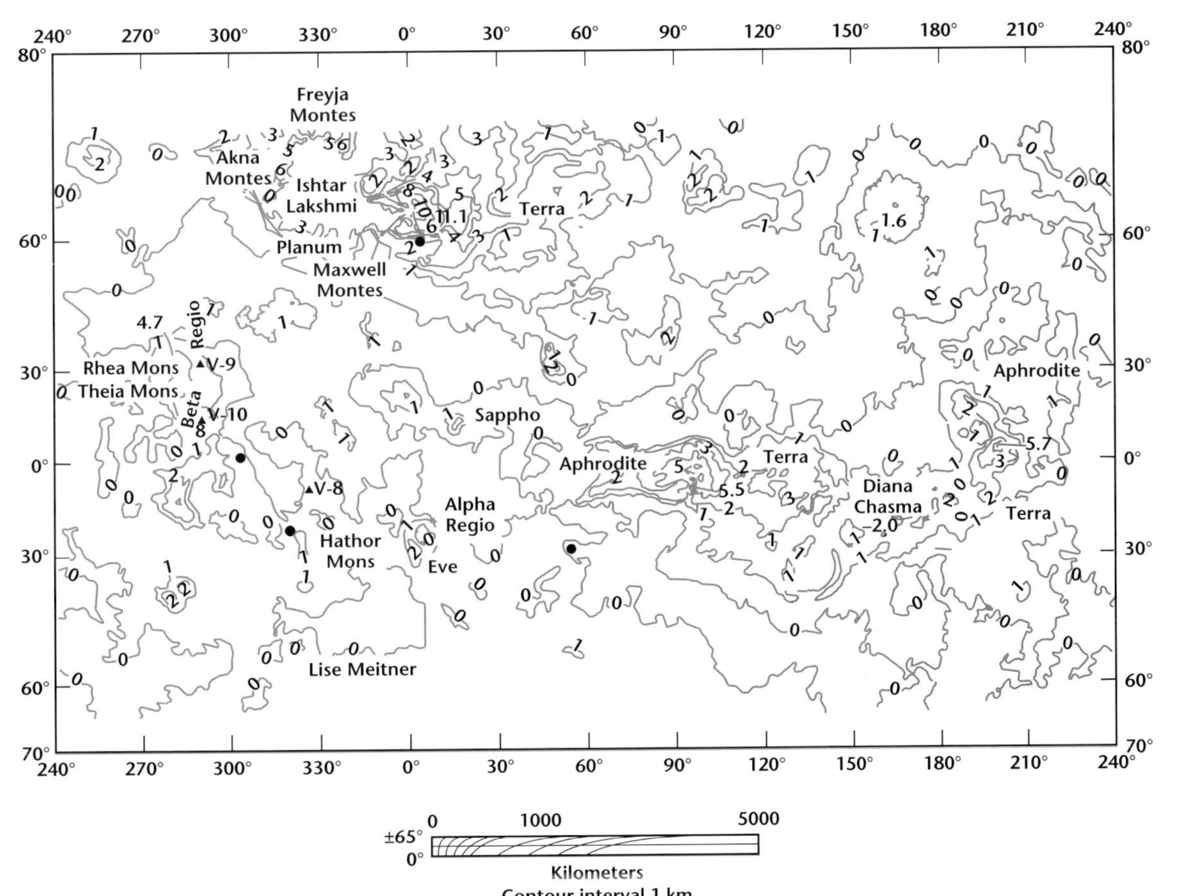

Figure 6-1. Cartographers made the first topographic maps of Venus in the late 1970s from Pioneer Orbiter data. (Top) Topographic map of the surface from radar data; dark gray is low, light gray is high. (Bottom) Contour map of the surface with contour intervals of 1.0 km (0.6 mile). The highest point is the summit of Maxwell Montes at about 350° east longitude and 60° north latitude. The lowest point is in the rift valley, Diana Chasma, at about 160° east longitude and 10° south latitude. The black triangles show the landing points of Veneras 8, 9, and 10. The black dots show the entry points of the four Pioneer probes. The map also shows the names and locations of major features that the text describes.

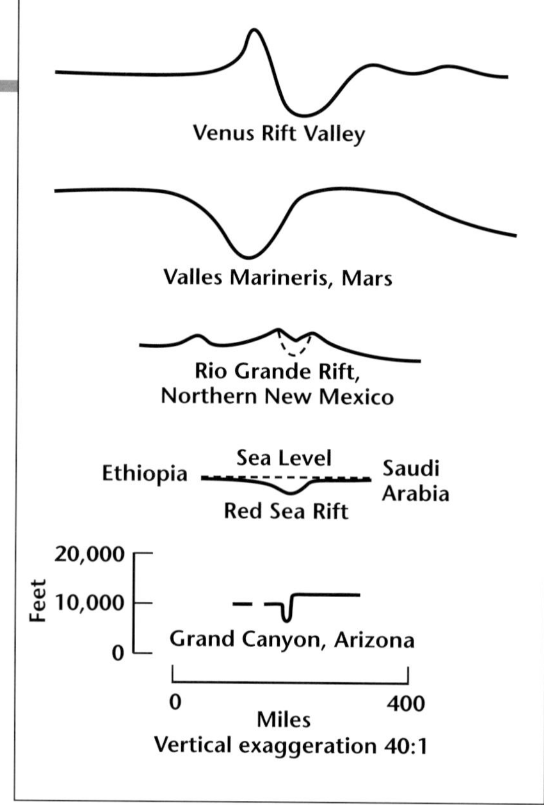

Figure 6-2. This drawing compares major valleys on Venus, Mars, and Earth. The vertical scale is exaggerated as you can see from the horizontal and vertical scales that represent Earth's Grand Canyon. These scales apply to all the valleys in the figure.

had derived estimates from the flybys of Mariner 2 (1969) and Mariner 10 (1974). Doppler radio tracking of Orbiter during its low altitude periapsis periods gave the first detailed gravity measurements. These allowed researchers to match gravity signatures with surface topography. Also, by processing long periodic variations of the Orbiter's mean orbital elements, estimates were obtained of the harmonic coefficients for Venus' gravity field. The data confirmed that the planet's oblateness is very small, as was expected from its slow rotation rate.

Detailed gravity measurements over a significant area of Venus revealed many anomalies (Figure 6-3). But unlike those on the Moon and Mars, Venus' gravity anomalies are relatively mild in amplitude and more like those of Earth. Geophysicists analyzed the spectrum of the harmonic model derived from these Pioneer data and found that topographic consequences of the anomalies in Venus' interior density are different from those of Earth. On Venus, the anomalies match up with topography; on Earth, most do not.

Scientists concluded that major adjustment to Venus' crust has taken place to reduce topographic effects. Also, partial isostasy or general equilibrium of crustal masses now prevails. Later, high resolution radar images obtained by NASA's Magellan spacecraft revealed lava flows over much of the planet's surface. Magellan also provided more gravity data that suggest isolated areas of upwelling among the networks of downwellings in a convective mantle. The uniform distribution of impact craters on Venus does not, however, support a terrestrial type of plate tectonics. There also appear to be zones of thick crust, associated with tesserae such as Alpha Regio. Large volcanic features such as Beta Regio are probably located over upwellings. Beta is most probably a young feature supported by the internal convective process.

Pioneer Orbiter investigators plotted the ratio of observed gravity to theoretical gravity from topography and found a definite trend of an increasing ratio eastward from Western Aphrodite around the planet to beyond Beta Regio. They suggested a possible explanation: a convection system moving eastward with its trailing topography more concentrated at larger distances from the most active region. Some investigators debated whether Aphrodite Terra was a spreading center like Earth's mid-Atlantic ridge. Later data from Orbiter and Magellan seem to rule out terrestrial-type plate tectonics from spreading centers. Even so, some topography is suggestive of subduction, for example, steep scarps on the northern boundary of Ishtar Terra and Ovda Regio.

Scientists used line-of-sight accelerations of Orbiter to deduce vertical gravity. They combined these observations with topography

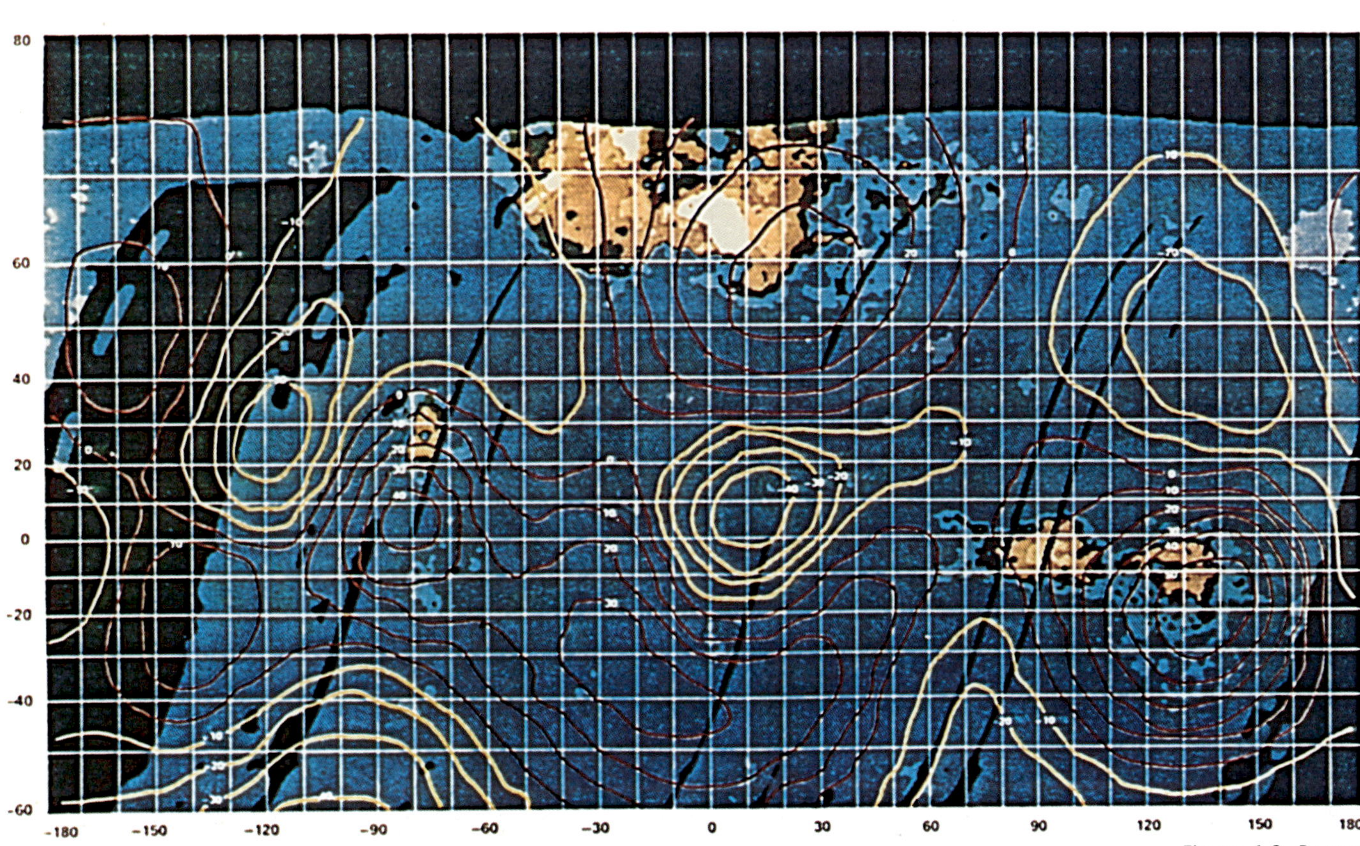

Figure 6-3. By measuring changes to the orbit of the Orbiter spacecraft, scientists determined Venus' gravity anomalies. This figure shows sixth degree and lower-order harmonic coefficients in milligals. Researchers calculated them at 100 km (60 miles) above the mean surface during an early phase of the Pioneer mission. The relationship to major surface features is clear. Later in the mission, researchers made even more precise gravity measurements, especially during the final phase when periapsis was at its lowest. These gravity maps showed that the relationship of gravity anomalies to surface features is different for Venus and Earth.

data to find mass anomalies on the crust-mantle boundary and in the upper levels of the mantle. Comparing these results with detailed radar maps from Magellan, scientists were able to link vigorous mantle upwelling with several "hot spots" on Venus' surface. They determined that mantle flow actively drives surface rifting. They also found that around some hot spots return flow is distributed asymmetrically. Geology of Venus appears to be more directly linked to mantle convection than is Earth's geology. Possibly because the upper mantle is very dry and less viscous than Earth's, stresses to Venus' lithosphere cause near-surface failure.

The Surface

Pioneer Venus Orbiter acquired radar data until its periapsis rose too high during Phase II. Researchers carefully adjusted Orbital tracks for the first part of Phase II (extended mission) so that new data points lay between those the spacecraft acquired during Phase I (nominal mission). To produce more complete morphological and geological maps of the planet (Figure 6-4) higher resolution data from Veneras 15 and 16 (1985-1986) were incorporated with the Pioneer Venus data. In the spirit of international cooperation, Soviet scientists made their data available before publication, so that U.S. scientists could plan more carefully NASA's Magellan radar mapper mission to Venus. Since the Magellan mission was so successful, Orbiter's radar mapper was not reactivated when periapsis returned to lower altitudes.

The Orbiter's "lifting of the veils of Venus" revealed a world of great mountains, expansive plateaus, enormous rift valleys, and shallow basins. Scientists had already deduced from Earth-based radar some of the features that Pioneer revealed. The wide range of Pioneer's

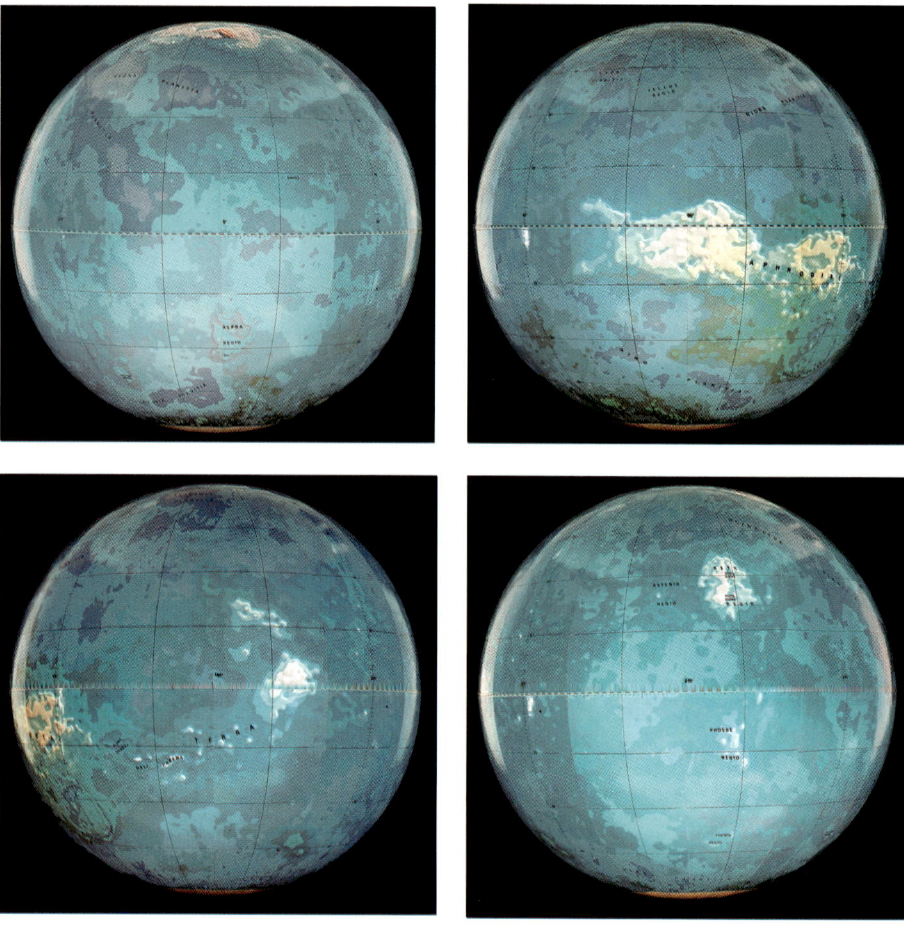

Figure 6-4. Geologists made use of Pioneer data to create a globe of Venus showing its surface features in detail. These photos show this globe from viewpoints centered on 0°, 90°, 180°, and 270° of longitude. The center of the crater Eve, located in Alpha Reggio, was selected as the 0° point of longitude on Venus.

data confirmed the existence of these features and expanded detailed coverage of the planet. However, the new Pioneer data caused scientists to revise many of their earlier theories. Venus' crust has three quite distinct regions. Relatively ancient crust seemed to be at intermediate elevations. Also, there were smooth lowland plains and highlands. Most of the planet's ancient crust—those parts of the planet between 0 and 2 km (0 and 1.25 miles) above the mean radius—may be preserved in Venus' upland plains. Venera 8 landed in these regions and its gamma-ray experiment showed that rocks there have uranium, thorium, and radioactive potassium contents that are consistent with a granitic composition. However, later data showed that these rocks may have a different composition (see also Chapter 7).

The Pioneer Venus data revealed that most of Venus (65% to 70%) consists of upland rolling plains on which circular dark features were identified as remains of large impact craters. The circular features were about 500 to 800 km (311 to 497 miles) in diameter but were very shallow—only 200 to 700 m (650 to 2300 ft) deep. Scientists attributed the shallowness to erosion or to flooding with lava or windblown deposits. Bright spots in radar images of the craters suggested central peaks, which were later confirmed by Magellan images with greater resolution. There also were small

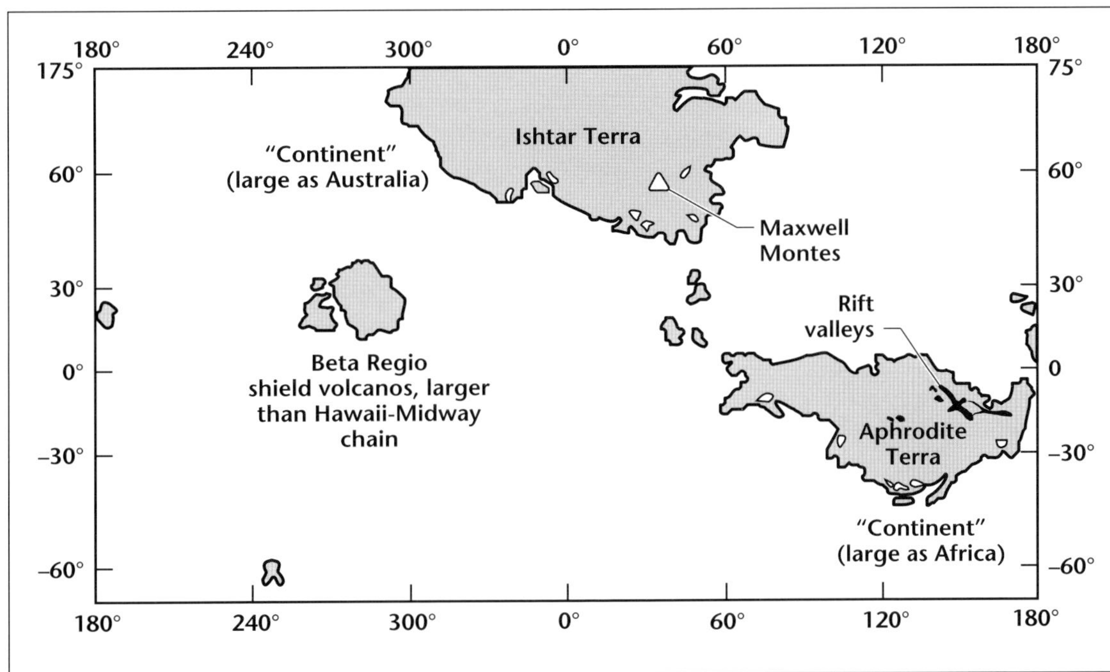

Figure 6-5. This map shows the major continental areas of Venus discovered by the Pioneer Venus spacecraft. Because of the Mercator projection, the size of Ishtar Terra, in relation to Aphrodite Terra, is exaggerated. This is similar to the exaggeration of North America's size compared with Africa or India on Mercator projections of Earth's surface.

circular features that looked much like young impact craters. Ejected material had produced a surrounding rough area, bright on the radar images. The existence of these young craters was also confirmed by Magellan images.

Orbiter discovered that Venus' lowlands cover about 25% of the surface. By contrast, terrestrial lowlands cover 70% of Earth. Plateaus and mountains on Venus are as high as or higher than those on Earth. The lowlands, however, are only one-fifth the greatest depth of Earth's lowlands.

A vast lowland basin, Atalanta Planitia, is centered at 170° east longitude and 65° north latitude. It is about the size of Earth's North Atlantic Ocean basin. The smooth surface of Atalanta Planitia, about 2 km (124 miles) below the mean elevation, resembles the mare basins of the Moon. Because there were few circular bright features that could be impact craters on these lowland areas, scientists thought that the surface may be young. The basin forms part of a large belt of irregular, unconnected lowlands encircling the planet, which were later discovered to be lava-flooded areas.

Precise observations of Pioneer's orbit around Venus allowed researchers to map the gravity field in detail. One theory for these gravity anomalies was that the plains have a thin crust of lower density than that below the upland plains. This is similar to conditions on the Moon and Mars. Some geologists also suggested that the low areas are depressions that later filled with basaltic lavas. This is similar to the mare surfaces of the Moon and some of the plains on Mars. Others theorized that they could be filled with now consolidated windblown sediments. Magellan images later showed that lava flows have globally modified Venus' surface.

There are only two highland or continental masses on Venus: Ishtar Terra and Aphrodite Terra (Figure 6-5). A much smaller elevated

Vertical exaggeration 200×

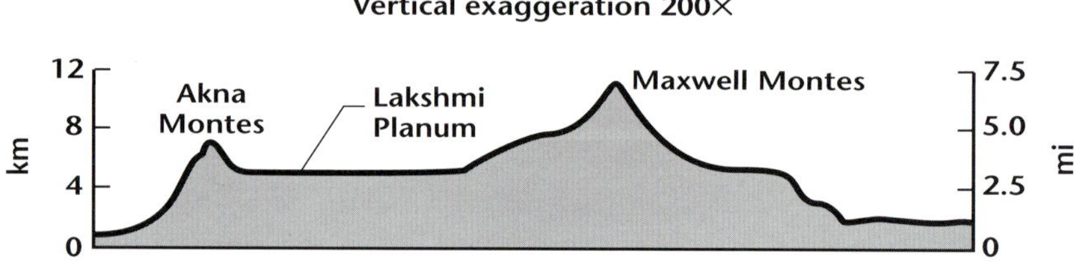

Figure 6-6. Ishtar Terra, the northern continental mass. (Top) Artist's concept of Ishtar Terra compared with an outline of the United States to show the relative sizes of the two land masses. (Bottom) A cross section of Ishtar Terra shows the relative heights of the mountains, the central plain, and the surrounding territory. Note that the vertical scale is exaggerated by 200 times the horizontal scale.

region, Beta Regio, appeared to be a volcanic area connected with a major rift valley system. It is now known that Beta Regio is a young region, and its volcanic mountains may still be forming. Ishtar Terra may be somewhat older, while the oldest region may be Aphrodite Terra. Atla Regio, the "Scorpion's Tail" at the east end of Aphrodite, also may be young.

Points on Ishtar rise to about 11 km (6.8 miles) and on Aphrodite to about 5 km (3 miles) above the mean radius of the planet. Only 5% or 6% of the surface in these "continental" regions is more than 1600 m (5200 ft) above the mean level. This measurement compares with 30% on terrestrial continents. The mass of these regions is about 80% compensated. Three possible causes explain this compensation: mantle convection is underplating the highland masses with silicic rocks, mantle plumes of upwelling magma are producing local differentiation to balance the thickness of the crust, or plate tectonics are causing continental growth. Continental growth by tectonics does not have supporting evidence of deep subduction troughs or midbasin ridges. These features are typical of terrestrial plate tectonics. The presence of some complex forms of troughs and ridges in many areas suggested large-scale motions of the crust, but terrestrial-style plate tectonics is unlikely on Venus.

Ishtar Terra, about the size of Australia or the continental United States (Figure 6-6), has the highest peaks on Venus. There are three geographic units: Maxwell Montes, Lakshmi Planum (with mountain ranges of Akna Montes and Freyja Montes on its northern and northwestern

Figure 6-7. An imaginary view of Maxwell Montes as you might see it across the plains of Venus. This is based on exploratory data from the Pioneer Venus mission. In the late 1980s, new computer techniques enhanced images from exploratory spacecraft. For example, such techniques were able to process higher resolution Magellan spacecraft images. These images allowed analysts to move a viewpoint sequentially over a three-dimensional image of Venus' surface. This gave the impression of viewing the mountains and valleys from a low-flying aircraft. At publication time, such programs for Mars and Venus are already available commercially for personal computer users and schools.

margins), and an extension of the Lakshmi Planum. Lakshmi is about 4 to 5 km (2.5 to 3.1 miles) above the mean level of Venus. This is about the same general elevation as the terrestrial Tibetan plateau is above Earth's mean sea level. It is twice the area of the largest terrestrial plateau. Researchers credited a bright scarp on the southern boundary to talus slopes of eroded debris along a fault zone. Such a rough surface could account for the strong radar reflection that observers noted.

If Ishtar consists of basaltic lava flows, scientists expected there would be a large gravity anomaly. But the data from Orbiter showed a rather mild positive anomaly. This suggested that Lakshmi Planum might consist of thin lavas overlying an uplifted segment of ancient crust. This would be similar to the Tharsis region of Mars.

On the eastern side of Ishtar, peaks of the towering Maxwell Montes thrust high into Venus' sky (Figure 6-7). Maxwell was first recognized on Earth-based radar images. It has a great circular feature that may be a caldera about 100 km (62 miles) across and 1 km (0.62 mile) deep. The caldera is offset toward the east flank of the mountain some 2 km (1.24 miles) below the summit. No bright flows radiate from this caldera. The assumption is that erosion has smoothed any lava flows. If so, the volcano must be much older than those in Beta Regio. Many slopes on Maxwell are, however, bright in the radar images. Such brightness suggests that they are covered with rocks that scatter the radar signal (probably because they are covered with debris). Alternatively, they could be bright because of extremely steep slopes. Magellan data suggest one is a 6-km (3.7-miles) high cliff.

Scorpion-shaped Aphrodite Terra (Figure 6-8) is about the size of Africa. It has two mountainous areas. On the east, mountains rise 5.7 km

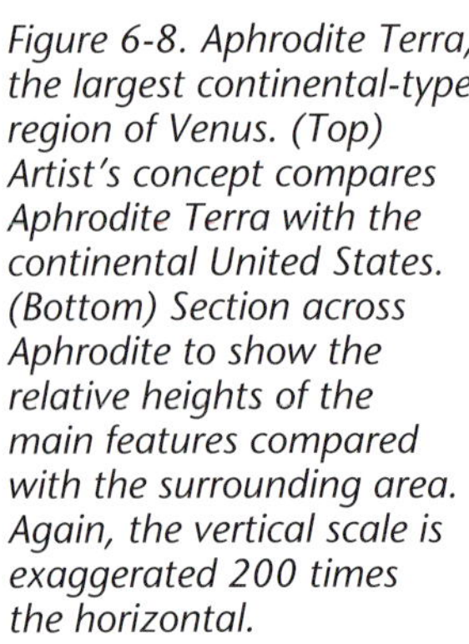

Figure 6-8. Aphrodite Terra, the largest continental-type region of Venus. (Top) Artist's concept compares Aphrodite Terra with the continental United States. (Bottom) Section across Aphrodite to show the relative heights of the main features compared with the surrounding area. Again, the vertical scale is exaggerated 200 times the horizontal.

(3.5 miles) above the mean radius of Venus. On the west, claw-shaped mountains are about 4 km (2.5 miles) high. Between them are rolling uplands. Also, there is a topographically complex mountain rising about 3 km (1.9 miles) above the uplands. The mountains have very rough surfaces like those of the Ishtar continent. South of Aphrodite is a large curved feature called Artemis Chasma.

Pioneer images of Venus' highland areas did not show circular features that could be craters. With radar, craters would be difficult to detect on rough terrain. The presence of these highlands may confirm lack of water in Venus' crust. This is because high surface temperatures would readily deform water-rich crustal rock, so highland areas could not persist.

The bright radar area of Beta Regio is also an interesting region dominated by a large complex shield volcano and a large trough (Figure 6-9). The trough is part of a fault zone that may extend far to the south where two small highland areas (Phoebe Regio and Themis Regio) are aligned. Other small highlands, including Asteria Regio, located west of Beta Regio, have a north-south trend. Lava flows extend radially from the volcanic centers. Two Soviet spacecraft landed directly east of Beta. They measured gamma-ray emanations from the surface that indicate the presence of "basalts." The highest mountainous features on Beta Regio, Theia Mons, and Rhea Mons, are 4 km (2.5 miles) high. They, too, have volcanoes. A large southward trending ridge has elevations up to 2 km (1.24 miles). The images

showed a flat terrain west of Beta Regio. On it was a linear tectonic feature extending 4500 km (2796 miles) to the south-southwest.

Geologists found the new information about this region very interesting. At first, from Earth-based radar data, Beta seemed to be a shield volcano with a central caldera. Data from Pioneer Venus suggested that it is part of an upland area of volcanics. A great rift valley splits this region of Beta Regio. The valley has high shoulders. Its nearest Earthly analogue is the Great African rift valley system. Bright radial streaks radiating from the shield volcanoes are suggestive of lava flows. Scientists suggested that their presence showed Beta Regio is a young geologic feature.

Alpha Regio is a plateau within the rolling plains. It is located at 25° south latitude and 5° east longitude (that is, near the origin of longitude coordinates on Venus). One of the brightest features on Venus, Alpha Regio is elevated about 0.4 km (1600 ft) above the mean level. Its rim is 2 km (1.24 miles) high. Many fractures cut its surface.

Orbiter's radar revealed many rift valleys on Venus (Figure 6-10). They appear to be straight, or gently curved, tectonic features. Some are 5000 km (3000 miles) long, and in some regions they form striking patterns. There are many valleys east of Aphrodite and east of Ishtar. Geologists suggest regional tectonic distortions probably caused them.

Figure 6-9. The great volcanic area of Beta Regio appears in this artist's drawing. It is based on data from Pioneer Orbiter's radar. The area has many calderas, and scientists believe it is one of the youngest areas of the planet, with the most recent volcanic activity.

▲

Figure 6-10. Artist's concept of one of many rift valleys that Pioneer Orbiter discovered on Venus. Often the valleys have lines of mountains associated with them.

The gravity field of Venus mapped by the Orbiter closely matches the topography. East of Ishtar, a large region extends from 14° to 40° longitude and from 50° to 75° north latitude. It consists of complex ridges and troughs, probably disrupted by extensive faulting. This appeared to be Venus' most tectonically disturbed region. Geophysicists theorized that this region could be where plate tectonics started or where a plume of hot magma rose through the mantle to produce a thickened low density crust. Other features on the radar images also suggested tectonic activity on Venus: vertical uplift at Lakshmi Planum, and the northern and western mountainous ridges marginal to Ishtar Terra. These ridges may be due to plate motion. However, scientists saw no evidence for integrated plate tectonics on Venus. Development of thin crusted lowlands and thick crusted highlands would imply a long period of widespread mantle convection early in Venus' history. Within the limits imposed by the resolution of Earth-based and Orbiter radars, scientists concluded that, if plate tectonics took place on Venus, they are grossly different in character from terrestrial plate tectonics. The more detailed Magellan radar images confirmed these conclusions.

Venus appears different from the other Earth-like planets. There are signs of regional placements, which may be evidence of incipient, rudimentary, or past plate tectonics. Development of plate tectonics may have stopped because Venus lacks water, but there is no proof that the presence of much water has anything to do with plate tectonics. Geophysicists speculated on why Venus should be so different from Earth when it is so similar in many respects. They suggested that the higher surface temperatures led to domination of tectonics by a thick layer of basaltic material. This could not be subducted. The global lava flows, which Magellan revealed, appear to confirm this. The global distribution of impact craters, with little evidence of a widespread

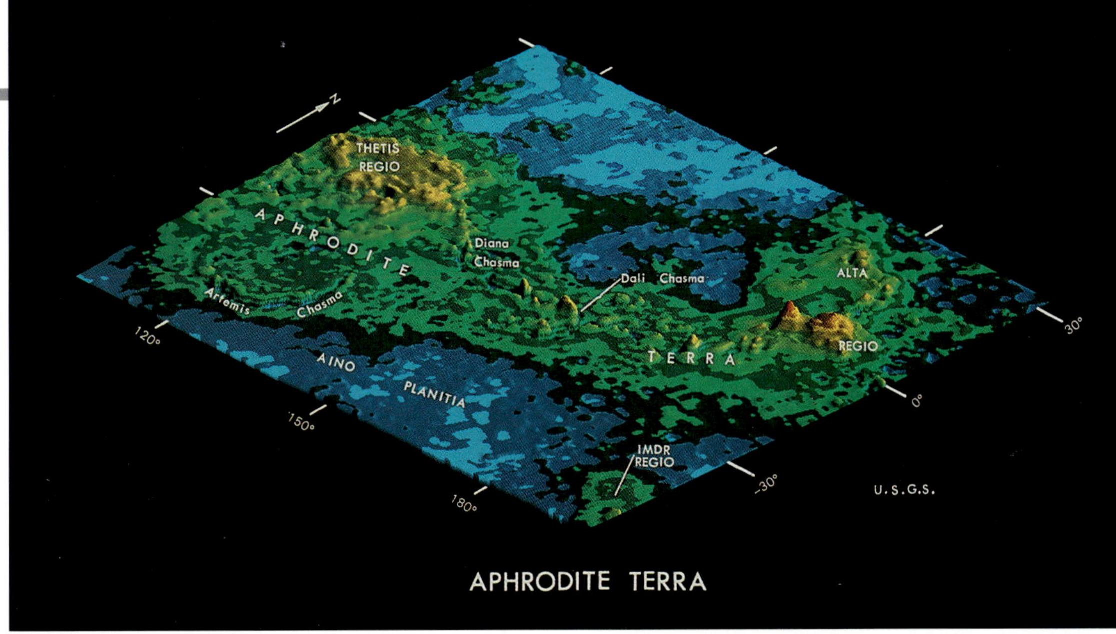

Figure 6-11. The USGS processed the radar data from Pioneer into perspective views of the surface showing surface relief in graphic detail. This computer enhanced image shows Aphrodite Terra from the southeast.

ancient crust like that on Mars and the Moon, also supports their suggestion.

Some computer enhanced surface relief images in Figures 6-11 through 6-15 are representative of the first spacecraft radar data on Venus. The United States Geological Survey (USGS) processed these Pioneer Venus data to create the first three-dimensional images of the planet's surface. These clearly show great depressions and mountains. The areas shown are Aphrodite Terra (Figure 6-11), Ishtar Terra looking toward the east (Figure 6-12), and Beta Regio (Figure 6-13).

Two Mercator projections based on the Pioneer Venus radar data provide the first detailed contour map of Venus' surface. Figure 6-14 is an annotated map showing the major topographical features discovered by the Pioneer Venus mission. On the map, the chart to the right shows the color scale indicating height in terms of the planet's radius, together with a kilometer scale above and below the mean radius.

The final image in this group (Figure 6-15) shows a detailed strip of the equatorial region of the planet in terms of radar brightness (see the scale on the right).

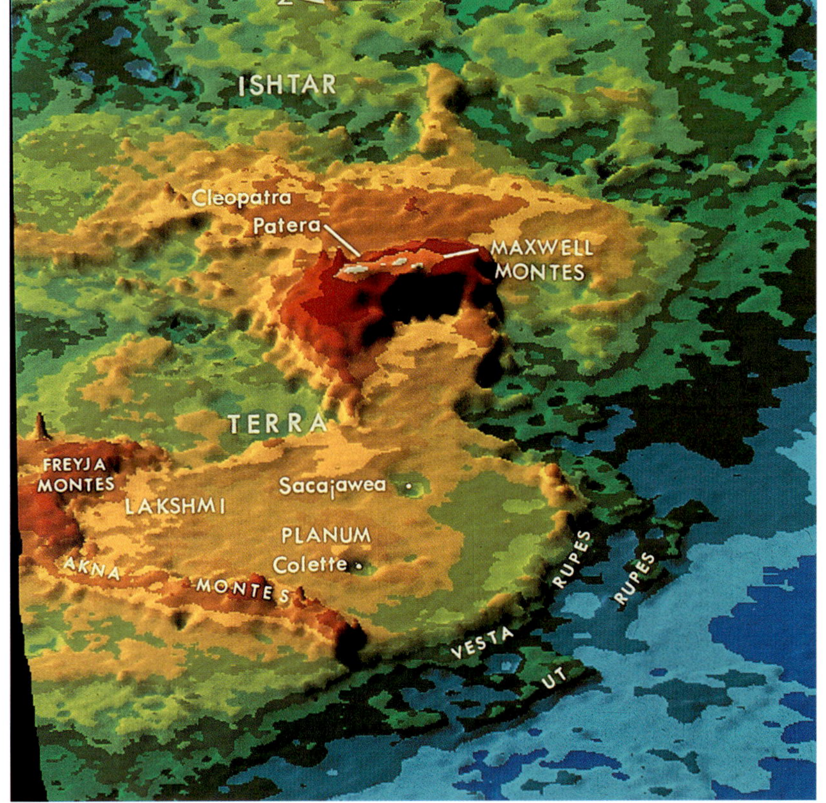

▲

Figure 6-12. Ishtar Terra in this computer enhanced relief image is from the west looking across Akna Montes and Lakshmi Planum toward Maxwell Montes (the red area just above center).

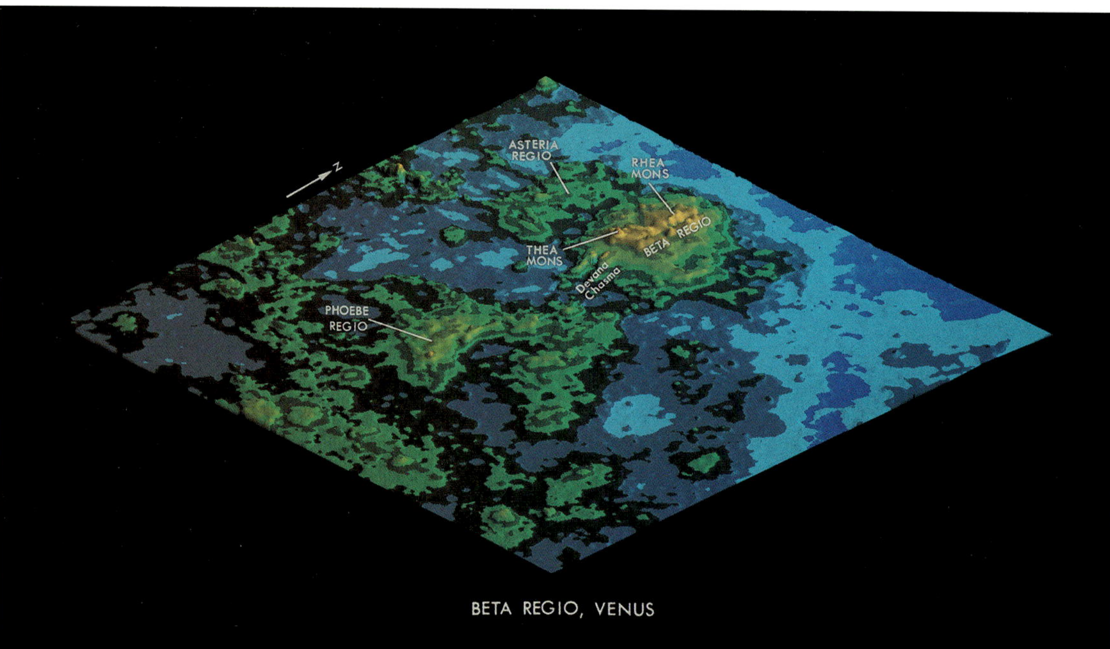

Figure 6-13. Beta Regio is the yellow area at top right of this relief image. Phoebe Regio is slightly below and to the left. The view is from the southeast.

The Atmosphere

Pioneer Venus Orbiter greatly extended observations of ultraviolet patterns in Venus' clouds. While Mariner 10 obtained eight days of pictures, Pioneer Venus obtained many hundreds of days of pictures. These images provide a greatly improved record of the bulk motions of the cloud tops. As Venus moved around the Sun once every 225 days, the cloud photopolarimeter was able to view Venus at all phases. It imaged the planet from waxing crescent phase to full phase and back to waning crescent phase.

Figure 6-16 is the first image that Pioneer Venus obtained. The low contrast is due to the oblique viewing conditions at crescent phase combined with high altitude haze in the atmosphere.

Figure 6-17 was obtained at the time the Soviet entry probe Venera 11 arrived. Venera descended at the equator near the bright limb (left edge) of this image.

Figure 6-18 shows Venus at full phase. Both poles have bright caps. An optically thick haze of small particles (radius about 0.5 micron) above the main cloud layer caused these caps.

The cloud photopolarimeter captured the data for Figure 6-19 on February 19, 1979, when the Sun illuminated almost the entire hemisphere visible from the spacecraft. Large-scale cloud patterns show a horizontal Y-pattern previously identified by lower resolution ultraviolet telescopic sightings from Earth. The mottled small-scale features in the center and left of center in the image probably represent convection cells driven by the Sun's heat.

A question arising from the Mariner 10 data was whether the features that move around in a four-day period are bulk movement of atmospheric masses or wave motions in the atmosphere. The Pioneer probe results suggested that the air is moving at about 100 m/sec (330 ft/sec). Probe data showed that the wind velocity starts to decrease below the clouds. At the surface, it is very small. Scientists regard the large ultraviolet features, especially the Y- and C-markings, as special kinds of waves that move around the planet at the same speed as the air. All four probes, and some Soviet probes, showed the same westward motion with little or no north-south motion.

The cloud photopolarimeter experiment on Orbiter obtained hundreds of images. These were four-color polarization maps and images of Venus in ultraviolet light. Orbiter obtained them when its orbit was farthest from the

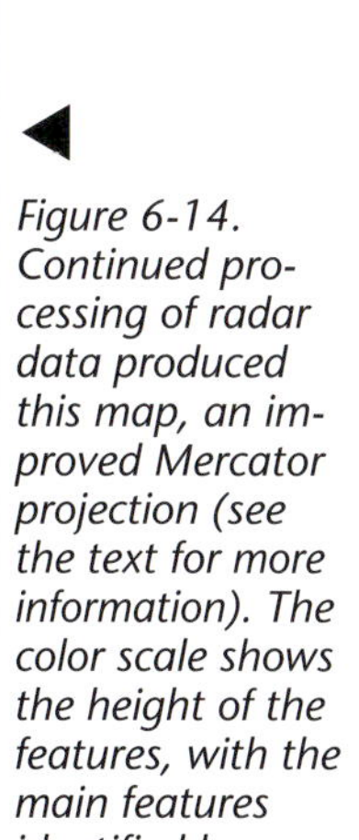

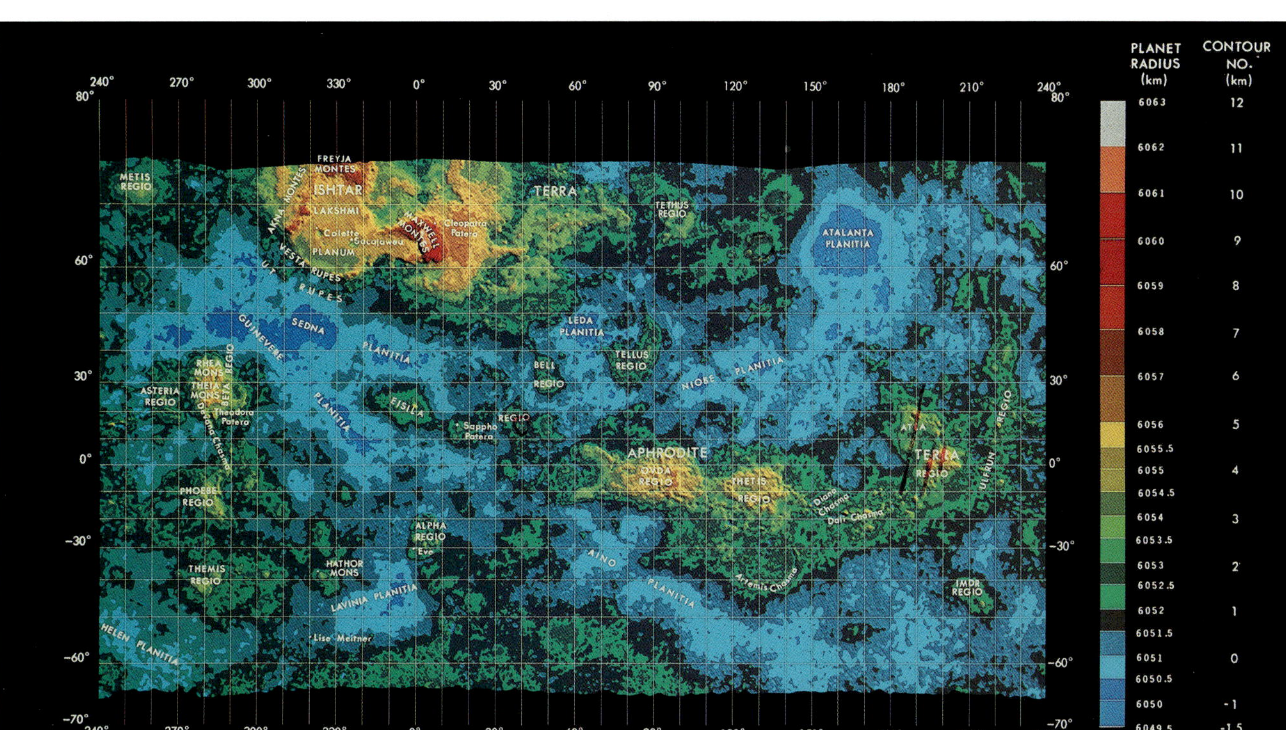

Figure 6-14. Continued processing of radar data produced this map, an improved Mercator projection (see the text for more information). The color scale shows the height of the features, with the main features identified by name. Note the increased detail compared with the first radar map in Figure 6-1.

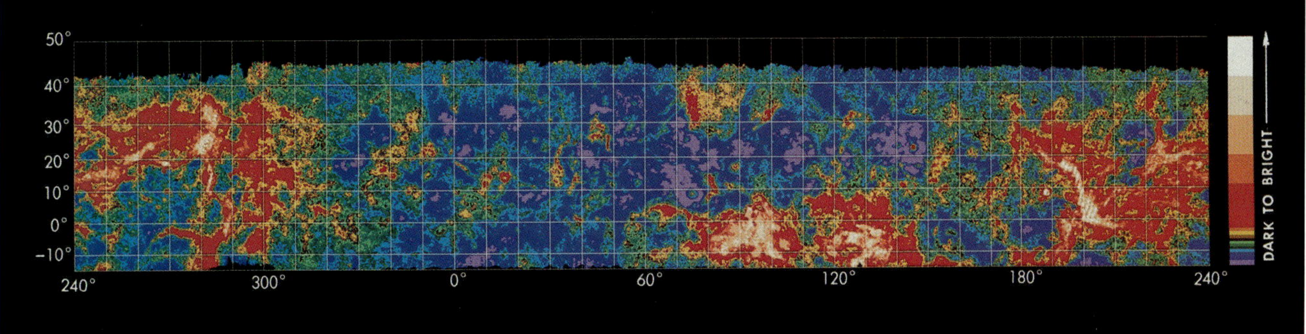

Figure 6-15. The advanced processing also produced very detailed maps of various regions of the planet. The example in this figure is one of Venus' equatorial regions. This map identifies radar brightness of the various features given by the color scale on the right of the image.

planet (apoapsis)—40,000 to 64,000 km (24,856 to 39,770 miles) away. At these times, the spacecraft was moving slowly. During the opposite portion of the orbit near periapsis, the spacecraft passed at high speed through Venus' tenuous upper atmosphere. During the nominal mission, it approached within 160 km (99 miles) of the surface. The spacecraft repeated its close approaches in 1992 at even lower altitudes before final entry. These were times when other instruments sampled atmospheric composition.

North polar regions of Venus were unusually bright in the images during the nominal mission. The polarimetry data show that a vast haze of submicron particles causes the bright polar caps. These particles are about 0.25 micron in radius. High-altitude haze also was present at lower latitudes, particularly in the morning sky. The haze extended vertically over at least 25 km (15.5 miles), reaching down into the main visible cloud layer where it coexisted with the larger (about 1 micron radius) sulfuric acid cloud droplets. The refractive index of the haze particles was 1.45 to 0.04, which suggested that their chemical composition could be the same as that of the main cloud deck. This was shown by the amount of haze above and

Figure 6-16. On December 5, 1978, Orbiter obtained this image of Venus appearing as a crescent. A high altitude haze obscured ultraviolet features.

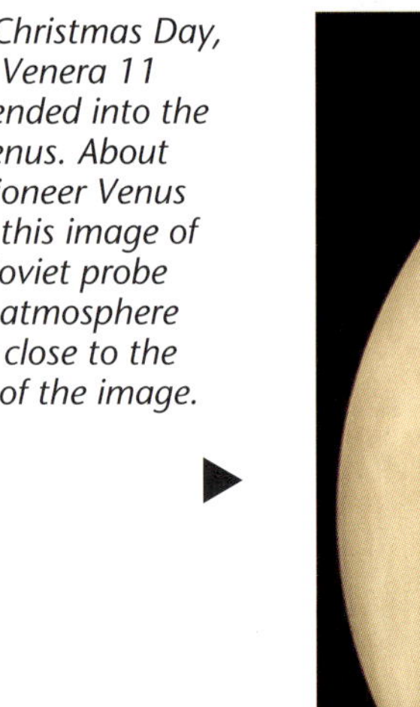

Figure 6-17. On Christmas Day, 1978, the Soviet Venera 11 entry probe descended into the atmosphere of Venus. About the same time, Pioneer Venus Orbiter obtained this image of the planet. The Soviet probe plunged into the atmosphere near the equator close to the bright limb (left) of the image.

within the main cloud deck in the polar regions, decreasing by more than one-half during the primary mission. Chemical and aerosol processes are at work on time scales of several months and longer.

Researchers used the images to study atmospheric circulation and its relationship to regional cloud patterns. The four-day rotation period of Venus' atmosphere was first determined from the reappearance at four-day intervals of a faint horizontal Y-shaped feature in ground-based ultraviolet images. Pioneer Venus images, taken at 24-hour intervals, show the planet's rotation in detail. As the Y-feature rotated around the planet, it confirmed the rotation period.

Wind speeds near the cloud tops were determined by tracking small cloud features. These measurements revealed nearly constant high-speed zonal winds, about 100 m/sec (330 ft/sec) at the equator. The winds decreased toward the poles. At cloud-top level, the atmosphere rotated almost like a solid body. This zonal circulation differed from that observed by the 1974 Mariner 10 flyby. (Mariner found strong midlatitude jet streams.) The planetary scale patterns of the clouds changed during the Pioneer Venus primary mission. For example, the dark horizontal Y-shaped feature disappeared for periods of a few weeks.

The cloud photopolarimeter functioned perfectly during Pioneer Venus' extended mission. Observations of aerosol evolution and atmospheric circulation over many years contributed valuable knowledge about processes that are important components in Earth's climate system.

Continued observations from orbit showed that the wind speed of 320 km/hr (199 mph) near the equator corresponds to the rotation

period between four and five days at most latitudes (Figure 6-20). At higher latitudes, however, the period decreases to three-and-a-half days, and under polar hazes an infrared dipole rotates in three days. Another phenomenon showed up near the equator. It was a wave of brightening that was, perhaps, a thickening of the haze layers. The wave passed through the clouds and circled the planet in four days. However, this wave was mysteriously absent in 1982 and 1983. The figure also shows that the distinct midlatitude "jet stream" obtained from 1974 Mariner 10 images was missing in the Pioneer Venus observations.

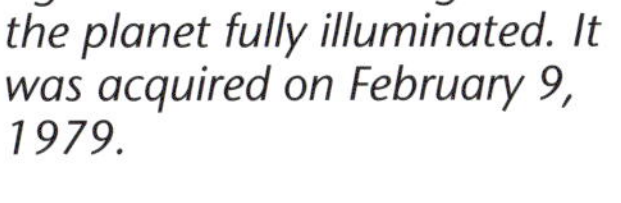

Figure 6-18. This image shows the planet fully illuminated. It was acquired on February 9, 1979.

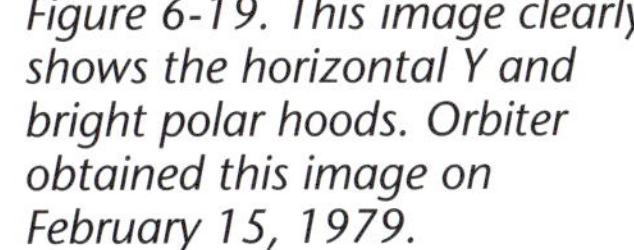

Figure 6-19. This image clearly shows the horizontal Y and bright polar hoods. Orbiter obtained this image on February 15, 1979.

Scientists have put forward a theory for Venus' rapid easterly winds. The theory suggests that the nature of the general circulation varies between wind profiles observed by Mariner 10 and those observed by Pioneer Venus. Such changes might be linked to long-term variations of clouds and aerosols, such as the appearance and disappearance of polar caps. Observations during the extended Pioneer Venus mission provided more data, and researchers continued to try to resolve these questions.

One day after the spacecraft took the image in Figure 6-19, high zonal winds changed the atmospheric pattern. They carried the clouds forming the prominent Y-feature from right to left by about 90° in longitude, leaving only the tail of the Y visible (Figure 6-21). The hemisphere of the planet opposite the Y (Figure 6-22) revealed a pattern of linear

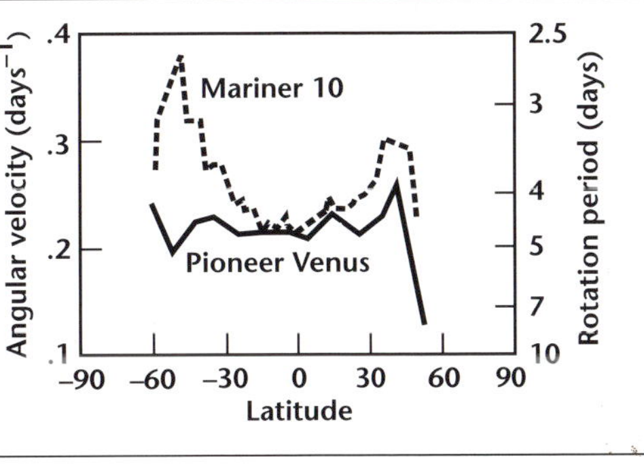

Figure 6-20. Imaging and polarimetry by Mariner 10 and Pioneer Orbiter show how wind velocities at various latitudes on Venus changed considerably between the two missions (except at equatorial regions).

features nearly parallel to the latitude circles. Curvilinear features predict reappearance of the Y-feature, which was recorded in Figure 6-23.

When Pioneer arrived at Venus, both poles were covered by bright cloud caps, which had been seen on one or both poles several times during earlier Earth-based sightings. The cloud photopolarimeter experimenters identified the "cloud" caps as a thick blanket of small haze particles, about 0.5 micron in radius. A series of parallel dark bands breaks the edge of the bright polar cap (Figure 6-24). Bright streamers of haze particles extend from the polar cap toward lower latitudes (Figure 6-25).

Figure 6-21. High zonal winds have moved the top of the horizontal Y-feature toward the left, leaving the tail a prominent feature in this image from February 16, 1979. Compare this image with that in Figure 6-19, which shows more of the Y-feature.

Sunlight becomes polarized when clouds reflect it. The nature of the polarization can provide information about the physical properties of cloud particles. Studies of ground-based polarization measurements of Venus had already revealed that the major cloud deck consists of spherical sulfuric acid droplets 1 micron (10^{-4} cm) in radius.

Figure 6-22. In this image (February 17, 1979), the hemisphere of the planet opposite the Y-feature has a pattern of linear features nearly parallel to latitude circles.

The droplets in this deck produce positive polarization at ultraviolet wavelengths. This pattern appears at the center of the disk in the Pioneer Venus polarization map (Figure 6-26). The map also indicates anomalous regions of negative polarization near both poles. Their location corresponds to the bright polar caps in Figure 6-27. This image was obtained just five hours before the polarization map. Polarization of the polar caps indicates a thick haze of very small particles (0.25 micron in radius) overlying the main cloud layer. Except for effects of their small

Figure 6-23. On this image (February 18, 1979), curvilinear features predict the reappearance of the Y-feature.

Figure 6-25. (Above) In this image (February 16, 1979), bright streamers of haze particles extend toward lower latitudes from the bright polar cap at the bottom of the image.

Figure 6-24. A series of parallel dark bands breaks the equatorial edge of the bright polar cap.

size, the polarization properties are similar to those of the droplets in the main cloud. This suggests that the haze also may be composed of sulfuric acid.

The polar haze began to partially disappear in mid-1979. The number of haze particles above each square centimeter of the main cloud deck, which was about 300 million in January and February, decreased to less than half of that over a period of several months. Continued observations during the extended Pioneer Venus mission were used to study cloud and haze variations and their possible link to long-term changes of atmospheric dynamics.

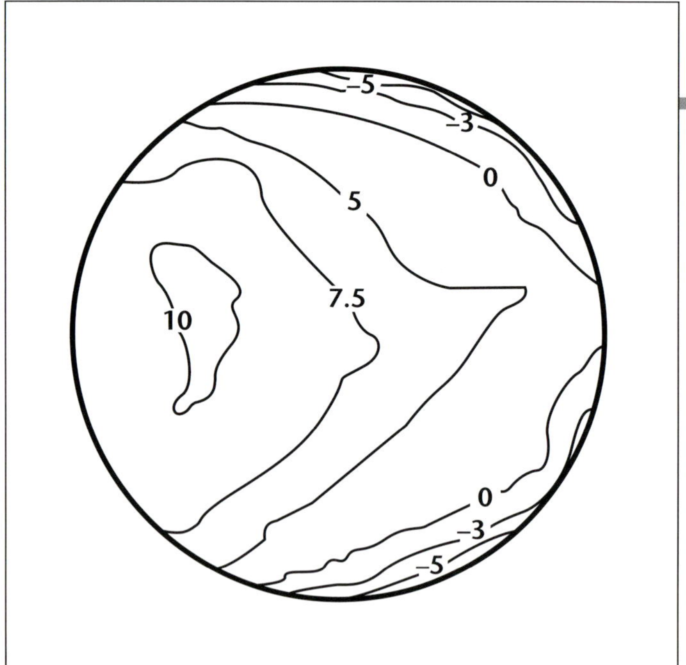

▲

Figure 6-26. (Top left) This drawing shows a polarization map that Orbiter obtained on February 25, 1979. The map shows anomalous regions of negative polarization caused by haze near the poles. It also shows positive polarization from sulfuric acid cloud drops at the center of the disk.

Figure 6-27. (Top right) This image was obtained several hours before the polarization map of Figure 6-26. The relation between ultraviolet features and map contours is clear.

The nature of these aerosols and their relationship with climate on Venus are of interest for studies of Earth's climate. Similar aerosols are produced in Earth's stratosphere following large volcanic eruptions. Some scientists believe that they may cause significant climate changes.

Venus' cloudy atmosphere reveals a rich spectrum of dynamic events, especially in the equatorial region. Some of these features are:

(a) Bright-rimmed cells appear as mottled cellular cloud patterns. Scientists believe these are convection cells driven by the Sun's heat. They may have some analogy to tropical cumulus cloud clusters on Earth.

(b) Wave-trains are series of short streaks cutting across background features. Their almost vertical lines are strongly suggestive of a wave phenomenon.

(c) Circumequatorial belts, vaguely visible in some images, appear as bright lines parallel to the equator, where they stretch several thousand miles from the limb across the disk.

The atmosphere was probed by many instruments from the Orbiter and sampled by others on the four probes and the Bus. Regions of the atmosphere are generally based on temperature, as shown in Figure 6-28. Solid lines represent data collected by the Probe Bus and the Orbiter. Dashed lines indicate limited data from the Small Probes. Direct probe measurements cover the range from the mesosphere to the surface. The Orbiter infrared radiometer provided almost global information for the stratosphere. More results came from the radio occultation experiment. All these data, from the surface to the ionosphere, provide an almost complete picture of the temperature, pressure, and density structure of Venus' atmosphere.

An exciting discovery was the enormous range of temperature between day and night in the upper atmosphere. Yet, even on the dayside of Venus, the upper atmosphere temperature is not as hot as Earth's upper atmosphere. On Earth, temperatures are 700 to 1000 K at sunspot minimum. Heat comes from formation of the ionosphere by very short wavelength solar ultraviolet radiation. Somehow, Venus manages to keep a lower temperature than Earth's upper atmosphere even with twice the flux of incoming solar radiation.

But the real surprise is the low temperature of the upper atmosphere on the nightside. This region cannot be called a thermosphere (hot sphere) like the equivalent region in Earth's atmosphere. The thermosphere is the atmospheric region where the incoming solar photons are absorbed and solar heat is transferred into the atmosphere. Scientists coined the name cryosphere (cold sphere) to describe this cold region of Venus' upper atmosphere. Although the Sun does not directly heat the nightside, heat must flow to the nightside from the dayside. It also must flow upward on the nightside from the warmer mesosphere. The gradient between day and night is rather sharp, occupying little more than the twilight zones, 20° to 30° of longitude. Theories developed to describe the behavior of Earth's thermosphere do not apply to Venus and leave unexplained many temperature features of Venus' atmosphere. Clearly, improvements in the theory were needed. Data from Phases II and III made it possible to refine models and test the theories against these new data.

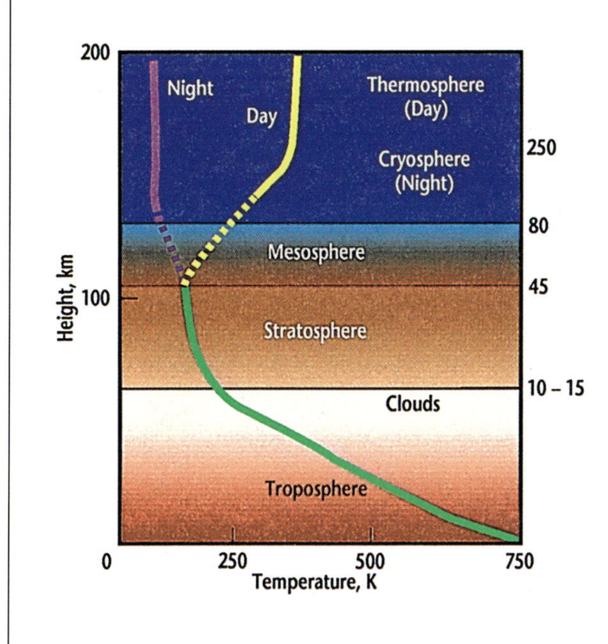

Figure 6-28. Typical temperatures for Venus' atmosphere relate to altitude of the various regions. Corresponding heights for Earth's atmosphere appear for comparison.

Scientists knew that the dayside thermosphere responds to short-term changes in solar activity, such as those caused by the Sun's 27-day rotation. They expected that it would also respond to changes in solar activity over the 11-year solar activity cycle. Despite solar heating, the temperature of the dayside thermosphere is only about 300 K. This was much colder than predicted. Researchers explained this low temperature on the basis of eddy and radiative cooling. Eddy cooling occurs when heat is transported down into the mesosphere. Radiative cooling is due to atomic oxygen exciting carbon dioxide into a strong emission at 15 microns. This radiative cooling appears to be the main mechanism keeping the thermosphere temperature down. On this basis, some researchers concluded that the response of the thermosphere to the 11-year solar activity cycle would be small. Also, they predicted that the atomic oxygen/carbon dioxide ratio would increase as increased solar radiation photodissociated more carbon dioxide. In turn, this would increase the cooling mechanism and help to weaken the effects of solar activity on the thermosphere's temperature.

New data provided important information about the atmosphere between 130 and 210 km (80 and 130 miles) on both nightside and dayside and at low solar activity. These were atmospheric drag measurements from the orbital decay of the Orbiter in 1992, coupled with drag data from the Magellan spacecraft. Researchers compared these data with earlier data they obtained at high solar activity in 1978-1980. The result has been an increased knowledge of the detailed response to solar variations of temperature, atomic oxygen, and carbon dioxide in Venus' thermosphere. For example, studies of images in light from carbon monoxide and oxygen, made by the ultraviolet spectrometer, showed that, as solar activity declined, the gases decreased relative to carbon dioxide (from which they are derived by photodissociation). The effects reversed after solar minimum.

Investigators found a weak but detectable temperature response on the dayside, which was in accord with the predicted response based on strong carbon dioxide radiative cooling. This was an important discovery because it highlighted a mechanism that might cause an otherwise unexpected strong cooling of Earth's thermosphere if terrestrial carbon dioxide builds up here in the future.

It now seems clear that with decreasing solar activity the oxygen/carbon dioxide ratio in the lower thermosphere decreases, as indicated by decreased photodissociation of carbon dioxide and a lower temperature. The decrease in this ratio results in less effective oxygen-carbon dioxide cooling and a partial cancellation of the decreased extreme ultraviolet heating at times of low solar activity.

Investigators found that the percentage decrease in atomic oxygen with decreasing solar activity on the dayside was about the same as that of atomic oxygen transported to the nightside. Also, they saw a weak response of temperature on the nightside to solar variations.

Scientists now conclude that there is evidence of photochemical, radiative, and dynamical responses of Venus' upper atmosphere to changes in solar activity. On the dayside, there is a weak temperature response to long-term solar activity variations. Also, carbon dioxide increases in the lower thermosphere with decreasing solar activity. This is due to reduced photodissociation of carbon dioxide. The consequence is a strong decrease in the ratio of atomic oxygen to carbon dioxide in the lower thermosphere when solar activity is low. In turn, this leads to a reduction in the amount of cooling by the atomic oxygen-carbon dioxide mechanism. Scientists consider that similar effects might operate in the Earth's upper atmosphere, and that further study is needed.

On Venus' nightside, researchers found an unexpected strong response of atomic oxygen to changes in solar activity. The atomic oxygen on the nightside reflects the changes in atomic oxygen on the dayside. Also, there is a decrease in the day-to-night flux across the terminator at low solar activity.

Because the nightside is so cold, atmospheric pressure falls very rapidly with increasing height. At each atmospheric level, pressure is much less than on the dayside. The nightside high atmosphere is exceptionally cold. Its temperature of 100 K is less than any other planetary atmosphere closer to the Sun than Saturn. This large difference between day and night temperature causes very strong winds to blow from day to night. The Orbiter observed this large temperature difference directly. Unfortunately, it did not carry an instrument to observe the winds directly. However, there were indirect confirmations of their presence.

Data gathered by Orbiter's instruments also showed that the high atmosphere, like the cloud top regions, "superrotates" much faster than the planet itself. Violent high-altitude winds are patterned by density waves that begin in the lower atmosphere. They appear in the data from the neutral mass spectrometer and as bright streaks in the images from the ultraviolet spectrometer.

The bottom 65 km (40 miles) of Venus' atmosphere is the troposphere. The boundary of this region, the tropopause, coincides with the cloud

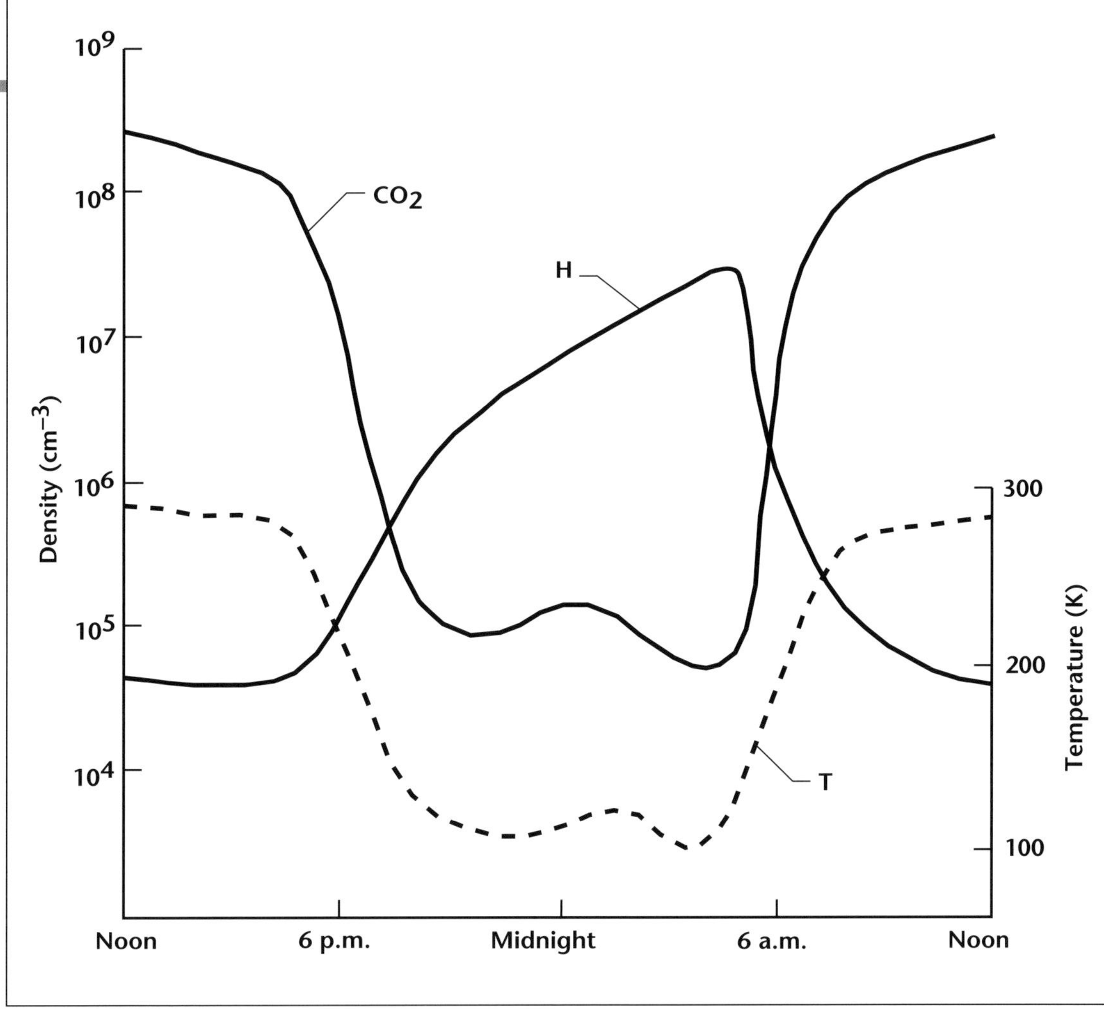

Figure 6-29. Instruments onboard Orbiter revealed a very cold nightside thermosphere, or cryosphere. Resulting pressure gradients drive strong winds from the dayside to the nightside of the planet. Atomic oxygen densities show a strong cold-trapping effect on the nightside similar to the CO_2 curve in the diagram. The displacement of the nightside atomic hydrogen peak toward the dawn terminator, as indicated by the H curve, suggests that the thermospheric winds have a super-rotating component.

tops. Pressures at the tropopauses of Earth and Venus are similar, but their heights are quite different because of different surface pressures.

Above the tropopause is the middle atmosphere. On Earth, this region consists of the stratosphere and mesosphere. The boundary between them is a temperature maximum caused by ozone absorbing solar ultraviolet radiation. Venus has no detectable ozone and no temperature maximum to divide its middle atmosphere. Scientists have not yet agreed upon a single name for this combined region in Venus' atmosphere. They believe the middle atmosphere has much chemical activity driven by solar ultraviolet radiation. Oxygen and ozone must continue to be released into the atmosphere by the breakdown of carbon dioxide. However, very scarce atoms, such as chlorine, probably reduce oxygen and ozone to levels that make their detection impossible.

Data from the neutral mass spectrometer and the atmospheric drag experiment confirmed a very cold nightside thermosphere, or cryosphere. The resulting pressure gradients drive strong winds from the planet's dayside to its nightside. Researchers deduced atomic hydrogen densities from ion mass spectrometer data on H^+ and O^+. They obtained densities of oxygen from neutral mass spectrometer data. The hydrogen peaked on the nightside, and the bulge was displaced toward the dawn terminator (Figure 6-29). The oxygen densities showed that there was a strong cold-trapping effect on the nightside. The displacement of the hydrogen peak showed that thermospheric winds have a superrotating component.

Thermal contrasts provide the driving mechanism for general atmospheric circulation. They set up pressure differences to drive the flow. Pioneer made a major discovery about the lower atmosphere below the clouds. There is little thermal contrast between night and day and from the equator to 60° latitude. Thus, temperature variations at and near Venus' surface are small.

Absence of large thermal contrast in the atmosphere led scientists to several other conclusions. At some levels, there must be an effective transport of heat from equator to poles and from the subsolar to the antisolar points by atmospheric circulation. The atmosphere must transport heat efficiently from the region below the Sun to the rest of the planet. Because the lower atmosphere is so dense, slow winds alone are sufficient. For the same reason, the rate at which temperature rises or falls due to varying inputs of solar heat is small.

A surprising discovery was that most of the deep atmosphere is stable and stratified like Earth's stratosphere. An analogy is the stagnant air layers in the Los Angeles basin on a smoggy day. From the clouds down to 30 km (18.6 miles) altitude, a layer 23 km (14 miles) deep, and in a lower layer between 15 and 20 km (9 and 12 miles) altitude, the atmosphere is stratified and free of convective activity. It does not rise and overturn in the way that air does on Earth over hot farm or desert lands, or in cumulus clouds. This was unexpected because scientists thought that the high temperatures in the deep atmosphere would be a source of hot, rising gas. If true, this would lead to deep convective cells and turbulence. Also, before Pioneer Venus, theoretical studies indicated that, at radiative equilibrium, much of the lower atmosphere would be unstable and overturning. The Pioneer Venus data quickly led scientists to revise these earlier models.

All four probes and several Soviet probes measured high surface temperatures. They were equal within uncertainties of a degree or so when corrected to a constant distance from the center of Venus. Earth-based instruments also have sensed surface temperatures at radio wavelengths, with comparable results. One thing that sets Venus apart from Mars and Earth is its very high surface temperature. One main objective of the Multiprobe mission was to test the belief that the "runaway greenhouse effect" caused the high surface temperature. This effect requires that only a small percentage of the solar energy reaching the surface be converted into heat, and be redistributed globally. Further, the atmosphere and clouds must form an insulating blanket that infrared radiation can penetrate only with difficulty. The heat cannot be reradiated into space.

Pioneer Venus data left no doubt that a strong greenhouse mechanism is at work. This mechanism describes the state of the atmosphere above about 35 to 50 km (22 to 31 miles) altitude. Below that, dynamics control the temperature. Radiative heating associated with the greenhouse mechanism drives the dynamics. About half the heating of the atmosphere by incoming solar radiation occurs near the top of the clouds. The rest of the energy is distributed at lower altitudes and at the surface.

Measured infrared fluxes on the probes showed several anomalies. These anomalies suggest that parts of the atmosphere transmit upward about twice the energy available from solar radiation at the same level. Instrument errors in this difficult measurement may be responsible. A possibility is that two of the probes entered regions that are unusually transparent

to thermal radiation. However, this is unlikely because much of the absorption is due to carbon dioxide. Scientists suggested that the heat balance oscillates around its average state. Also, the anomalous measurements occurred during the cooling phase. Despite these problems in interpreting some observations, the runaway greenhouse effect, coupled with global dynamics, is accepted as explaining the high surface temperature.

The Atmosphere—Clouds

When viewed from the Earth in visible light, the disk of Venus appears to be completely covered with a bright veil of unchanging, featureless, yellowish clouds. Before Pioneer Venus, astronomers had diligently observed these clouds from Earth without adding much to our knowledge. Some *in situ* data through the cloud depths were available from Soviet missions to Venus, especially from Veneras 9 and 10. Earlier, Mariners 5 and 10 flyby spacecraft experiments also yielded some information, primarily about regions near the cloud tops. An objective of Pioneer was to determine the nature and composition of the clouds.

Earth-based observations first revealed the clouds' featureless, global nature. These sightings were at both visible and infrared wavelengths. However, astronomers also had discovered features at near ultraviolet wavelengths, hinting at some form of horizontal cloud structure. Further, these features appeared to circulate around the planet about every four days. Mariner 10 obtained detailed imaging of Venus to confirm this four-day rotation period. It also obtained detailed measurements of the circulation near the cloud tops. The images showed that the motions are generally zonal; that is, parallel to Venus' equator.

Observations from Earth also provided evidence about the detailed properties of the particles composing the uppermost clouds. Scientists measured scattered sunlight that had interacted with the uppermost layers. Particularly useful were measurements of how scattered sunlight polarized at various angles of observation of the clouds relative to the solar illumination. Researchers compared these measurements with calculations based on models that considered particles with various properties. Best agreement was found when the particles were all assumed to be spherical and about the same size. Their effective radius was about 1.05 microns, and their index of refraction was 1.44 for visible light.

These conclusions, coupled with spectroscopic data obtained from Earth, suggested that the upper cloud particles were principally concentrated sulfuric acid.

Optical experiments aboard the Veneras 9 and 10 probes as they fell through the atmosphere obtained data consistent with these conclusions. Analyses of data from light scattering (nephelometer) experiments on the Soviet probes showed that the vertical cloud structure had three main layers. The data also yielded information about variations in particle sizes and indices of refraction in each of these layers and the regions between them.

Data from these experiments also suggested that larger particles with large indices of refraction were present at lower altitudes. Scientists tentatively identified the particles as large sulfur droplets. Furthermore, since sulfur seemed a likely candidate, they also proposed sulfur crystals as a high-altitude absorber responsible for the ultraviolet contrasts.

Although invisible from Earth, a very tenuous haze was revealed on the Mariner 10 images. The haze layers were above the cloud tops at altitudes of 70 to 80 km (43 to 50 miles). Also,

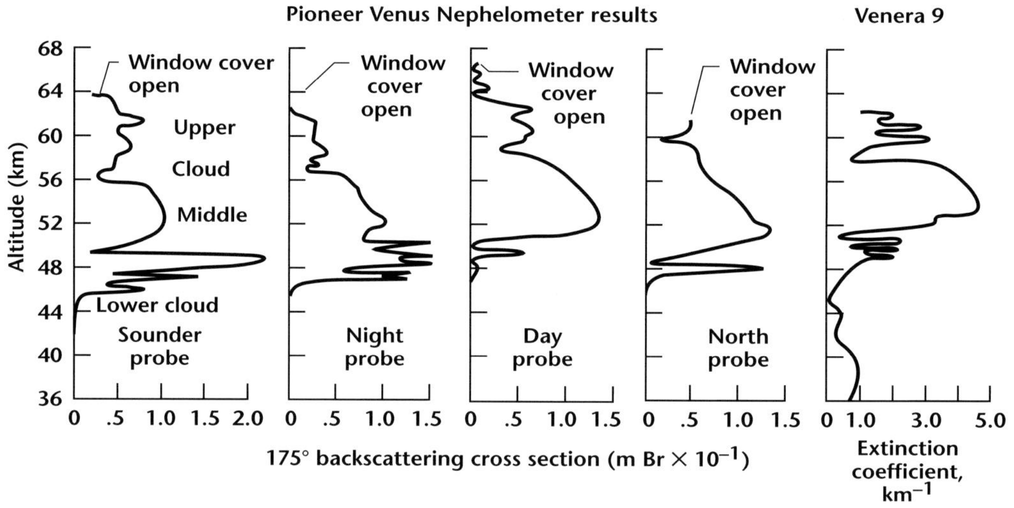

Figure 6-30. Comparison of results from Pioneer Venus probes and Venera 9 show a remarkable similarity in profiles of the cloud layers plotted against altitude. These results suggest that the cloud system is global, even though similarities in features of the vertical structure do vary because of what may be changes in large-scale dynamics of the cloud system.

bright transitory polar caps or bands, lasting from weeks to months, were observed from Earth.

Based on the above background, scientists chose experiments for Pioneer to investigate, in detail, cloud properties at depth and temporal "weather-related" features at cloud tops. For example, experiments on the probes were selected to detail the vertical cloud structure at each of the four entry sites. Orbiter experiments provided many years of cloud-top observations.

Primary experiments selected specifically to examine clouds included equipment such as Large Probe and Small Probe nephelometers, Large Probe cloud-particle size spectrometer, and Orbiter cloud photopolarimeter/imager. Cloud-related experiments that provided information from which scientists could infer cloud properties included several pieces of equipment. These were the Large Probe solar net flux radiometer, the Large Probe neutral mass spectrometer, the Large Probe gas chromatograph, Orbiter infrared radiometer, and Orbiter ultraviolet spectrometer. Further supporting information was obtained from the Large Probe infrared radiometer, the Small Probe net flux radiometer, and the Large and Small Probe atmospheric structure experiments.

Investigators combined data from in-depth measurements from the four probe locations with the Orbiter's planetwide observations. Data from the probes served as "ground truth" for the Orbiter's data. This led to a more complete general understanding of the clouds, their morphology, the microphysical description of their particles, and their physical and chemical composition. It also led to increased understanding of their optical properties, their role in planetary energy processes, and their interaction with atmospheric motions.

Cloud Morphology

From Pioneer data, scientists identified several particle-bearing regions in Venus' atmosphere:

(a) A haze region extends upward from 70 to 90 km (43 to 56 miles). It is composed of very small particles observed by the Orbiter cloud photopolarimeter, ultraviolet spectrometer, and infrared radiometer experiments.

(b) The main cloud deck consists of three more-or-less distinctly separate regions. An upper cloud region is at 56.5 to 70 km (35 to 43 miles), a middle cloud region at 50.5 to 56.5 km (31.4 to 35 miles), and a lower cloud region at 47.5 to 50.5 km (29.5 to 31.4 miles). Each has varying microphysical properties observed by the probe nephelometer and cloud-particle size spectrometer experiments.

Table 6-1. Summary of Characteristics of Venus Clouds

Region	Altitude, km	Temperature, °C	Refraction index	Composition	Diameter, µm
Upper haze	90.0–70.0	–83 to –48	1.45	Sulfuric acid + contaminants	0.4
Upper cloud	70.0–56.5	–48 to 13	1.44	Sulfuric acid + contaminants	0.4, 2.0 (bimodal)
Middle cloud	56.5–50.5	13 to 72	1.42 1.38	Sulfuric acid + contaminants + crystals	0.3, 2.5, 7.0 (trimodal)
Lower cloud	50.5–47.5	72 to 94	1.32	Sulfuric acid + contaminants + crystals	0.4, 2.0, 8.0 (trimodal)
Layers	47.5–46.0	94 to 105	1.46 1.50	Sulfuric acid + contaminants	0.3, 2.0 (bimodal)
Lower haze	47.5–31.0	94 to 209	—	Sulfuric acid + contaminants	0.2

(c) A lower haze extends from 47.5 km to about 31 km (29.5 to 19 miles). It was observed by the probe cloud-particle size spectrometer. Also, investigators saw evidence of matter suspended in the atmosphere at lower altitudes. Some of the probes' nephelometers provided this evidence.

(d) Additional thin-layered structures, or pre-cloud layers, exist as transitory clouds in the upper part of the lower haze region.

Figure 6-30 shows the results of the nephelometer measurements of the clouds' vertical structure at four Pioneer Venus sites and one Venera site. Table 6-1 summarizes properties of the hazes and main clouds assembled from Pioneer Venus experiments.

The upper and lower cloud regions are much more variable in structure than the middle cloud region. By analogy with Earth clouds, all clouds are stratiform. They consist of fairly large scale, uniformly layered structures. With the possible exception of the middle cloud region, the cloud regions are remarkably stable against vertical overturning. For such cloud structures, there may be a possibility of light mist or drizzle. Heavy precipitation typical of cumulus-scale convection in an unstable atmosphere is unlikely. Futhermore, similarities in the main cloud deck profiles and in stability properties (measures of the atmosphere's tendency to overturn by convection) at each of their four probe sites strongly suggest that the major features of the cloud systems are global. They are not very dependent on local longitude or latitude, except perhaps at high latitudes and at the equator.

Scientists suggest that the features observed at ultraviolet wavelengths are principally identified with the motion of an ultraviolet absorber in the atmosphere. This is because changes in the concentration of sulfuric acid particles cannot account for these patterns. Also, the haze is not dense enough to provide the observed contrasts. The ultraviolet absorbing species, other than sulfur dioxide, which investigators have identified as one of the absorbing species, remains unknown. Since solar energy is absorbed mainly in and above the upper levels of the main cloud deck, vertical motions of this unknown species from below the haze may be responsible for the observed dark regions. However, such an assumption would imply a dark region of upwelling ultraviolet radiation absorber at the subsolar point on the planet. Yet, the subsolar point is a bright region.

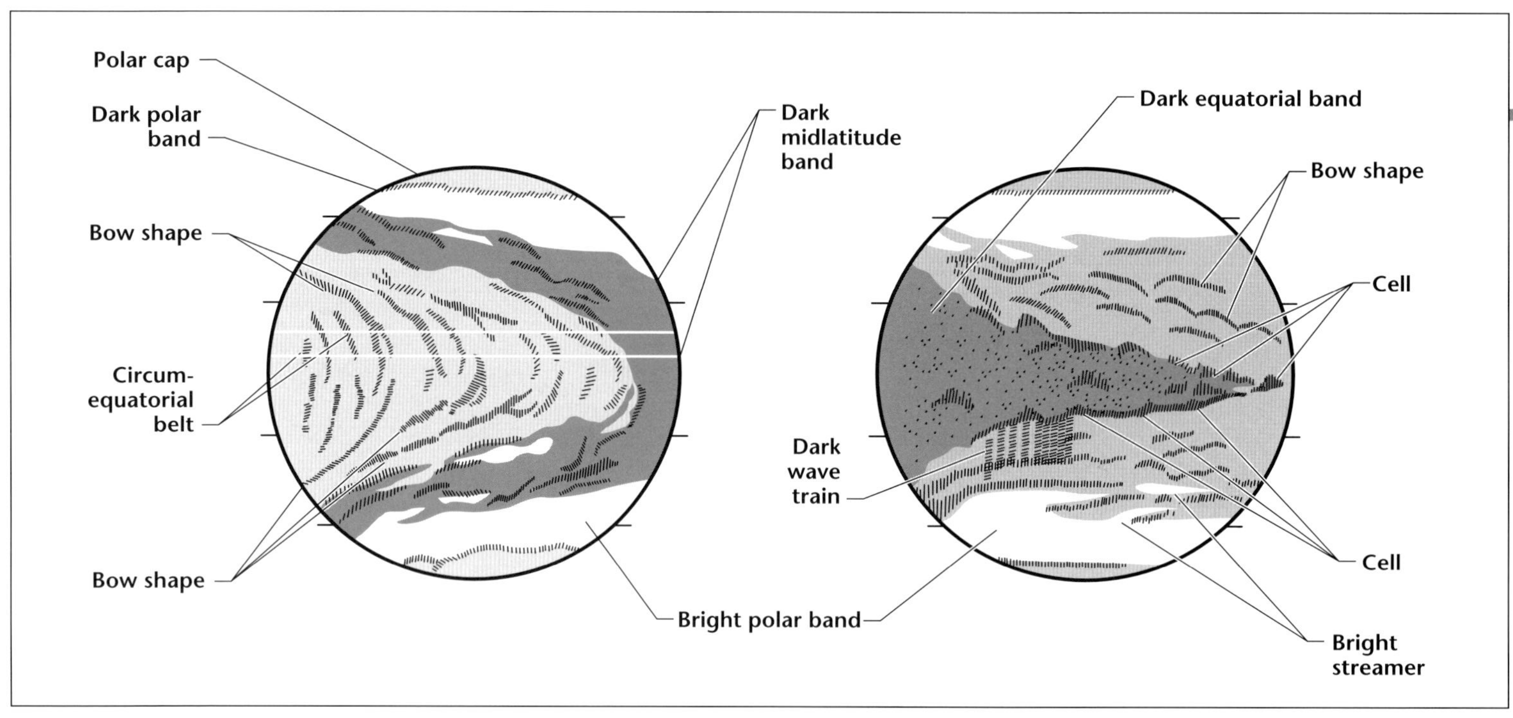

Figure 6-31. Basic types of cloud features observed on Venus in ultraviolet images. These views typically occur two Earth-days apart.

Nonetheless, researchers think that the absorber masks motions in the atmosphere by indicating regions of horizontal variation or of vertical displacement of the absorber. This is presumably from below the cloud tops to higher altitudes where regions of ultraviolet absorption would appear darker. The ultraviolet absorber acts as a marker of motion. Also, since it absorbs appreciable amounts of energy, it may play a role in cloud layer dynamics.

Ultraviolet features may be categorized into those associated with three distinct regions of the planet. A polar zone is above 50° latitude, a midlatitude zone between 20° and 50°, and an equatorial zone extends about 20° north and south of the equator. A small-particle haze covers the planet, varying in density with latitude. A polar haze collar, bright in ultraviolet light, encircles the polar regions at about 55° latitude. However, even at lower latitudes, there are significant amounts of haze above the cloud tops. Also, there is evidence of increased amounts of haze at the morning and evening terminators. The haze even covers the polar regions where it obscures, in ultraviolet images, features that can be seen in infrared images. Changes in the general haze features appear to occur in times ranging from months to years.

The large variety of dark features in the ultraviolet images of midlatitudes and equatorial regions is composed of three types of features: bow shapes, dark midlatitude bands, and a dark equatorial band (Figure 6-31). The dark equatorial band forms a tail that, together with a bow feature, produces the characteristic Y-feature astronomers can see from Earth. This feature appears clearly in the images from Mariner 10 and Pioneer Venus. At times, it keeps its structure as it moves around the planet, showing a four- or five-day periodicity. At other times, the Y-feature is absent from the ultraviolet cloud patterns. Even when it is present, many detailed features change with time. The variability of the Y-feature suggests that its smaller features change independently.

Cellular features with either dark or bright surroundings are common at low latitudes. Most have dark centers. They are, on the average, about 200 to 300 km (124 to 186 miles) in diameter and are present in bright and dark regions. They are more numerous in the dark equatorial region and during the afternoon on Venus.

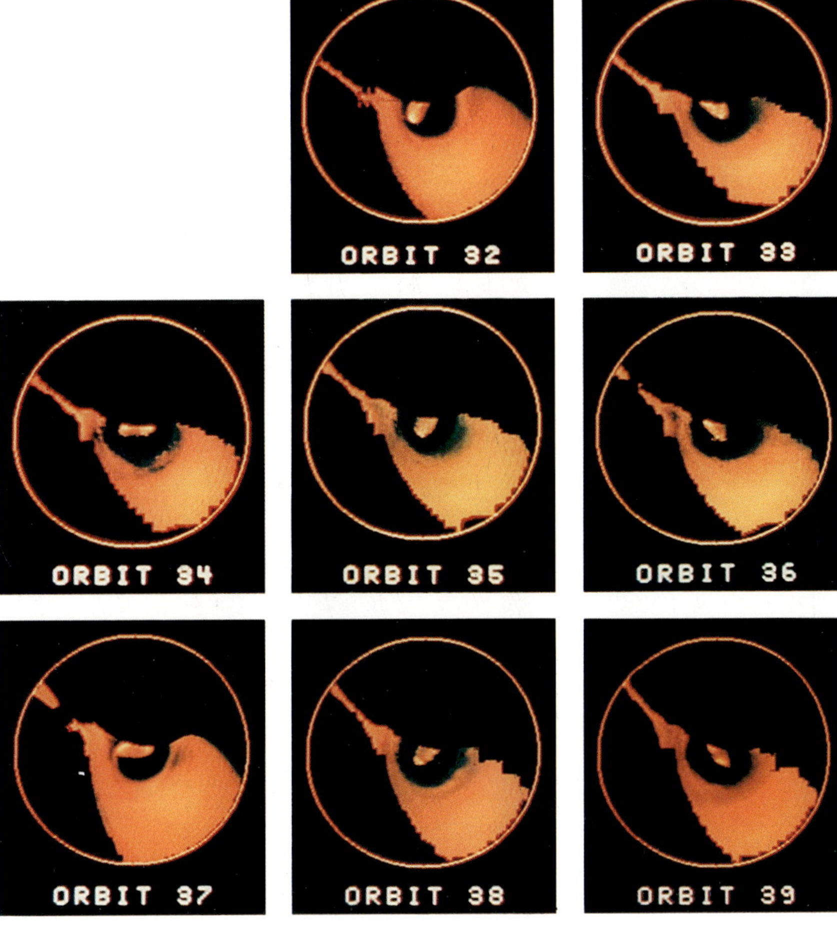

Figure 6-32. Pioneer spacecraft took these eight consecutive polar stereographs of Venus' northern hemisphere at 11.5 microns in the infrared region of the spectrum. The images were taken one each day in orbits 32 through 39 (January 5 through 12, 1979). The north pole is at the center of each image, and the equator is the outer boundary with the noon point at the bottom. Note that the polar dipole returns to about the same position in three days.

Ultraviolet images from Orbiter also showed wave-like features about 1000 km (621 miles) long and 200 km (124 miles) apart. They made large angles with the equator and cut across other features, thereby showing that they were at different altitudes from the other features.

The Orbiter's infrared radiometer data showed a dark polar band at about 65° to 75° north latitude. This broad, cold feature formed a collar around the pole. It was most likely an unusually cold region near the base of a temperature inversion. Its coldest part seemed to follow the antisolar point around the planet. Earth-based observations indicate that polar collars usually persist for weeks or months. They are most pronounced near only one pole throughout the period when the planet is suitably positioned for observation from Earth.

A localized polar brightening at very high latitudes is generally associated with collars in ground-based observations. Pioneer Venus infrared images resolved this pattern into a pair of "hot spots" that straddle the pole. These hot spots were at about 85° north latitude. They appeared as a dramatic "dipole" in images and maps.

Infrared images revealed structure on the nightside as well as polar regions. Near the pole, thin hazes and an unfavorable angle of solar illumination made observations at other wavelengths difficult. The hot spots of the polar dipole were probably clearings in the polar cloud deck. This feature rotated about the pole in approximately 2.7 days (Figure 6-32). Brightness temperatures within the hot spots approached 260 K at 11.5 microns. But the temperature could have been as high as 280 K if the spacecraft had viewed the region from directly above. Bright filamentary streaks emerging from one eye of the dipole and dividing the collar are visible in several images.

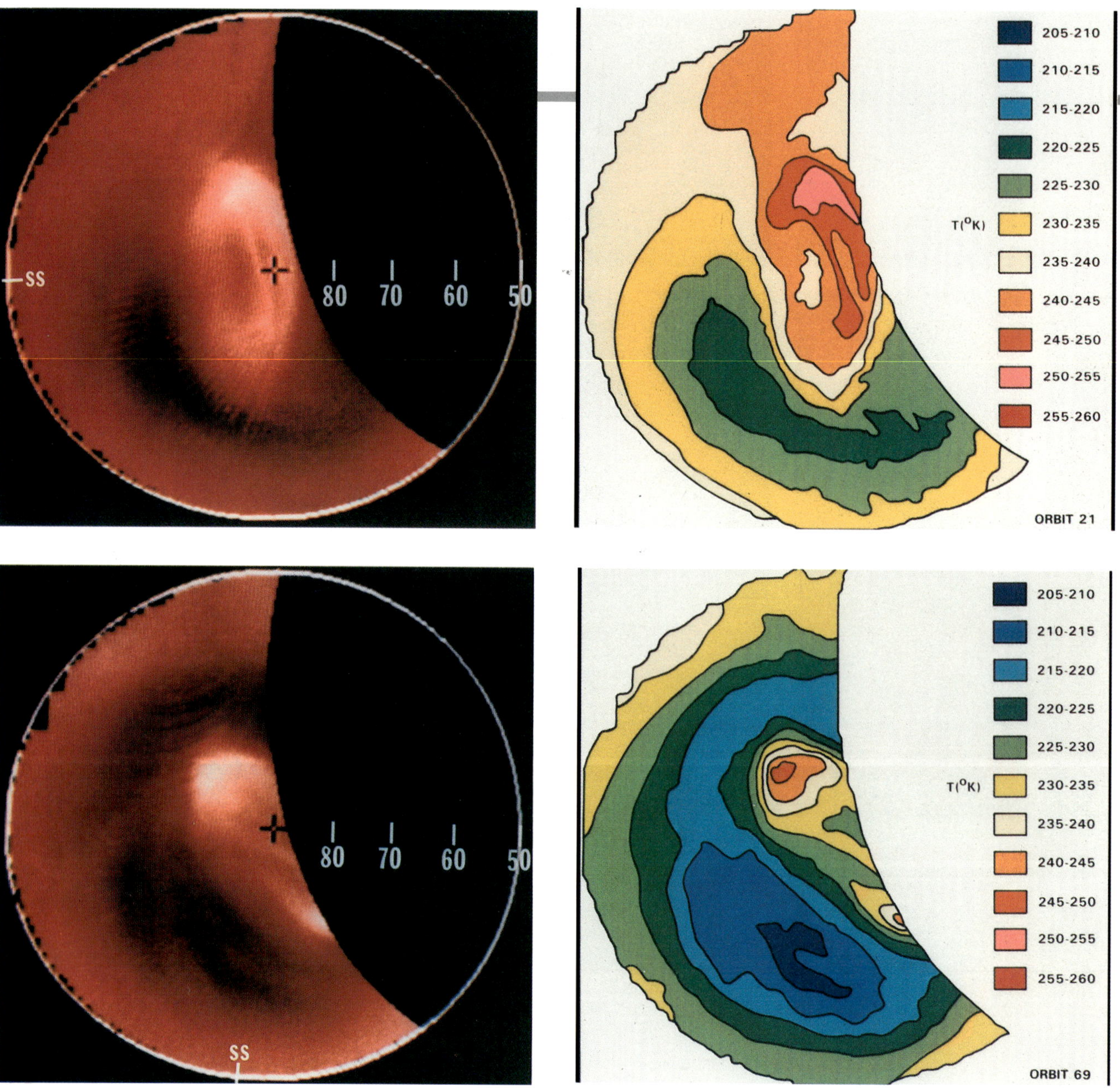

▲

Figure 6-33. Infrared images and plots of Venus' north polar regions. The pole is at the center, and the outer boundary is 50° north latitude. The noon point is at the left. The blacked out region indicates lack of data because of orbit geometry. The Orbiter obtained the top pair on orbit 21 (December 26, 1978), the bottom pair on orbit 69 (February 11, 1979).

The dipole was about 2000 km (1243 miles) long and about 1000 km (621 miles) across (Figure 6-33). These polar hot spots may be evidence of atmospheric subsidence at the center of the polar vortex. Because descending motions are not observed elsewhere in the northern hemisphere of Venus, the evidence points to a single large circulation cell filling the hemisphere at the level of the cloud tops.

Particle Microphysics

Size groupings distinguish the particles in the main decks of the upper, middle, and lower cloud regions. These groupings have more than one maximum and so are multimodal. By contrast, haze particles seem to group around one maximum value and are unimodal.

Data from Pioneer showed that, in the lower and middle cloud regions, the size distribution is trimodal, with modal diameters of 0.1 to 0.5,

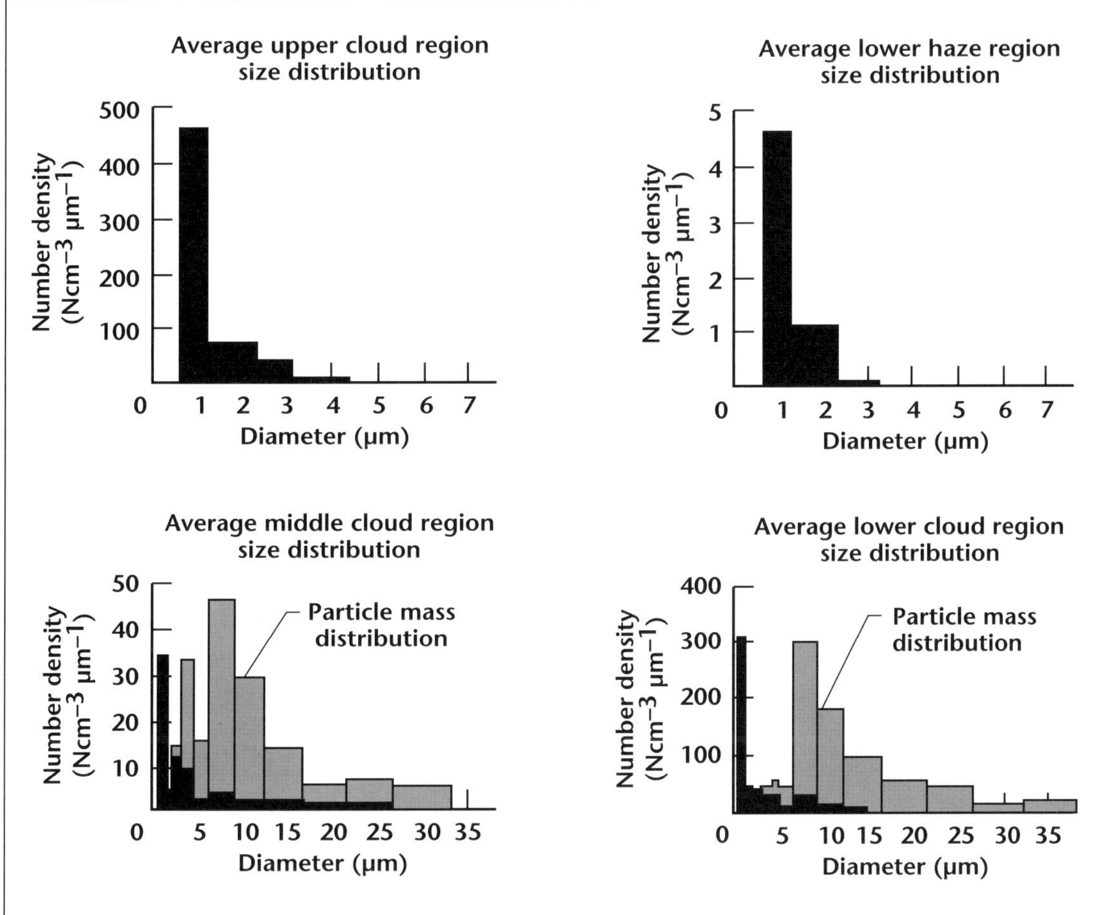

Figure 6-34. Average size distribution of cloud particles from Large Probe data. A multimodal distribution is evident, especially in the middle and lower cloud regions. You also can see it in the upper cloud layer and the haze layers. The precloud region, part of the lower haze, accounts for nearly all the particles larger than 1.2 microns. The mass-relative distribution assumes that all particles are spherical.

1.8 to 2.8, and 6 to 9 microns. The upper cloud region is bimodal, consisting of particles from the first two ranges of modal diameters.

The smallest-size mode is a widespread aerosol population extending throughout the main cloud deck and 15 to 20 km (9.3 to 12.4 miles) above and below it. Its number density varied greatly with height but was enough for the particles to act as centers for the growth of larger particles. This probably occurs by heterogeneous nucleation from parent vapors in the atmosphere. The second-size mode consists of droplets of sulfuric acid with primary growth taking place in the upper cloud region. There was a gradual increase in number density descending through the main cloud deck. This mode's size distribution was extremely narrow at any one altitude, and it was the dominant-size mode of the planet.

The largest-size mode for best agreement with all Pioneer Venus and Venera data consists of thin plate-like crystals. Their high aspect ratio prevented accurate determination of their mass. Aspect ratio is the ratio of maximum to minimum projected area.

Figure 6-34 shows the average size distribution within each cloud region, as measured by the Large Probe's cloud-particle size spectrometer.

The particle-size distributions were not unusual except for the second-size mode, which appeared to be monodisperse. Such narrow distribution can be explained by assuming competitive diffusional growth. However, the uniformity of the distribution width over the planet, as hinted from the probe and Orbiter data, is mysterious. It is highly unlikely that droplets grow by coalescing because there is low probability of their colliding with each other.

Particle Composition

Multimodal size distributions usually indicate several different chemical components of a population of particles. Scientists easily identified the second-size mode particles, mode 2, as sulfuric-acid droplets. This identification was primarily from their optical properties. These particles are traced throughout the main cloud deck. While the concentration of sulfuric acid in the droplets may decrease from 90% at 60 km (37 miles) to 80% at 50 km (31 miles), concentration has little effect on drop size.

The mode-1 aerosol is of variable composition as inferred from its optical properties and from considerations of particle growth. The aerosol is mainly sulfuric acid in upper and lower cloud layers, precloud layers, and upper haze regions. Sulfuric acid forms in the region above the boundary between the upper and middle clouds. The mode-1 aerosol apparently contains other chemical species or direct condensates as scavenged contaminants. These could account for most of the particle mass remaining in the lower haze and perhaps the middle cloud regions.

Composition of mode-3 particles is uncertain. They may well be chlorides, but, if so, scientists have yet to identify the cation. Except for any particulate matter in the lower atmosphere, essentially all particle mass is volatile at temperatures above 20°C (68°F). Venera 11 instruments detected chlorine in large amounts, but its role in cloud chemistry is uncertain.

Optical Properties

The major absorption of solar energy in Venus' atmosphere takes place at high altitudes corresponding to the locations of the high hazes down through the upper cloud regions. The actual role of the cloud particles in the absorption process is not clear, but they certainly play an important role in several ways. They redirect incident solar energy by scattering processes. They increase the actual absorption of incident photons in a horizontal layer of the atmosphere. They redirect most of the incident light into space.

Much of the absorption observed at far ultraviolet wavelengths is attributable to sulfur-dioxide vapor. Measurements showed that this gas wells upward in quantities that match the rate at which acid droplets of the clouds settle downward. Sulfur dioxide is oxidized at the cloud tops by ultraviolet photochemistry. It then dissolves in water droplets to form sulfuric acid. Infrared absorption is also attributed primarily to other gaseous constituents such as carbon dioxide and sulfuric acid. However, the absorber of an important part of the solar spectrum extending from about 3200 angstroms into the visible, which is also, in large part, responsible for the presence of the ultraviolet markings observed remotely, has yet to be identified.

Since they are transparent at the wavelengths involved, particles of pure sulfuric acid do not qualify as candidates for this absorption. Therefore, if the missing absorber is in the particulate matter, it must be in the form of a contaminant or aerosol core to the sulfuric-acid particles. Two factors point to the location of the absorber. First, the contrast of the ultraviolet features as observed by Orbiter's cloud

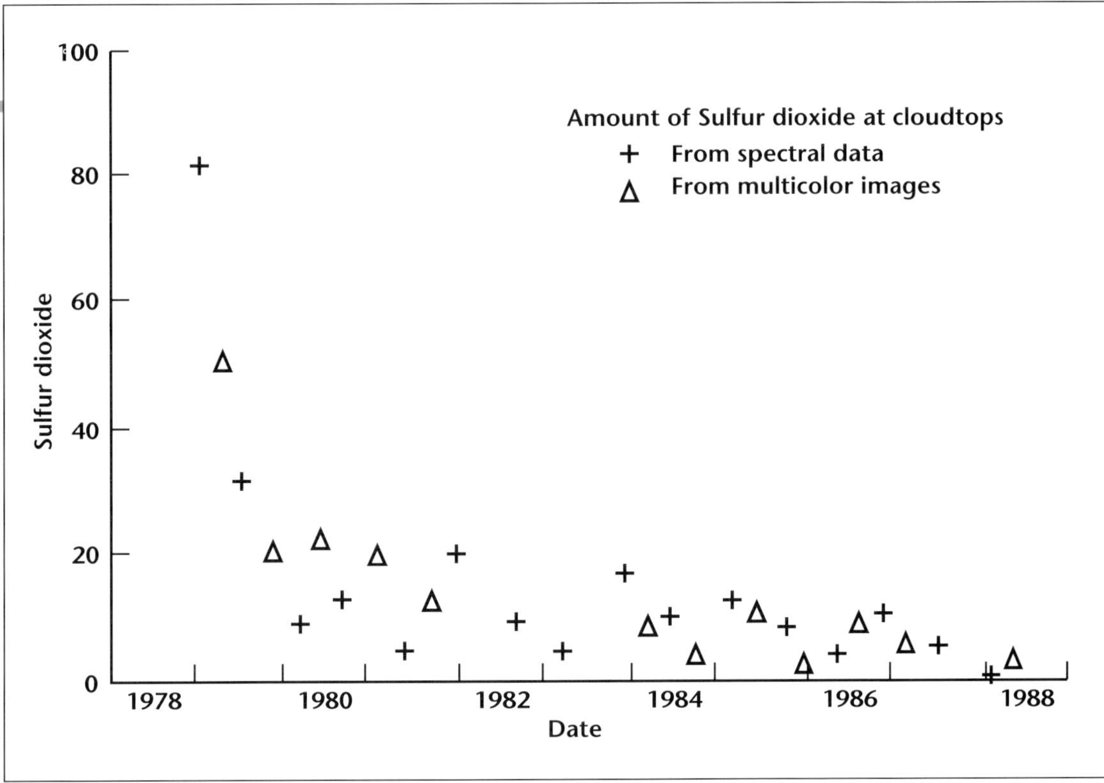

Figure 6-35. Investigators observed a dramatic decrease in sulfur dioxide at the cloud tops during the Pioneer Venus mission. In 1978, the ultraviolet spectrometer easily detected the gas. Scientists had not detected it before, even though they searched for decades from Earth. During Orbiter's mission, the amount of gas declined steadily as the figure shows. Scientists suggested that a major volcanic event had occurred on Venus just before Orbiter's discovery of the gas. It would have injected large amounts of sulfur dioxide into Venus' atmosphere.

photopolarimeter decreases as the phase angle of observation increases. Second, the greatest contrasts appear when viewing normal to the clouds. Therefore, the ultraviolet absorber must lie much deeper than the overlying haze.

However, data from the Large Probe's solar net flux radiometer indicated that absorption of solar energy takes place at altitudes above optical depths of 6 or 7. That is, most absorption is in or above the upper cloud region, with little absorption in the middle or lower clouds. In addition, Orbiter ultraviolet spectrometer measurements suggest that the unknown absorber's location is connected with the location of the sulfur-dioxide absorber.

Before Pioneer Venus, scientists had searched in vain for decades for the signature of sulfur dioxide. In 1978, sulfur dioxide was easily detected in Venus' atmosphere (Figure 6-35). However, subsequent measurements from orbit showed a steady decline in the amount of sulfur dioxide. Scientists suggested that a major volcanic episode occurred early in 1978 before Orbiter's arrival at Venus. The episode injected large amounts of sulfur dioxide into Venus' atmosphere.

Several years of observation by Orbiter showed that sulfur dioxide in Venus' atmosphere decreased to between 10% and 29% of its value in late 1978. One possible explanation is that Venus is still volcanically active. Volcanoes periodically inject massive amounts of gas into the atmosphere, and quantities of the gas later decrease by various processes. Venus' lack of water makes explosive volcanic episodes capable of pushing large amounts of sulfur dioxide to the cloud level unlikely. There might, nevertheless, be episodes of low-level volcanic activity with the gases carried upward by normal atmospheric mixing.

Fits of models to data from the Large Probe's solar flux radiometer experiment suggest that the imaginary index of refraction (the absorption portion of the index of refraction) could reach 0.05 for the mode 1 aerosol. However, correlating bright polar regions with large amounts of cloud above the sulfuric-acid main cloud at high latitudes argues for a small

Table 6-2. **Radiative Properties of Cloud Layers at Sounder Probe Location**

Cloud layer and range, km	Pressure, atm		Optical depth	Fractional contribution to optical depth			
	Top	Bottom		Mode 1	Mode 2	Mode 3	Mode 4
Upper haze above 75	0	0.015	0.04				1.0
Upper cloud 75–56.2	0.015	0.025	0.425	0.2	0.8		
	0.025	0.035	0.5	0.2	0.8		
	0.035	0.050	0.65	0.55	0.45		
	0.050	0.067	0.8	0.55	0.45		
	0.067	0.083	0.8	0.55	0.45		
	0.083	0.1	0.8	0.55	0.45		
	0.1	0.132	2.49	0.7	0.3		
	0.132	0.187	3.01	0.72	0.28		
	0.187	0.25	2.96	0.63	0.37		
	0.25	0.402	2.49	0.56	0.44		
Middle cloud 56.2–50	0.402	0.661	3.82	0.14	0.5	0.36	
	0.661	0.77	1.41	0.17	0.48	0.35	
	0.77	0.991	2.42	0.24	0.5	0.26	
Lower cloud 50–48.3	0.991	1.102	2.5	0.21	0.25	0.54	
	1.102	1.225	2.5	0.2	0.29	0.51	
Sub-cloud 48.3–46.8	1.225	1.501	0.8	0.43	0.43	0.14	
Lower haze 46.8–31	1.501	8.56	0.21	1.0			

amount of absorption. Single scattering albedos (the ratio of the probability of scattering to the sum of the probabilities of scattering and absorption for a single particle) range from 0.95 (high absorption) in the upper cloud region to 0.999 (low absorption) in the lower cloud region. The larger mode 3 particles are essentially nonabsorbing. However, an unknown mechanism appears to generate some absorption near the boundary between the upper and middle clouds.

The entire Venus cloud system has an optical depth of 25 to 35 at visible wavelengths. That is, the probability of a single normally incident photon passing through the cloud system without a single interaction with a cloud particle is e^{-25} to e^{-35}. The relative contributions of each cloud region to the total optical depth appear in Table 6-2. Figure 6-36 shows a plot of the clouds' measured optical properties.

The radiometric albedo, essentially the reflection coefficient weighted over the solar spectrum, is 0.77 to 0.82. It increases from equator to poles. The particle real refractive indices at visible wavelengths for modes 1 and 2 are approximately 1.40 to 1.46. While consistent with sulfuric acid, the indices permit the presence of many other species. The real refractive index of mode 3 particles is unknown but probably ranges from 1.5 to 1.7. The imaginary index for these particles must be less than 10^{-3}.

Dynamical Processes

The cloud system is embedded in the general circulation of the atmosphere at altitudes of greatest wind velocity and vertical wind shear. Atmospheric motions consist mainly of a zonal circulation. The atmosphere moves from east to west with velocities increasing from a few meters per second at the surface to sometimes as high as 150 m/sec (490 ft/sec) at cloud

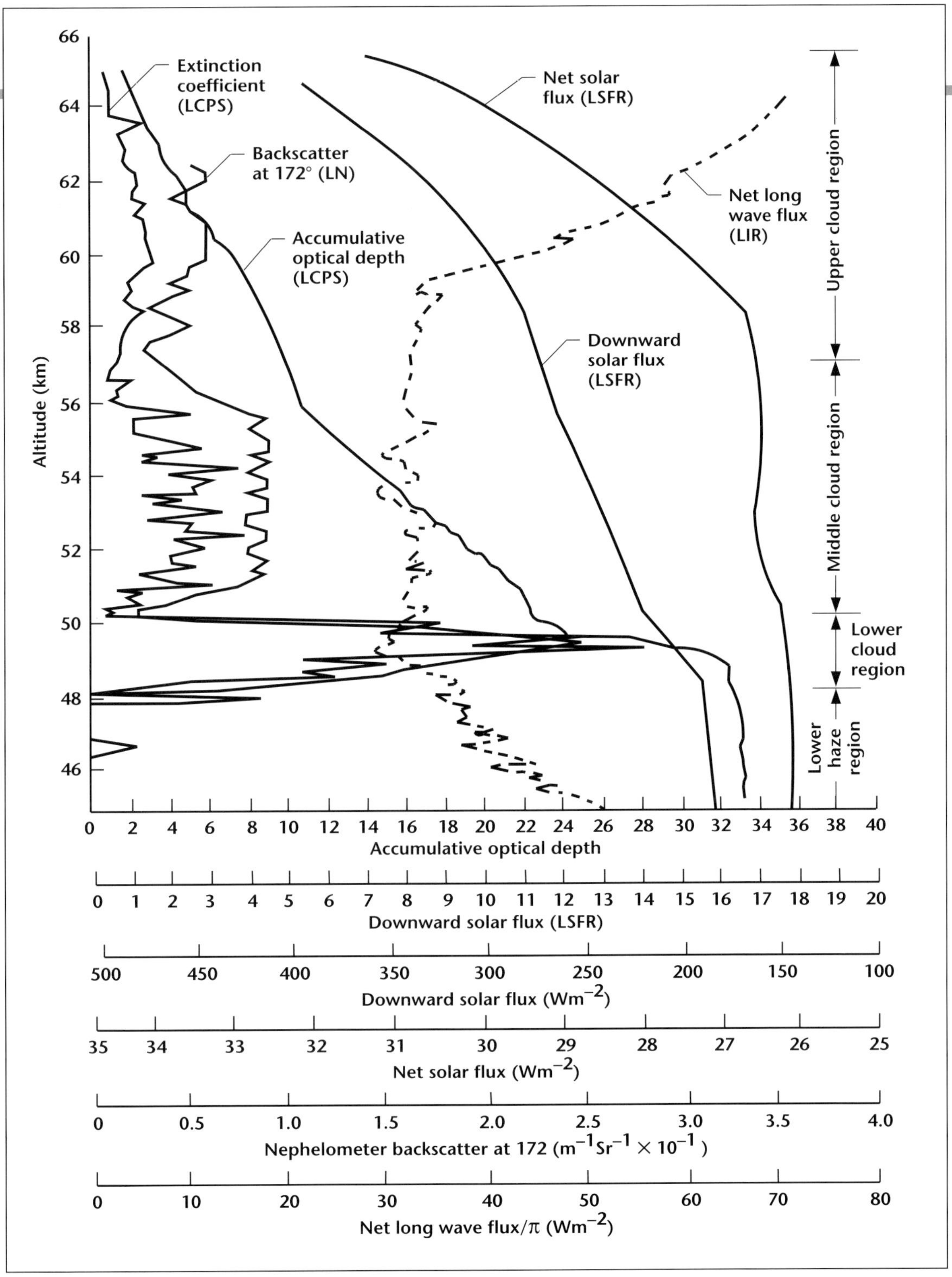

Altitude (km)
66 64 62 60 58 56 54 52 50 48 46

Extinction coefficient (LCPS)

Backscatter at 172° (LN)

Accumulative optical depth (LCPS)

Net solar flux (LSFR)

Downward solar flux (LSFR)

Net long wave flux (LIR)

Upper cloud region

Middle cloud region

Lower cloud region

Lower haze region

0 2 4 6 8 10 12 14 16 18 20 22 24 26 28 30 32 34 36 38 40
Accumulative optical depth

0 1 2 3 4 5 6 7 8 9 10 11 12 13 14 15 16 17 18 19 20
Downward solar flux (LSFR)

500 450 400 350 300 250 200 150 100
Downward solar flux (Wm^{-2})

35 34 33 32 31 30 29 28 27 26 25
Net solar flux (Wm^{-2})

0 0.5 1.0 1.5 2.0 2.5 3.0 3.5 4.0
Nephelometer backscatter at 172 ($m^{-1}Sr^{-1} \times 10^{-1}$)

0 10 20 30 40 50 60 70 80
Net long wave flux/π (Wm^{-2})

◀ *Figure 6-36. This figure summarizes optical properties of Venus' cloud systems. The scale on the right identifies the various regions of the atmosphere. Scales for the various plotted curves appear below the graph. When studied in detail, these plots reveal a wealth of detail about the planet's atmosphere.*

tops. The average cloud top velocity corresponds roughly to the four-day circulation.

Also, the data suggest a major, although much slower, north-south circulation at several meters per second. It occurs at altitudes corresponding to the cloud region. There seems to be atmospheric movement from equator to poles at altitudes corresponding to the tops of the clouds. The movement subsides at the poles. Return flow toward the equator is at altitudes that match the lower part of the

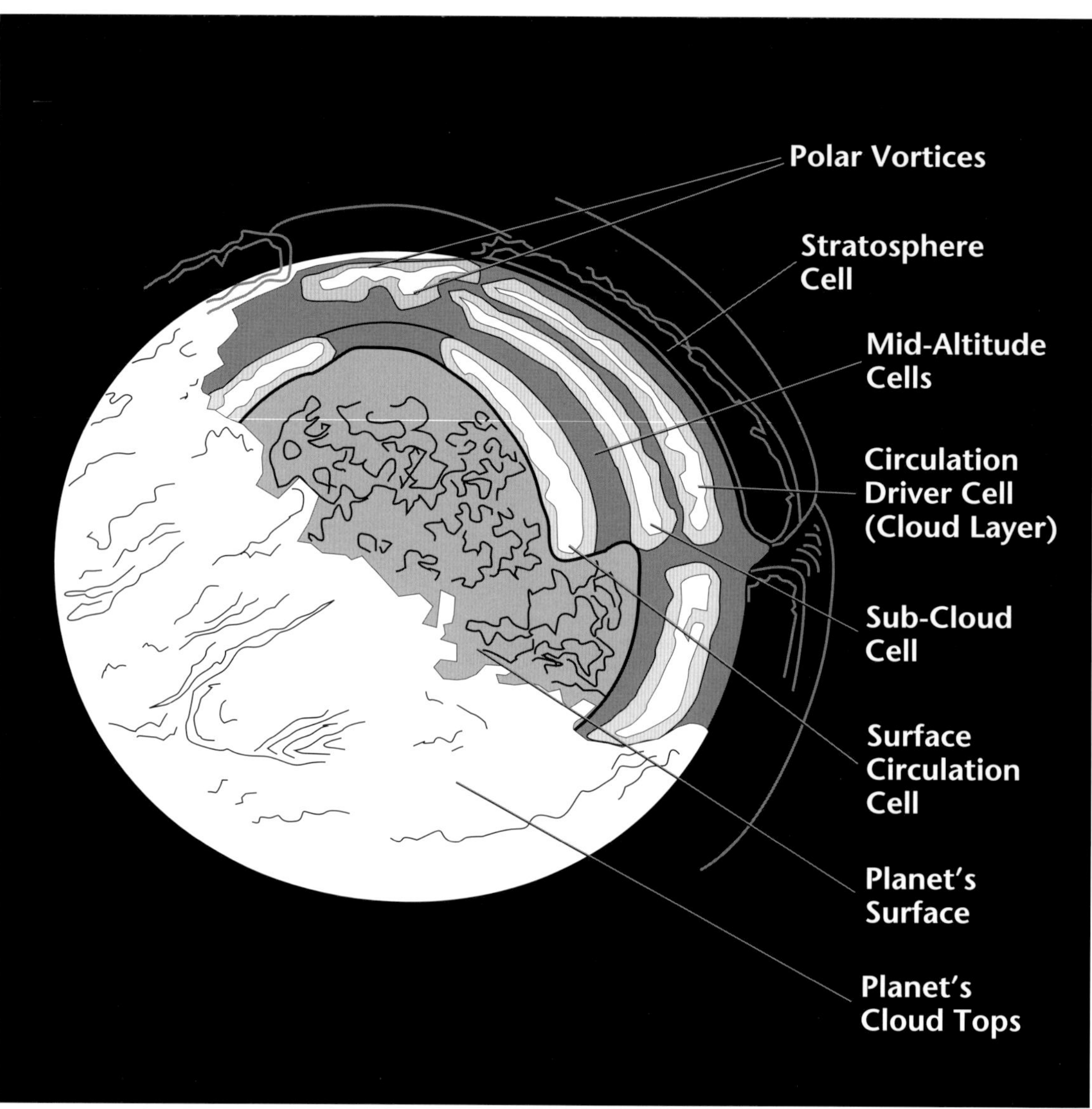

Figure 6-37. A possible pattern for the meridional circulation in the atmosphere of Venus.

main cloud region. The atmosphere rises again near the equatorial region. Such north-south cellular motions are called Hadley cells. The combination of east-west and north-south motions produces vortices in the polar region. These affect the haze layer and produce an apparent cloud top depression in the vortices. They also might be the reason for the "pileup" of high latitude hazes and the even higher latitude "cold ring" observed by the Orbiter's instruments. Figure 6-37 is a schematic drawing of the suggested circulation pattern.

The detailed ultraviolet and infrared features observed from Earth, and from flyby and orbiting vehicles, may thus be in accord with the general behavior predicted from the *in situ* probe measurements. Features involving the four-day zonal rotation are evident in the ultraviolet images. Most other features result from wave motions and convection cells disturbing the level of the upper-altitude ultraviolet absorber. Thus, some features, such as the large-scale Y-shaped structures, prominent at lower altitudes, may propagate slowly with respect to the atmosphere. They may appear and disappear as the wave motion dictates. The east-west wind moves their major features around the planet. Smaller convection-type features, suggesting rising atmosphere motion, also are evident. Finally, the suggested circulation pattern may plausibly

describe the bright polar collar, the cold ring, polar hot spots, and infrared holes.

Cloud particle growth is not strongly influenced by the large-scale planetary circulation. Acid particles go along for the ride, simply adjusting their acid concentration to each new equilibrium the circulation offers. Rapid circulation together with particle volatility produces the planetary cloud structures.

Growth of sulfuric-acid droplets appears to be a very slow process except in the lowest cloud regions. Recondensation of sulfuric acid might be quite rapid there. A large range of particle lifetimes extends from years in the upper hazes to hours in the lower cloud region. Mode-3 particles provide much of the middle and lower cloud structure. Their growth starts near the top of the middle cloud. The particles evaporate at the bottom of the lower cloud.

As a result of the Pioneer mission, scientists now have a much better understanding of the chemistry in Venus' clouds. The clouds are basically the product of a cyclical chemical process involving elemental sulfur. Sulfur originates from surface rocks through mineral buffering between the surface and the atmospheric gases of carbon dioxide and carbon monoxide. Reaction with these gases produces carbonyl sulfide and elemental sulfur. The carbonyl sulfide then interacts with oxygen in a hot layer above the surface to form sulfur dioxide and carbon monoxide. High in the atmosphere above the clouds, the sulfur dioxide reacts with water under the influence of solar ultraviolet radiation to produce sulfuric-acid droplets. After they form, the droplets sink slowly toward the planet's surface. They grow as they collide with each other and condense sulfuric acid and water from the atmosphere in a condensation zone. Finally, as they fall toward the hot surface, the lower atmosphere's high temperature causes the droplets to vaporize and break up into sulfur dioxide and water vapor. These chemicals then circulate in the atmosphere and continue the process of sulfuric-acid droplet formation above the cloud layers.

Although knowledge about the clouds of Venus has been enormously increased by the successful missions to the planet, there are still unanswered questions. The identity of the remaining ultraviolet absorber still eludes us. Scientists must know what this absorber is to fully understand upper atmosphere motions and cloud details. Also, this information is vital to understanding the planet's energetics and atmospheric chemistry. The detailed composition of mode-3 particles and the nature of contaminants in other cloud particles are still in question. The role of chlorine in cloud chemistry is unknown. There also are questions about precipitation within the atmosphere. Finally, we know little about particles suspended in the atmosphere at low altitudes, the presence of which is hinted at by data from several probe instruments and the possible occurrence of lightning discharges.

Lightning

On Earth, most lightning occurs in strongly convective clouds, but it can also be produced in volcanic clouds, dust storms, and snowstorms. Some scientists have suggested that the energy of lightning strikes on Earth may have played an important part in producing complex molecules for the evolution of terrestrial life. Volcanic eruptions are not major producers of terrestrial lightning, but individually are very active producers. Over the past decade, lightning has been recorded optically on Jupiter, Saturn, Uranus, and Neptune. These strikes have occurred mainly in nightside hemispheres where production rates would be expected to be lower than on

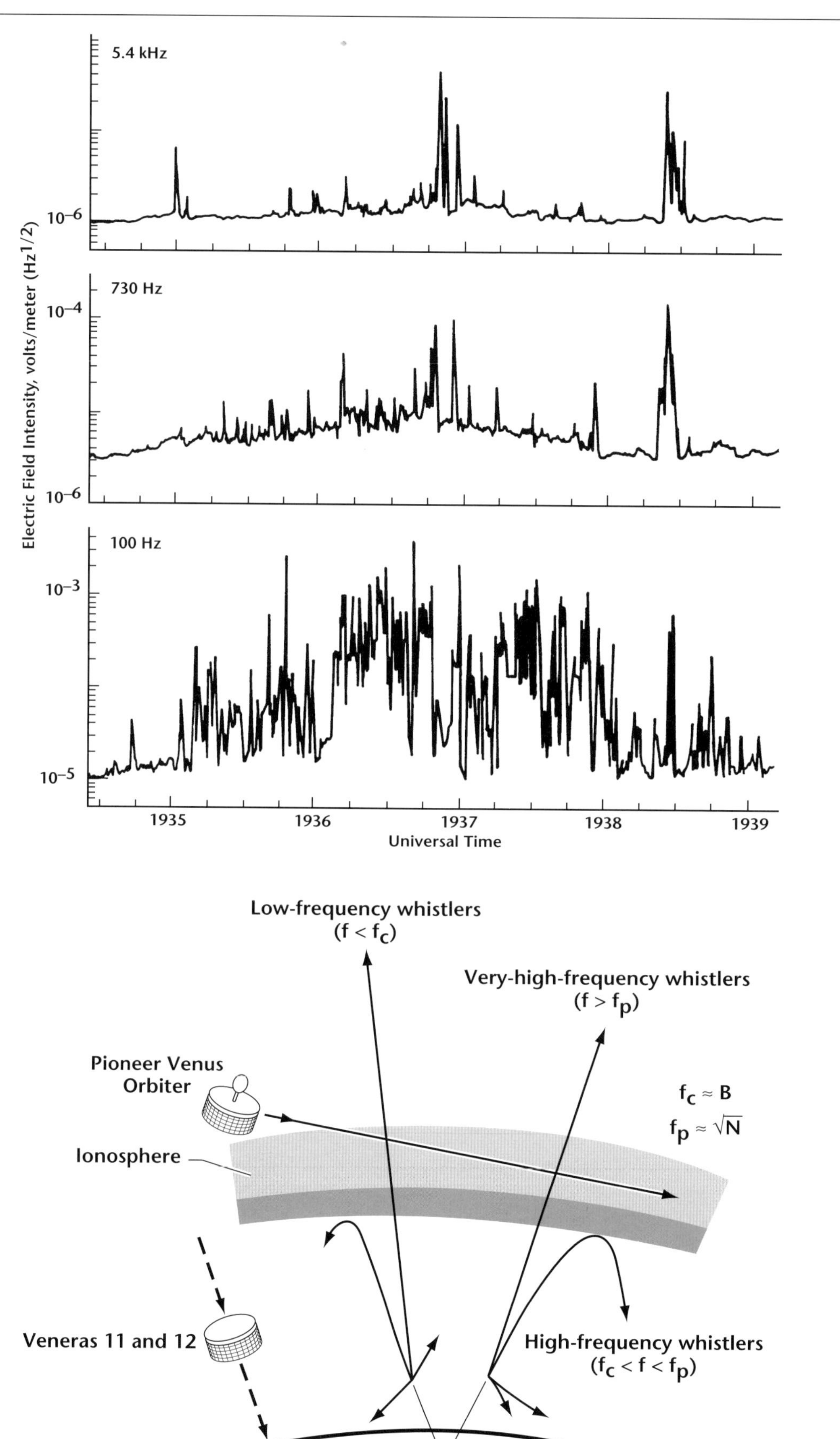

Figure 6-38. Lightning on Venus? Investigators interpreted signals from Orbiter's electric-field detector as originating from lightning in Venus' clouds. The conceptual drawing shows how the signals at 100 Hz and higher frequencies can be interpreted.

the dayside. Dayside strikes can only be "seen" as they are recorded by electromagnetic emissions. At Venus, Pioneer Orbiter sought evidence of lightning electromagnetically and optically. The former method was successful; the latter was not.

Instruments on Veneras 11 and 12 observed electrical signals attributed to lightning on Venus (see Chapter 7). The Pioneer Venus orbiting electric-field detector also observed signals suggestive of lightning. It recorded these signals on its 100 Hz channel (Figure 6-38). Orbiter first recorded these whistler-mode electromagnetic noise bursts in December 1978 when its periapsis moved from sunlight into darkness. It obtained more detailed observations during Phase III.

An important observation was that lightning-associated signals depend strongly on the local time. As on Earth, high-frequency signals do not usually propagate far into the ionosphere, and they clearly mark the surface region of origin. The source of the waves on Venus was at local times before 10:30 p.m. The decrease in occurrence as dusk approached was an effect of the increasingly dense ionosphere, which altered propagation of the waves. Some investigators suggested that lightning probably occurs in the afternoon and early evening. A strong local time dependence suggests that lightning is generated by weather conditions on Venus as on Earth. At one time, however, early in the Pioneer Venus mission, scientists speculated that the lightning discharges were related to volcanic activity, while other scientists did not acknowledge that lightning was the source of the electromagnetic signals.

Many scientists now believe the signals do originate from lightning. They cite four reasons: (1) the signals are intense and highly impulsive, (2) they occur near periapsis, (3) their spectral characteristics are consistent with whistler-mode propagation, and (4) they often appear when low and variable electron densities are present.

Known processes for the formation of lightning on Earth require large particles and strong updrafts in cloud regions. Potential latent instability (the difference between the rate at which the temperature would vary with altitude in an idealized atmosphere and the actual lapse rate) is a measure of the tendency of the atmosphere to overturn and undergo convective motion. Scientists found evidence of planet-wide instability in Venus' middle cloud region. There, updrafts probably occur over a limited altitude range from 50 to 56 km (30 to 35 miles). However, there is no direct evidence for large precipitative-type particles similar to rain or hail. Thus, if cloud processes generate lightning, then large undetected particles may exist in Venus' atmosphere. Lightning could be the result of local, large-scale events such as volcanic eruptions or strong and still undetected convective motions at the subsolar point. Also, because the cloud base is roughly 45 to 50 km (28 to 31 miles) above the surface, lightning flashes on Venus would most likely be from cloud to cloud rather than from clouds to ground.

One goal during Orbiter's entry phase was to find out if plasma wave signals could be detected at low altitudes with the spacecraft below the ionospheric density peak. The results were positive. Bursts of 100 Hz were detected before final entry. They were recorded at an altitude of about 130 km (81 miles) around 4:00 a.m. local time. The wave activity lasted for tens of seconds, and the bursts were not symmetric about periapsis. Their vertical attenuation scale height of about 1 km was consistent with whistler-mode waves propagating through the ionosphere. Researchers credited the waves to signals of electromagnetic radiation entering the bottom of the ionosphere from several discrete sources. This would be expected of lightning flashes occurring frequently within Venus' atmosphere.

The data were gathered in the predawn hours, but other researchers suggested that lightning would be an afternoon or early evening phenomenon. One possibility is that entry phase observations gathered ambient wave noise caused by lightning flashes occurring at locations remote from the spacecraft. The intense signals might have propagated for considerable distances from their source because the planet's surface and ionosphere acted as a giant waveguide. Investigators concluded that the intense bursts they observed during the final two periapsis passages were most probably direct subionospheric detection of atmospheric discharges. In other words, they were lightning. Some scientists, however, still linked the wave bursts to local plasma instabilities and not to lightning flashes.

Several experiments attempted to observe lightning optically using the Pioneer Venus Orbiter's star sensor. They showed no statistically significant difference in signals from the planet's dark hemisphere compared with control signals from pointing the star sensor into deep space. These experiments thus implied that most lightning occurs on Venus' dayside and, except for early evening, lightning would be relatively rare on the nightside. The Venera orbiters did, however, obtain somewhat inconclusive data that investigators attributed to optical detection of lightning flashes. (It is important to note that Pioneer Orbiter's star sensor was not originally designed for optical lightning detection.)

It is now generally accepted that nearly 14 years of observations of electromagnetic signals from lightning at Venus indicate that the flash rate is similar to or greater than that of Earth. More information, however, is necessary before we can speak with certainty about the lightning's origin and any atmospheric composition changes it may cause.

Atmospheric Gases—the Neutral Atmosphere

An important source of information for the way the terrestrial planets—Mercury, Venus, Earth, and Mars—formed is an analysis of their atmospheric gases. Scientists generally accept the chemical composition of gases that formed the primitive atmospheres of these planets as resembling that of the Sun and the giant planets. These gases were lost during the early stages of the Solar System's formation because of the high temperatures prevailing at that time. Scientists believe the present atmospheres consist of volatiles that were originally incorporated in the solids that combined to form the planets. Probably during the first few million years in the lives of these planets, high internal temperatures and tectonic activity drove the volatiles from their crusts and mantles. Some of the volatiles make up the present atmospheric gases. Others, such as water vapor, have condensed or otherwise been transformed. On Earth, water constitutes the oceans. On Mars, water is hidden below the surface in some form such as permafrost. On Earth, carbon dioxide has been converted chemically to carbonate rocks such as limestone. On Mars, carbon dioxide remains in the atmosphere and in polar caps.

Based on this scenario, the amount of each kind of gas in the atmosphere of a terrestrial planet should depend mostly on the mass of that planet. Studies of Mars by Mariner and Viking probes showed this is not true. Even allowing for its smaller size, Mars seems to be deficient in volatiles compared with Earth.

Table 6-3. Comparison of Atmospheres of Venus and Earth

Gas	Venus at surface, % or ppm[a]	Earth at sea level, % or ppm[a]
Argon	70 +50 −30	0.93%
36	20 +20 −10	31
38	6[b]	6
40	31[b]	0.93%
Carbon dioxide	96%	0.02–0.04%
Carbonyl sulfide	<3	0.5
Chlorine	<10	
Hydrogen	500	
Krypton	0.05	0.5
Neon	10	18
20	9	16
22	1	2
Nitrogen	4%	78%
Oxygen	<30	21%
Sulfur dioxide[c]		

[a]1 ppm = 0.0001%
[b]Derived from ^{36}Ar
[c]<10 in clouds; <300 near surface

These volatiles include carbon, oxygen, nitrogen, and the noble gases (neon, krypton, and argon). Deficiency factors are as large as 100 to 200. After the Viking mission, an explanation for these results was that the material from which Mars formed lacked volatiles compared with Earth. Also, a smaller percentage of volatiles escaped from the Martian interior. The reason for the deficiencies remains unknown.

Because Earth and Venus are so similar in size, mass, and distance from the Sun, the volatile inventories of these two planets were expected to be very similar. An exception was water since astronomers knew Venus has no ocean. Hence, the stage was set for the Pioneer Venus mission to conduct a crucial test of models of planetary formation.

Before Pioneer Venus, scientists generally agreed that Venus' atmosphere was mostly carbon dioxide gas. Estimates of the fraction varied between about 95% and 98%. Also, scientists believed that most of the rest of the atmosphere was nitrogen. Atmospheric pressure on Earth is about 1% of that on Venus, and carbon dioxide makes up about 0.03% of Earth's atmosphere (Table 6-3). In stark contrast, Venus' atmosphere contains about 300,000 times as much carbon dioxide as Earth's. This does not necessarily mean that more carbon dioxide has escaped into the atmosphere of Venus from its interior. The supply of carbon in limestone rocks and elsewhere in the Earth's crust suggests that most of the carbon dioxide produced on Earth has been converted to carbonates. A rough comparison shows that Venus produced no more than about twice as much carbon dioxide as Earth. On Earth, carbon dioxide has been incorporated in rocks. On Venus, it has remained in the atmosphere because that planet lacks an ocean to mediate the transformation.

Many of the instruments carried by Orbiter were specially chosen to form a synergistic package for exploring Venus' atmosphere and ionosphere. During the years of Orbiter's mission, the spacecraft gathered vast amounts of data on the basis of which scientists began to understand the aeronomy of Venus better than that of any other planet except Earth.

A major problem faces scientists trying to understand the divergent evolutionary paths of the two planets. How can we account for the present-day absence of water on Venus? Was water never present? Or had large amounts of water evolved from the interior at an early stage only to be lost later—hydrogen to space and oxygen to the crust and interior?

There is another basic question of importance. Can some climatic change on Earth, man-made or natural, cause an increase in carbon dioxide and water in Earth's atmosphere that results in a runaway greenhouse? Because carbon dioxide and water inhibit the escape of heat radiation, an increase in their concentration would probably lead to a rise in atmospheric temperature. This, in turn, would lead to the release of more carbon dioxide and water into the atmosphere, and the temperature would rise further, and so on. The result could be an atmosphere like Venus'. All available carbon dioxide might be in the atmosphere and the temperature near the ground would approach 700 K as on Venus. However, recent studies have thrown doubts on the role played by burning fossil fuels in raising carbon dioxide levels in our atmosphere. Natural processes linked to solar activity appear to play a more dominant role.

One of the major tasks of the instruments on the Large Probe, the Orbiter, and the Multiprobe Bus was to confirm that carbon dioxide and nitrogen are, indeed, the main atmospheric constituents on Venus, and to determine their precise concentrations. The instruments also had to identify other atmospheric components, even if these were only one part per billion (1 ppb). Instruments on the Large Probe that were assigned to these tasks were the neutral mass spectrometer and the gas chromatograph. The mass spectrometer covered altitudes from 62 km (39 miles) to the surface. The gas chromatograph sampled the atmosphere at 52, 42, and 22 km (32, 26, and 13.7 miles). On the Bus, a mass spectrometer obtained data above 130 km (81 miles). On the Orbiter, another mass spectrometer sampled the atmosphere above 145 km (90 miles). Important data about atmospheric composition above the clouds also came from the Orbiter's ultraviolet spectrometer.

Scientists have reached a consensus over measurements from the Pioneer instruments and from those on the Veneras 11 and 12 landers. They now agree Venus' atmosphere is 96% carbon dioxide and 4% nitrogen. Its surface pressure is 94.5 times that of Earth and its temperature is 732 K. These figures mean that Venus has outgassed 1.8 times as much carbon dioxide as Earth and 2.3 to 4 times as much nitrogen. The nitrogen, however, depends on how much is still in the Earth's crust. Thus, the expectation was confirmed of a rough equality in the volatiles of Earth and Venus for carbon dioxide and nitrogen.

However, an assay of the remaining volatiles in Venus' atmosphere delivered a rude shock to the planetary science community. The case of argon is an example. Two isotopes of argon are of interest to scientists studying planetary atmospheres. Radiogenic argon-40, the most abundant kind of argon in Earth's atmosphere, is produced by the radioactive decay of potassium. Its abundance tells us about the primitive concentration of potassium and about outgassing conditions throughout the planet's 4.5-billion-year history. On the other hand, argon-36 and argon-38 are primordial gases which tell us about the early volatile content of planetary interiors and how they outgassed.

Table 6-4. Mixing Ratios in the Lower Atmosphere

Gas	Amount, ppm
Argon	40–120
40/36	1.03–1.19
38/36	0.18
Carbon dioxide	96%
Carbon monoxide	20–28
Krypton	0.05–0.5
Neon	4.3–15
Nitrogen (percentages)	3.41% (at 24 km)[a]; 4%[b] 3.54% (at 44 km)[a] 4.60% (at 54 km)[a]
Oxygen	16 (at 44 km)[a]; <30[b] 43 (at 55 km)[a]
Sulfur dioxide	185 (at 24 km) <10 (at 55 km)
Water	20 (at surface) 60–1350 (at 24 km) 150–5200 (at 44 km) 200–<600 (at 54 km)

[a]Large Probe Gas Chromatograph
[b]Large Probe Mass Spectrometer
Based on six different instruments—four mass spectrometers and two gas chromatographs.

On the basis of carbon and nitrogen results, scientists expected that there would be about as much argon-36 and argon-38 in Venus' atmosphere as in Earth's atmosphere. Instead, the mass spectrometers on the Pioneer and Venera landers found about equal concentrations of radiogenic argon-40 and nonradiogenic argon. About 30 atoms in every million atmospheric molecules (30 ppm) were argon-36. The gas chromatograph, which could not distinguish among argon's various isotopes, supported the mass spectrometer results (Table 6-4). Their data showed a total concentration between 50 and 70 ppm. Since Venus' atmosphere contains about 75 times as many molecules as Earth's, it contains 75 times as much argon-36 as Earth's atmosphere. Yet, the ratio of argon-38 to argon-36 is almost identical to the terrestrial ratio.

One discordant note came from the Bus' neutral mass spectrometer. It could not detect argon at 130 km (81 miles). By extrapolation to the lower atmosphere, this result would imply that there is less than 10 ppm of argon-36 in Venus' atmosphere. Even this upper limit, however, does not exclude the possibility of 25 times as much argon-36 as in Earth's atmosphere.

Examination for neon, another primordial rare gas, confirmed the argon story. Pioneer instruments and Venera's neutral mass spectrometers placed the abundance of neon between about 4 and 13 ppm compared with 18.2 ppm for Earth. The ratio of neon-22 to neon-20 was 0.07. Compared with the argon isotopes, this ratio is lower than the value on Earth (about 0.1), but is close to the solar ratio.

The notion that Venus, Earth, and Mars formed from materials containing the same endowment of volatiles, already shaken by the Viking results, was completely refuted by the data from Pioneer Venus. Why should Venus have received about twice as much carbon dioxide and nitrogen as Earth? And why does it have about 50 to 100 times as much neon and nonradiogenic argon?

After a review of early data from Pioneer and Venera missions, researchers suggested a possible reason. The planets, they hypothesized, formed from dust grains in the solar nebula. These grains were surrounded by gas at a pressure that diminished rapidly with increasing distance from the center of the nebula. Reactive volatiles such as carbon, nitrogen, and oxygen would be chemically combined within the grains. Rare gases would be adsorbed from the surrounding gas in amounts depending on the pressure. As a result, grains forming the three planets would possess about the same reactive volatile content, while the rare gas concentration would decrease rapidly with increasing distance from the Sun. This model required that the nebula's gas temperature should be fairly constant. Also, early outgassing from Mars should be less efficient by a factor of 20 than from the other two planets.

Analysis of the Large Probe's neutral mass spectrometer data produced another surprise. Although Venus' atmosphere contains a large excess of neon and primordial argon, this is not so with two other rare gases. The absolute abundance of krypton is only about three times larger in Venus' atmosphere than in Earth's. There is much less than 30 times more xenon. In the grain accretion model, there is no reason to expect enrichment of one rare gas to be greater than another. In fact, a close look at Mars data shows that, from Mars to Earth, the enrichment decreases from a factor of about 220 for neon, through 165 for argon and 110 for krypton, to 30 for xenon.

Another way to look at these results is to compare the ratio of primordial argon to krypton on the terrestrial planets with the ratio on the Sun. The ratio is 4000 in the Sun's atmosphere, 1000 on Venus, 50 on Earth, and 40 on Mars. So, the ratio gets more solar-like the closer the planet is to the Sun. This suggests that perhaps the material that accreted to form the planets was exposed to a strong irradiation by gas of solar composition flowing away from the Sun as the Solar System formed. If so, the grains and small bodies that formed the planets would have volatiles from the Sun in addition to those from the nebular gas in their neighborhood. The material forming Venus may have received a larger share of solar gases than the other planets. In intercepting much solar gas, the material forming Venus would have shielded the outer regions of the Solar System from this gas.

Another possibility is that Mars formed earlier than Earth, and Earth much earlier than Venus. This would explain why Mars lost most of its volatiles. The planet may have originated early enough to have retained such highly radioactive substances as aluminum-26 left over from a nearby supernova explosion believed to have triggered the formation of the solar nebula. The heat produced by the decay of this radioactive aluminum might have driven off many of the Martian volatiles very early.

Two important noble gases are produced by radioactive decay of heavy elements such as uranium. One is argon-40, the other is helium-4. The consensus regarding Pioneer Venus and Venera measurements is that argon-40 and argon-36 are about equal in abundance on Venus. On Earth, argon-40 is about 400 times as abundant as argon-36. Since there is 75 times as much argon-36 on Venus as on Earth, this means there is only about one-fourth as much argon-40 on Venus as on Earth. Venus either started with much less potassium than Earth or is yielding up its argon from the interior more slowly than is Earth. Several factors may account for a slow escape of gases during Venus' 4.5 billion year lifetime. These factors include lack of widespread tectonics, a thicker

and relatively plastic unfractured lithosphere, and absence of surface erosion by water.

Measurement of helium in the upper atmosphere by the Bus' neutral mass spectrometer agrees with this picture. Extrapolation to the lower atmosphere suggests that there are about 12 helium atoms per million molecules in the planet's atmosphere. This works out to an absolute abundance of helium on Venus 250 times greater than on Earth. Yet, we cannot conclude that Venus has vented that much more helium-4. Scientists know that the present atmospheric amount of helium would be produced by radioactivity in Earth's interior in about one million years. Earth's atmosphere is losing helium at a great rate. The amount actually produced, vented, and lost is at least 10,000 times what now remains. The best estimate is that 5 to 10 times as much helium has been produced and escaped from Earth's atmosphere compared with Venus. Hence, inefficient present release of gas from Venus' interior may account for the difference between the radiogenic gas inventories of the two planets, if they contain equivalent amounts of potassium and uranium.

The amount of water vapor present in an atmosphere has important implications for the atmosphere's temperature structure. Water vapor plays a major role in the greenhouse mechanism invoked to account for the very high temperature of the atmosphere near Venus' surface. It also has an important bearing on the chemical composition of the atmosphere.

Unfortunately, accurately measuring the amount of water vapor in an atmosphere is very difficult. Scientists are even uncertain about the exact amount of water in Earth's stratosphere. After the Venus Multiprobe mission of 1979, a similar state of confusion developed about the amount of water vapor in Venus' atmosphere. Data from the Large Probe's neutral mass spectrometer showed less than 0.1% water in the atmosphere. A special optical device on the Venera probes found a small amount, too. Its measurements suggested that water decreases from 200 ppm at 50 km (31 miles) to 20 ppm at the surface. On the other hand, the probe's gas chromatograph data showed 0.52% of water at 42 km (26 miles) and 0.13% at 22 km (13.7 miles). These were much greater amounts.

The amount of carbon monoxide gas in Venus' atmosphere is minute. According to data from the gas chromatograph, its concentration is about 20 ppm at 22 km (13.7 miles). At the cloud tops, it is about 50 ppm as deduced from Earth-based observations. If carbon monoxide is produced by photodissociation of carbon dioxide above the clouds and subsequently diffuses downward, this kind of distribution would result. However, the amount of carbon monoxide expected to accompany carbon dioxide as it vents from a planet's interior is far greater than the amount observed on Venus. At least a thousand times as much carbon monoxide should have been produced. It is conceivable that carbon monoxide may have reacted with water to form hydrogen and carbon dioxide early in the planet's history. This explanation could account for the lack of water on Venus. Hydrogen might have escaped into space. However, it is most unlikely that the initial amounts of water and carbon monoxide were so nearly equal that they would have mutually reduced each other to such minor quantities as are now on the planet.

Oxygen is one of the other constituents found by various instruments. This gas increases from 16 ppm to 43 ppm between 42 and 52 km (26 and 32 miles), according to data from the gas chromatograph. The Large Probe's neutral mass spectrometer produced data that show the amount of oxygen as less than 30 ppm. Earth-based measurements find less than 1 ppm at the cloud tops. The coexistence of carbon monoxide and molecular oxygen in the atmosphere is difficult to understand thermodynamically. Photolysis of carbon dioxide above the clouds would form oxygen along with carbon monoxide. These should decrease in abundance downward. However, the amounts researchers found below 52 km (32 miles) were quite inconsistent with the small amount they observed above the clouds. Thus, the oxygen measurements presented an enigma.

Carbon dioxide, which makes up the bulk of Venus' atmosphere, is mysteriously stable. Orbiter found the reason. In the highest part of the atmosphere, rapid decomposition of carbon dioxide by sunlight is an ongoing process, releasing heat that drives the planetwide system of winds. These winds blow from dayside to nightside. On the nightside, winds descend to lower altitudes carrying the carbon monoxide and oxygen down with them. A striking confirmation of the process came from several ultraviolet spectrometer images that showed atoms of nitrogen from the dayside "burning" on the nightside to produce an ultraviolet "flame." This occurred in regions low enough for sufficient oxygen pressure to support the reaction. Calculations show that once the dissociated gases reach lower altitudes, they reform carbon dioxide under the influence of chlorine-catalyzed photochemistry above the cloud tops. The carbon dioxide is recycled with the result that the bulk of the atmosphere remains stable.

Among sulfur compounds, the measurements would allow no more than 3 ppm of the interesting molecule carbonyl sulfide. Yet, sulfur dioxide appears to be present near 22 km (13.7 miles) in the fairly large amounts of 130 to 185 ppm. Above the clouds, the amount is only 0.1 ppm. Finally, the neutral mass spectrometer detected hydrogen sulfide gas with a mixing ratio decreasing from about 3 ppm at the surface to 1 ppm in the clouds. These results have an important bearing on the question of how Venus' clouds form. We know the clouds contain large amounts of sulfuric acid. Before Pioneer Venus, scientists suggested a cycle of chemical reactions similar to one responsible for formation of sulfate aerosol layers on Earth. In this cycle, carbonyl sulfide plays a key role. Failure to find carbonyl sulfide in Venus' atmosphere was a major surprise. Now scientists are considering mechanisms that use a sulfur dioxide and water source to produce the sulfuric acid.

Upper limits for other important species have been set by data from the gas chromatograph. These are 10 ppm for hydrogen, 1 ppm for methane, and 1 ppm for ethylene.

The neutral mass spectrometer made atmospheric composition measurements during the final entry phase of the Orbiter's mission. The entry data at lower solar activity filled a gap in the midnight to 5:00 a.m. local solar time range of the earlier data gathering period of 1978-1980. The earlier data extended to 140 km (87 miles) from midnight to 1:00 a.m. and were above 155 km (96 miles) from 2:00 to 4:30 a.m. The entry data extended down to 130 km (81 miles) during the same local times. On the last orbit, the spacecraft obtained data down to 128.8 km (80 miles). In Phase III, data were gathered about helium above 170 km (106 miles) from 6:00 p.m. to midnight local solar time. Also, from midnight to 4:30 a.m. below 200 km (124 miles), Orbiter gathered data on helium, atomic nitrogen, atomic oxygen, carbon monoxide, molecular nitrogen, and carbon dioxide.

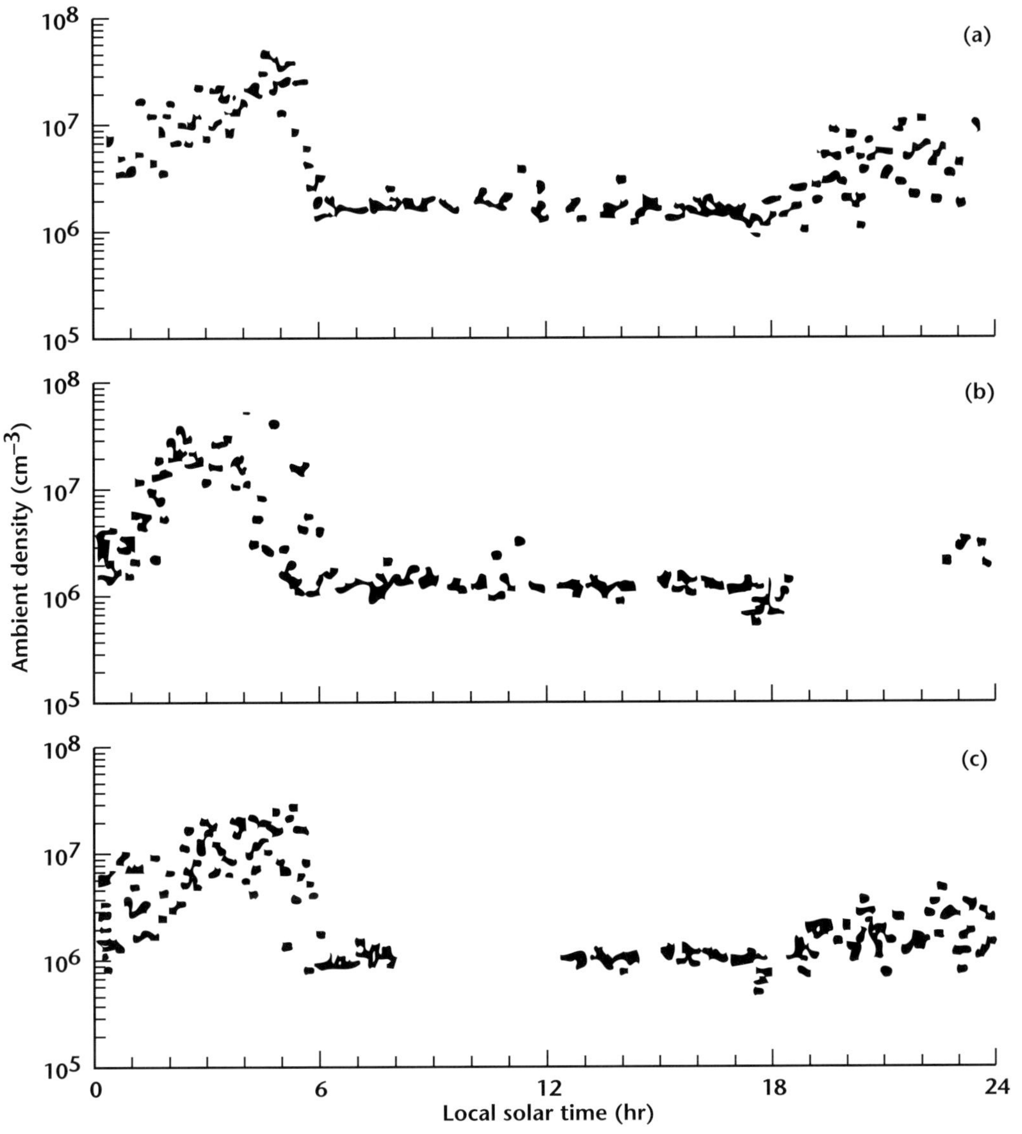

Figure 6-39. Measurements of helium number densities from 170 km (105 miles) altitude over three of Venus' diurnal cycles (each of 225 Earth days). The letters a, b, and c identify these cycles. Note how they are very much the same over the cycle of solar activity.

During Phase III of the mission, helium was the dominant species in the postmidnight sector above 170 km (106 miles). The number densities of helium at an altitude of 170 km (106 miles) over three diurnal cycles of the Pioneer mission appear in Figure 6-39. The figure shows the three cycles separately and identifies them as a, b, and c. Very little change is apparent over the solar activity cycle.

Also in Phase III, oxygen was the dominant species from 140 to 170 km (87 to 106 miles). Carbon dioxide was dominant below 140 km (87 miles). Estimated scale height temperatures for helium, oxygen, and carbon dioxide were about 105 to 120 K. This was similar to temperatures researchers observed in 1978-1980 at a period of higher solar activity. The diurnal variation of exospheric temperature, based on number densities and scale heights over one sidereal period of 225 Earth days early in the mission, appears in Figure 6-40.

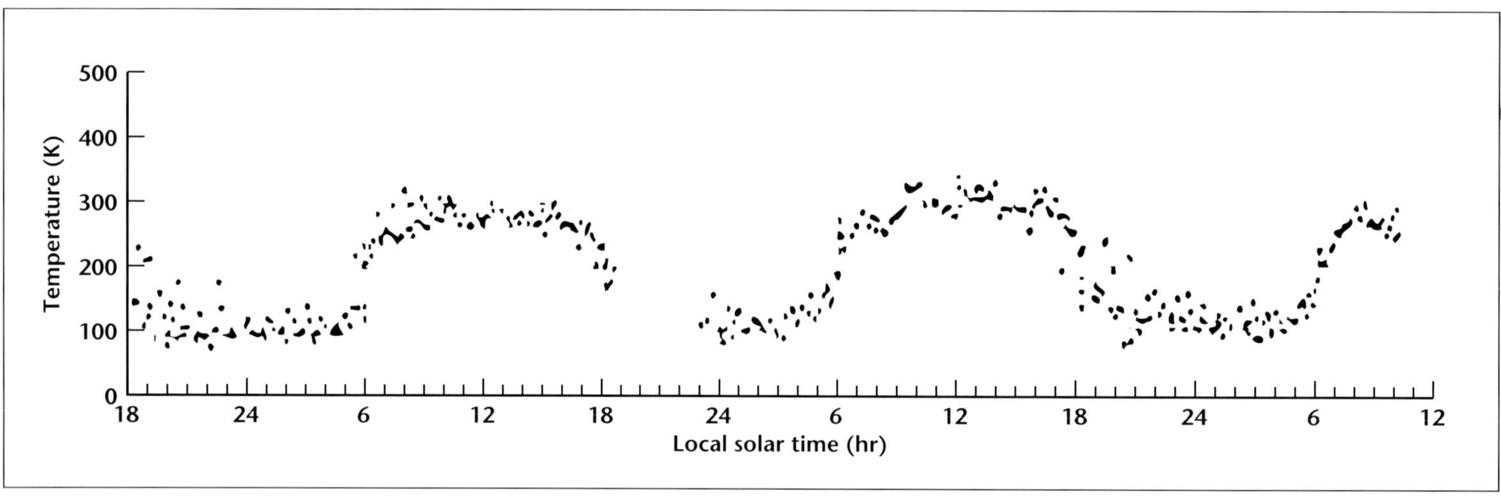

Figure 6-40. Diurnal variation of exospheric temperature. Researchers derived these data from number densities and scale heights that Orbiter's neutral mass spectrometer measured. The data cover one diurnal cycle of Venus that lasted 225 Earth days.

The densities at 1:00 a.m. local solar time and at 150 km (93 miles) altitude were within 35% of earlier measurements. These measurements occurred during the concluding phase of the mission. Also, the helium bulge was similar to that in 1978-1980. This confirmed that superrotation of the thermosphere was still occurring. It appeared that small changes in the dayside thermosphere arising from changes in solar activity have little impact on the nightside thermosphere. The densities at an altitude of 170 km (106 miles) for carbon dioxide and atomic oxygen plotted against local solar time appear in Figure 6-41. Orbiter obtained these over almost three sidereal days, and they showed little change over the solar cycle.

Water-vapor measurements presented major theoretical problems. Use of the high value obtained from the Pioneer Venus gas chromatograph in a thermodynamic calculation created an anomaly. It predicted amounts of hydrogen sulfide and carbonyl sulfide somewhat larger than the gas chromatograph itself would allow, but consistent with the mass spectrometer measurements. The smaller amount that the Venera photometer found would not allow nearly so much hydrogen sulfide as the mass spectrometer found. On the other hand, an elementary conservation law states that the ratio of hydrogen atoms to the total number of gas molecules of all kinds must remain constant in the atmosphere below the clouds. Whether the gas chromatograph measurement of 0.52% water at 52 km (32 miles) or the photometer value of 200 ppm is correct, compounds with equivalent amounts of hydrogen atoms must exist at the surface. Their concentrations must vary to keep the hydrogen mixing ratio constant. Scientists could not find these hydrogen compounds. Thus, hydrogen presents a continuing dilemma as it generally does in studies of planetary atmospheres.

An important question for many reasons is whether the atmosphere is reducing or oxidizing. Scientists are sure that it is very close to the dividing line between these two states but are still unsure as to which side it is on. The amount of carbon monoxide detected seems to be slightly greater than the amount of molecular oxygen. Some scientists doubt the presence of the latter. Thus, a case can be made that Venus' atmosphere is in a reducing state.

Orbiter's instruments recorded wave-like perturbations in the nightside neutral atmosphere. These were interpreted as being due to gravity waves penetrating upward from the lower thermosphere. Gravity waves couple the upper atmosphere to the lower thermosphere and modify the circulation of the lower thermosphere. Researchers suggest that these gravity waves couple the lower atmosphere superrotation at the cloud tops to the superrotation in the thermosphere. The latter was inferred from measurements of the neutral composition.

The neutral atmosphere, unlike the ionosphere, showed very little variability from solar maximum to solar intermediate conditions. This was especially true on the nightside, probably because the nightside neutral atmosphere is insulated from solar cycle dependent changes in the dayside thermosphere.

Aurora and Airglow

An unexpected discovery by Orbiter was the presence of ultraviolet emissions from oxygen in the high nightside atmosphere. Researchers explained these emissions as being due to energetic particles, either electrons or ions, entering the atmosphere from space. Such emissions are common on planets that have magnetic fields. The particles originate in the solar wind and travel along magnetic field lines toward the planet's magnetic polar regions. On Earth, we call the emissions aurora borealis and aurora australis, or northern and southern lights.

However, Venus has no intrinsic magnetic field to trap and direct the solar wind's charged particles. The origin of the particles responsible for Venus' aurora is uncertain. Because the brightness of emissions from Venus' atmosphere is related to solar activity, scientists suggested one possible reason. The particles may be photoelectrons that extreme ultraviolet solar radiation produces on the dayside. Weak but turbulent magnetic fields produced by the action of solar wind on Venus' ionosphere could carry them into the night hemisphere.

Observation of high-altitude airglow at the limb was an important discovery by Mariners 5 and 10. Pioneer Venus Orbiter added information about this phenomenon. All three spacecraft observed Lyman alpha radiation, which is a tracer for hydrogen atoms. Also, Mariner 10 provided data for helium that allowed the first unambiguous determination of the temperature of Venus' exosphere. The exosphere was found to have a probable temperature not of 700 K as previously supposed, but one of only 350 K.

Among the light atmospheric elements that travel to the nightside are oxygen and nitrogen atoms, produced on the dayside by solar ultraviolet radiation. There is a nightside bulge of atomic oxygen (Figure 6-41). When the atoms are carried down again into lower atmospheric regions, they recombine into oxygen molecules and nitric oxide. In so doing they emit airglows. Mapping of the nitric oxide airglow as observed by the Orbiter reveals a concentration near 2:00 a.m. local time.

Oxygen glows in the infrared and visual regions of the spectrum. The infrared glow had been observed but not mapped from Earth. Venera 15 mapped the glow in the visual region, but its variations are more subdued than the nitric oxide glow.

The Ionosphere

Data from Pioneer Venus greatly increased our knowledge of Venus' ionosphere. The ionosphere of a planet is a region of the upper atmosphere with a high density of electrically charged particles—electrons and ions. These charged particles are usually a product of extreme ultraviolet solar radiation interacting with neutral molecules and atoms of the upper atmosphere. The types and densities of ions in an ionosphere depend on the neutral composition, the chemical reactions that occur, and how the ions move from place to place within the ionosphere. Magnetic fields affect the behavior of a gas consisting of charged particles (known as a plasma).

Measurements of the delay time in the arrival of a radio wave passing from a spacecraft to receiving stations on Earth provides

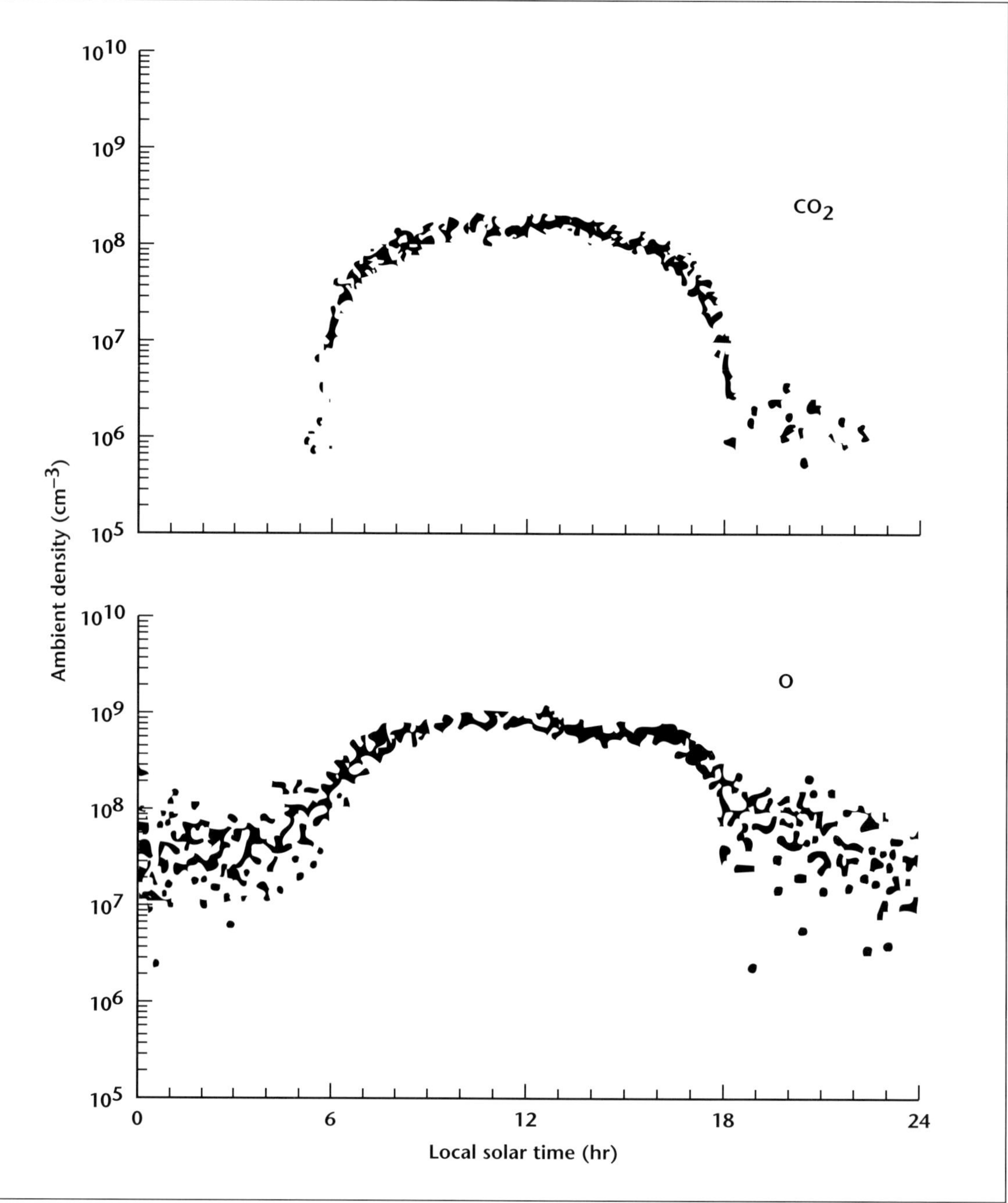

Figure 6-41. Measurements of carbon dioxide and atomic oxygen number densities at 170 km (105 miles) altitude. Orbiter's mass spectrometer recorded them over nearly three diurnal cycles (about 675 Earth days). Note the smaller peak of oxygen to the right of the dayside curve.

information on the electron densities encountered along the way. Experimenters used this technique to obtain information concerning the ionosphere. They arranged for the radio waves to pass through Venus' atmosphere on their way to Earth as the spacecraft went into and emerged from occultation by Venus. On earlier flyby and orbital missions before Pioneer, this technique obtained some limited data on the total electron densities. Pioneer Venus Orbiter not only employed this technique but also made the first *in situ* measurements of Venus' ionosphere. The spacecraft used the following instruments: an ion mass spectrometer, a Langmuir probe, a retarding potential analyzer, and a fluxgate magnetometer. Information from these instruments helped develop a picture of global composition and dynamics.

On Venus, the ionospheric electron density reaches a maximum at altitudes near 140 km (87 miles). This occurs on both the dayside and the nightside. Generally, Orbiter could not directly access this level because it is slightly below the lowest periapsis altitude the spacecraft reached. Scientists, however, were able to examine this density maximum with the radio occultation technique. Above this density peak, the electron density decreases gradually with increasing height. In regions directly accessible to Orbiter's instruments, the Langmuir probe made high time-resolution measurements of both the electron density and temperature that revealed many unusual ionospheric events (Figure 6-42). These included ionospheric density depletions ("holes") and detached plasma clouds. Also, Orbiter's ion mass spectrometer measured plasma composition and its total density for the first time. More data on plasma composition came from the retarding potential analyzer and from measurements of ion temperature, photoelectron fluxes, and plasma drifts.

Earth's ionosphere reaches heights of many thousand kilometers, gradually tapering off with increasing altitude. This high altitude extension is possible because a strong intrinsic dipole magnetic field shields Earth's ionosphere from the solar wind. By contrast, Venus' intrinsic magnetic field is negligible and the solar wind interacts directly with the ionosphere. Venus' ionosphere is an obstacle to the solar wind and deflects it around the planet. As a result, the ionosphere ends rather abruptly at an altitude of only a few hundred kilometers. The boundary where the ionosphere ends and the region of decelerated solar wind (ionosheath) begins is the ionopause. This boundary altitude is variable.

Just outside the ionopause is a large horizontal magnetic field. It contains some ionosheath plasma and some rapidly moving "superthermal" plasma of ionospheric origin. This large magnetic field, induced by the interaction of solar wind with the ionosphere, transmits the solar wind pressure and acts as a "piston" on the ionosphere. When the pressure is high, the magnetic field is enhanced. The piston moves in and pushes the ionopause to a lower altitude. When the pressure is lower, the ionopause moves up (Figure 6-43). As a result, the ionopause height is quite variable, ranging from 200 km (124 miles) to over 1000 km (621 miles) on the dayside. On the nightside, there is no direct interaction of solar wind with the ionosphere because the solar wind is deflected around the planet. However, there must be indirect interactions that we do not yet fully understand because even on the nightside the height of the ionopause is usually less than 1000 km (621 miles). Also, there are many variations over the 11-year solar activity cycle (Figure 6-44).

Unlike the magnetic field just outside the ionosphere, the field within it is small. However, Orbiter's magnetometer detected unique magnetic structures. These structures, or flux ropes, are long, narrow, rope-like regions of strong magnetic field in which the field lines are twisted. One suggestion is that these regions form from the large magnetic field piled up just outside the ionopause. The solar wind, "pulling" on the "ends" of the ropes, draws them down into and through the ionosphere. Another explanation is that magnetic flux ropes form in a region of large ionospheric magnetic fields near the subsolar point.

On the nightside, the ionosphere's magnetic field is most often larger and more regular than on the dayside. The average field has the type of global symmetry expected from a "draping" of solar-wind field lines around the planet.

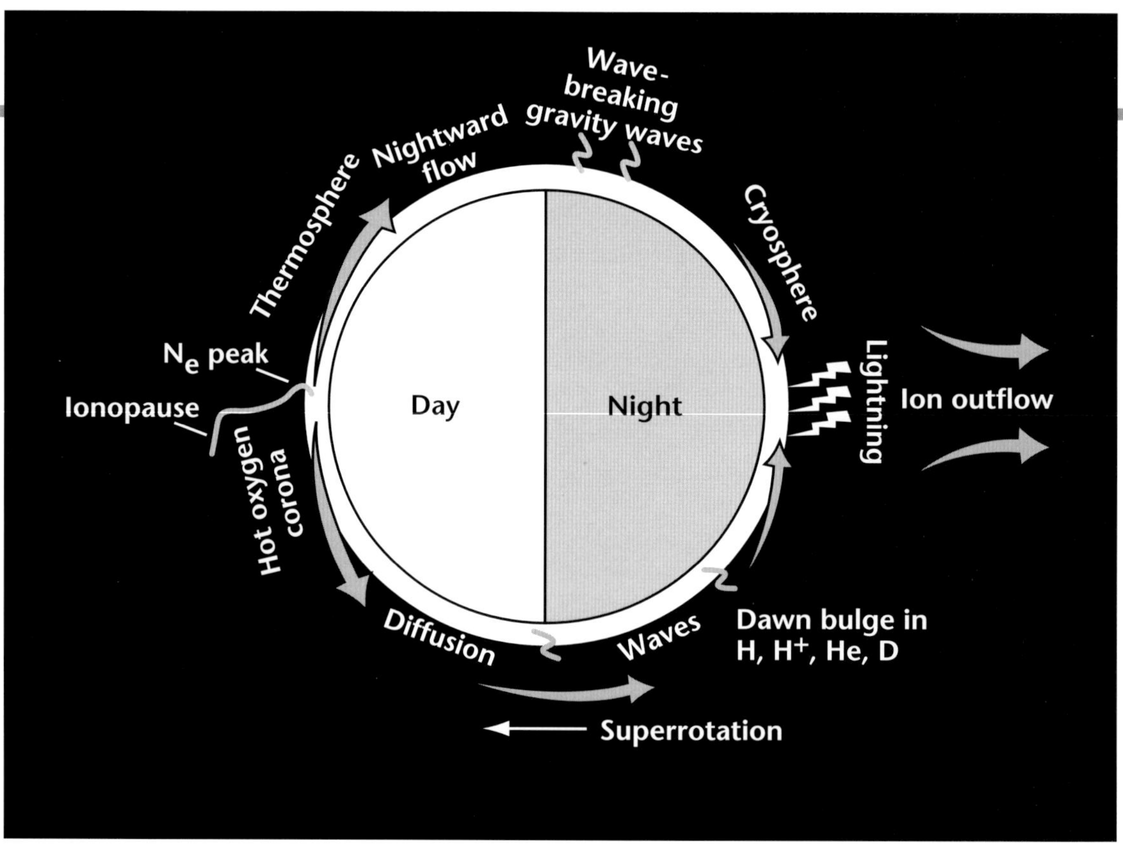

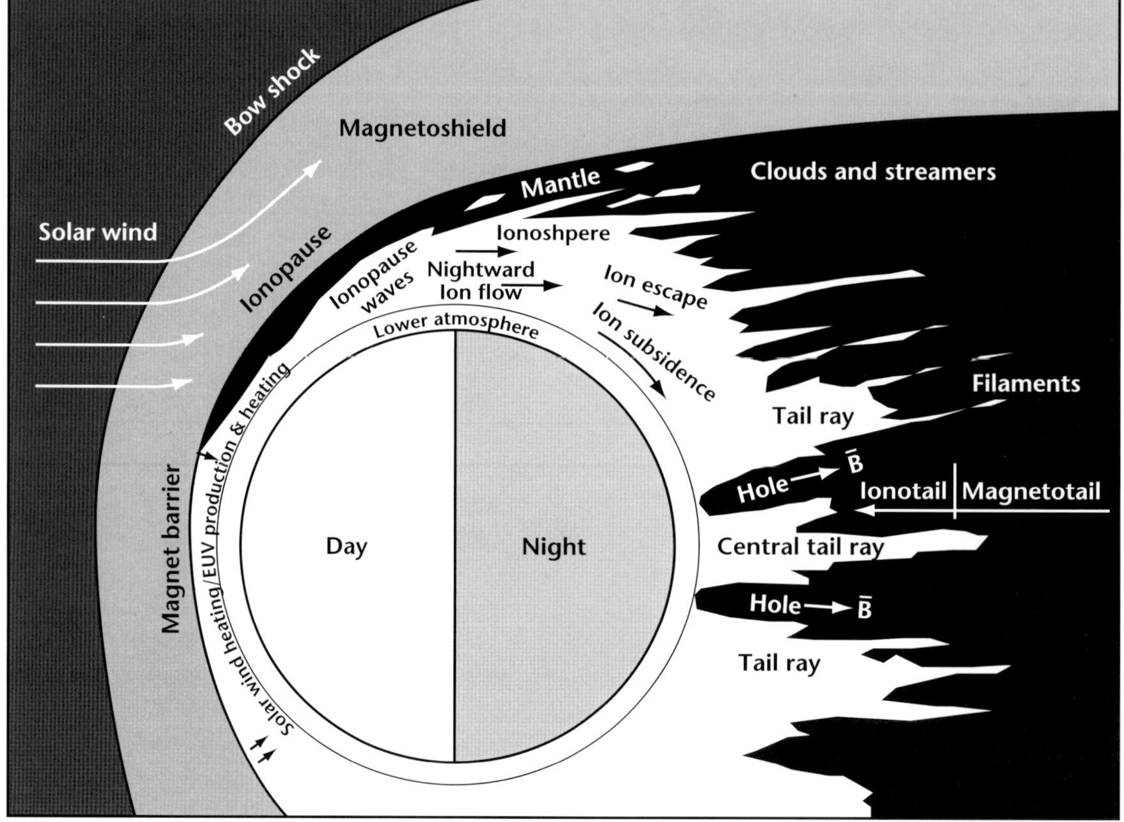

Figure 6-42. The complex environments of Venus' thermosphere and ionosphere and the planet's interaction with the solar wind. (Top) This diagram highlights major discoveries by Pioneer Venus. There is the extremely cold nightside upper atmosphere, gravity waves at predawn and early-dusk sides, and a dawn bulge in lighter constituents of the atmosphere. A large cloud of atomic oxygen extends over the cold dayside thermosphere. Low-frequency radio bursts during nightside passages of Orbiter suggest lightning flashes in the lower atmosphere. (Bottom) This diagram highlights major discoveries about the ionosphere and solar-wind interaction. On the sunlit side of Venus, the atmosphere ionizes to form a dense ionosphere. The planet has no intrinsic magnetic field. So ions and electrons flow at high speed to the nightside and form a strong ionosphere there. The solar wind interacts with the top of the ionosphere and forms a bow shock that moves in and out from the planet as the strength of the solar wind changes. There is a complex of plasma clouds, tail rays, filaments, and ionospheric holes on the planet's nightside. As a result of the Pioneer Venus mission, scientists have examined the ionosphere of Venus in more detail than any other planet besides Earth.

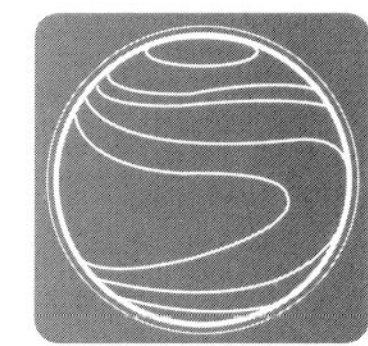

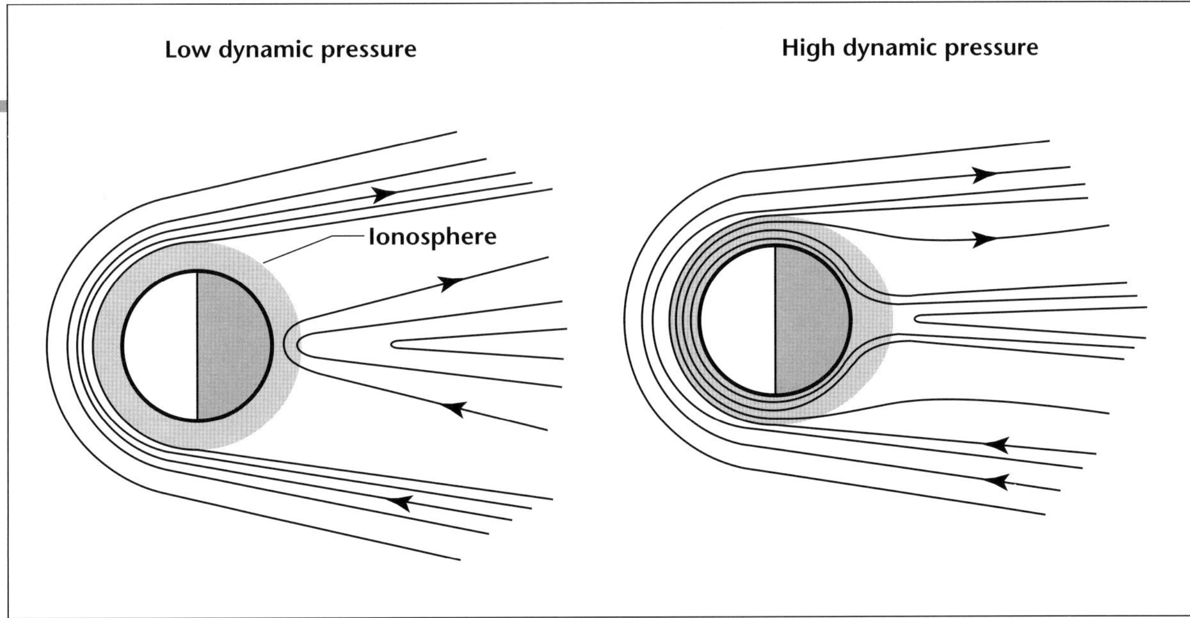

Figure 6-43. The two sketches show the effects of low and high dynamic pressure of solar wind on Venus' ionosphere and the changes in its shape on the nightside.

Heat conduction and transport of electrically charged particles is constrained along magnetic field lines rather than at right angles to them. The flux ropes may affect electron and ion temperatures in Venus' ionosphere. The electron temperature is a few thousand kelvin on both the dayside and nightside of the planet. This is much hotter than the neutral gas in the thermosphere, which has a temperature of only a few hundred kelvin. Another reason for high temperatures is that heat from the solar wind is "pumped" into ionospheric electrons at the ionopause. The temperature of the ions is also high, about 2000 K on the dayside and more than 4000 K on the nightside. Interactions, such as friction between the neutral gas and the ions, produce heat that helps keep the ionosphere hotter than the neutrals. On the nightside, some of the energy from rapid motions or horizontal drifts of the ions converts into heat and makes the nightside ions hotter than those on the dayside.

Orbiter's ion mass spectrometer established the presence of many different ions. From theoretical studies, scientists expected to find O_2^+, O^+, CO_2^+, He^+, and H^+ ions. Other ions they found in Venus' ionosphere were unexpected. These were C^+, N^+, NO^+, O^{++}, H_2^+, and N_2^+.

Molecular oxygen is the most common ion below 200 km (124 miles) on the dayside and below 160 km (100 miles) on the nightside. Above an altitude of about 160 to 200 km (99 to 124 miles), atomic oxygen becomes the most common ion. In the predawn region of the nightside, atomic hydrogen ions are just as abundant as atomic oxygen ions.

There is a strong day/night asymmetry, or local time variation, in the total plasma density. Each ion species has its own day/night asymmetry. That is, composition and total plasma density depends on local time (Figure 6-29). At 200 km (124 miles), atomic oxygen ion concentration gradually decreases by a factor of 10 from the dayside to the nightside. Molecular oxygen ion density decreases rapidly at the terminator and is almost one thousand times less on the nightside than on the dayside. Atomic hydrogen and helium ions behave quite differently from oxygen ions and are greater on the nightside than on the dayside. Yet, the nightside distributions are not uniform. There are more hydrogen than helium ions in the predawn region, no doubt reflecting the predawn bulges in neutral hydrogen and helium.

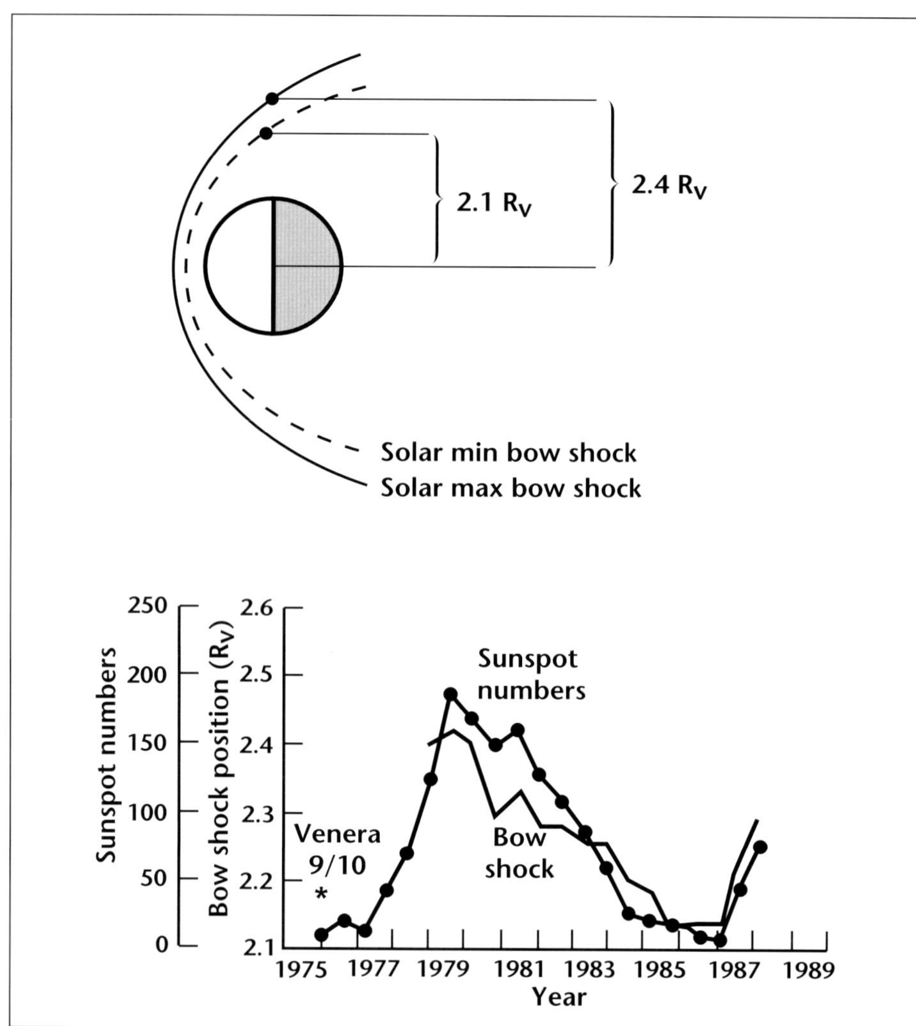

▲

Figure 6-44. The top diagram shows how the position of the bow shock changes between solar maximum and solar minimum. The bottom diagram shows the position of Venus' bow shock compared with sunspot numbers during the whole of the Pioneer Venus mission.

Data from Pioneer Venus led to great progress in understanding the mechanisms that maintained the nightside ionosphere. The problem is the length of the Venusian night. It is about 58 Earth days and much longer than the lifetime of the ions. Therefore, scientists did not expect a significant ionosphere of the type Pioneer found on the nightside. Two sources of ionization were identified, largely with the help of data from Orbiter's instruments. One source, first supported by data from the Soviet Venera spacecraft, is the bombardment of the nightside atmosphere by fast electrons that are energetic enough to ionize the neutral gas. This is much like the electron flux that gives rise to terrestrial auroras. The electrons originate in the wake of the planet outside the ionopause. This mechanism can account for a large fraction of the ionization in the lower part of the nightside ionosphere. However, another source of ions is required to account for conditions at higher altitudes and to supplement the "auroral" source at lower altitudes.

Instruments on Orbiter detected large horizontal flows, or drifts, of plasma from day to night hemispheres. Drift velocities were very large at high altitudes and near the terminator, up to 10 km/sec (23,000 mph). Plasma motions like these are more than enough to maintain the observed nighttime ionosphere at higher altitudes. A significant contribution also can be made to maintaining the lower ionosphere, since ions, as they flow to the nightside, also sink to lower altitudes. We do not yet fully understand the mechanism for the plasma drifts themselves.

At lower altitudes, day-to-night neutral winds help drag the ions along to the nightside. At very high altitudes near the ionopause, the antisunward flow of plasma on the high side of the ionopause can induce ionosphere flow below it. At middle altitudes, the day-to-night gradients in the ion densities seen by Orbiter can generate ion drifts.

Concentrations of all ions show pronounced fluctuations from orbit to orbit on the nightside as well as near the terminators. Usually there is an ordinary nightside ionosphere, but sometimes the nightside ionosphere disappears entirely. Perhaps solar wind (when its pressure is large) sweeps it downstream of Venus. The nightside magnetic field plays an important role in this. At other times, the nightside ionosphere looks normal except for localized holes in the plasma where the electron density is very low and the electron temperature is very high. The magnetic field in these holes aligns vertically, indicating that

these holes may be associated with the large-scale structure of the field on the nightside. Another phenomenon that is frequently observed on night and day hemispheres, mostly near the terminators, is the presence of detached layers of clouds of ionospheric plasma that lie outside the ionosphere, beyond the ionopause. It is likely that the solar wind, or an ionosheath flow, removes chunks of plasma from the ionopause region and carries this plasma downstream (Figure 6-42).

The radio occultation experiment clearly showed a major change in the scale height of the ionosphere between solar maximum and minimum. This may be due to a drop in exosphere temperature to about 200 K at solar minimum. It also could be due to a reduction in the amount of atomic oxygen in the thermosphere. This would have the same effect in reducing the average scale height. Dayside electron density profiles at solar maximum in 1980 and at solar minimum in 1986 show the effects of the solar cycle on ionospheric density (Figure 6-45). Scientists believe the depletion of the upper ionosphere at solar minimum is an important factor in the reduction of nightward ion flow, which, in turn, causes the generally less robust nightside ionosphere.

During the entry phase, Orbiter investigated the atmosphere and ionosphere under different conditions of solar activity from those during Phase I. Researchers expected variations in solar activity to have a strong effect on the nightside ionosphere. This ionosphere has two sources of ionization at solar maximum. One is ionization by electron precipitation. The other is ion transport from dayside to nightside. Scientists expected ion transport to be reduced during solar minimum because of a lower altitude of the ionopause at the terminator. This would result from reduced pressure of the ionospheric plasma. At solar maximum, on the other hand, solar wind has a high dynamic pressure. This also would restrict the day-to-night transport of plasma. It was suggested that the loss of ionosphere at solar maximum arose from a decreased transport of ions from day to night.

During Phase III of Pioneer's mission, researchers found that the nightside ionosphere was greatly reduced from conditions at solar maximum. Modeling ionospheric processes suggested that some ion transport should still occur at that time. Since entry took place at a period of intermediate solar activity, day-to-night ion transport might still occur. This agreed with observations.

When Orbiter penetrated below 140 km (87 miles) during Phase III, scientists had an opportunity to study the low-altitude ionosphere with repeated sequential observations of the nightside ion peak. When they compared the ion peak with what they observed during Phase I of the mission, researchers saw interesting similarities and challenging differences. Earlier in the solar cycle, details of the peak were similar to those during the final encounter. There were no noticeable differences in either the altitude of the peak or the maximum ion concentrations at the peak. However, the data from Phase I showed a much better developed high-altitude ionosphere, with higher concentrations extending to higher altitudes than in the final encounter data. Composition differences were seen in the ion mass spectrometer's data between the earlier mission and the final encounter. During Phase I, the nightside ionosphere at high altitudes was more extensive. It had a large concentration of ions extending to high altitudes. This well developed ionosphere was maintained, researchers presumed, by transport from the dayside. As they expected, the dominant ion at high altitudes was singly ionized atomic

Figure 6-45. The two curves on this diagram show the variation of electron density with altitude at two parts of the solar activity cycle: 1980, close to maximum, and 1986, close to minimum, activity. The peak density remains about the same over the solar cycle, but the density in the upper atmosphere is markedly reduced at solar minimum.

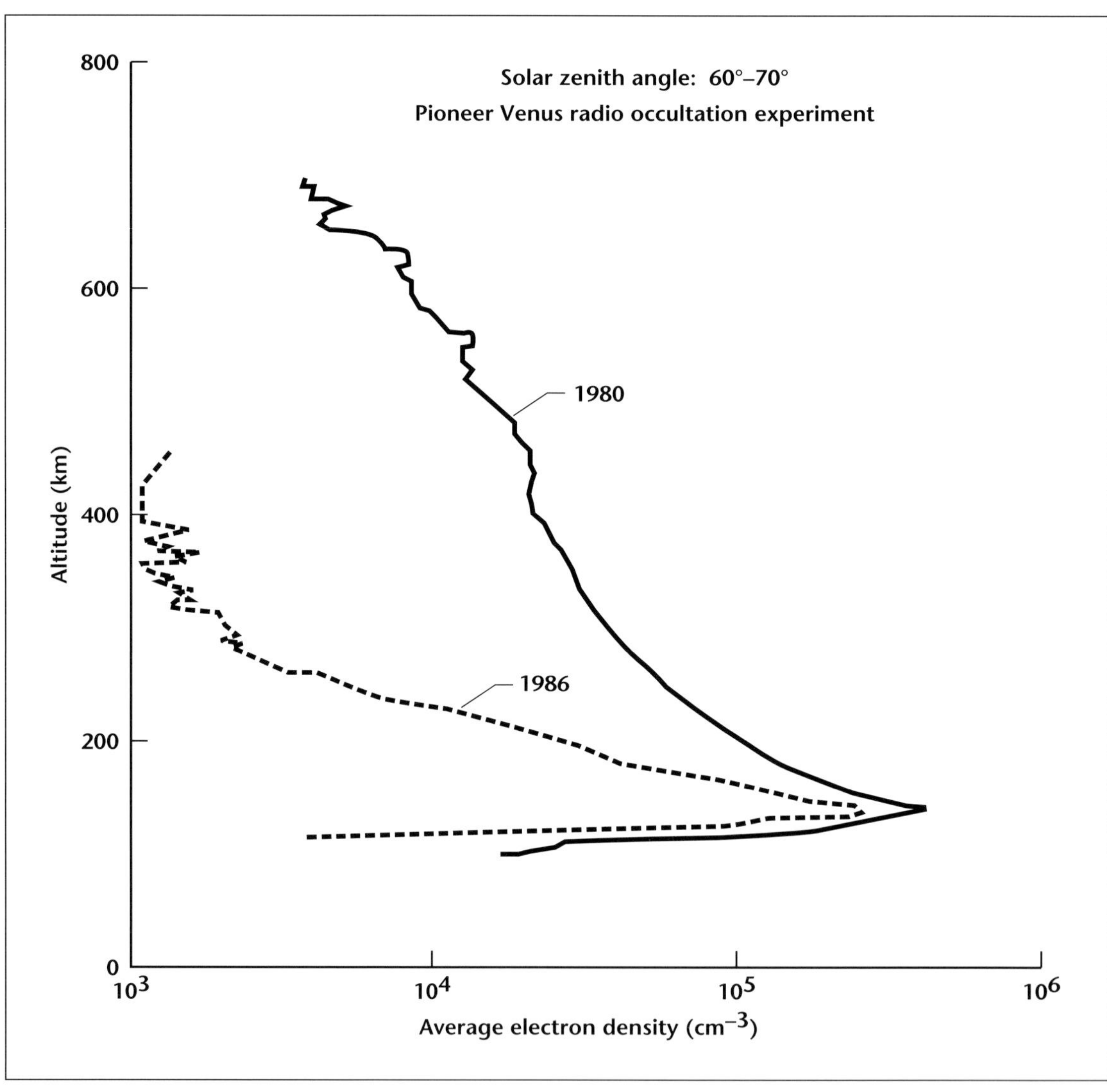

oxygen. However, singly ionized hydrogen was severely depleted relative to the oxygen ions.

Then, during the final encounter, concentrations of oxygen ions were much lower while hydrogen ions dominated on the nightside at high altitudes. Did the change reflect differences in composition of ions transported from the dayside to the nightside? Or are hydrogen ions produced in some other way and persist when oxygen ions are not being transported from the dayside? Researchers were not certain.

Several scientists have examined the day-to-night ion transport at low solar activity. By combining analysis of data from the ion mass spectrometer with mathematical modeling, several things were learned about the importance of plasma transport in the predawn ionosphere, especially in comparison with electron precipitation. Scientists computed the average peak density of oxygen ions as a function of solar zenith angle. Then they determined the fluxes of atomic ions or precipitating electrons needed to produce the

observed values. Calculations were compared with observations. The comparison showed that there must be significant day-to-night plasma transport at low solar activity. This refuted earlier suggestions that day-to-night transport would stop under conditions of low solar activity. These assumptions had been based on a decreased solar flux leading to dayside ion densities too low for efficient ion transport to the nightside. However, calculations showed that electron precipitation cannot reproduce the observed helium ion densities. As a result, researchers concluded that there are significant day-to-night fluxes of ions, at least in the predawn bulge region. This occurs even when solar activity is low.

Scientists used a one-dimensional magneto-hydrodynamic model to study the dayside ionosphere. If solar wind magnetizes the ionosphere at solar minimum, model results compared fairly well with observed electron density profiles. The model also could reproduce the layer of increased electron density at 170 to 200 km (106 to 124 miles) that appeared in Orbiter data. The layer structure was more apparent in the model results if it was assumed that electron temperatures below about 200 km (124 miles) are much lower at solar minimum than at solar maximum. Although there are still uncertainties about the upper atmosphere at solar minimum, the small scale height of the electron density can be reproduced under magnetized conditions. The mechanism for structure formation is much the same as at solar maximum. Also, the ledge structure is more apparent if low altitude electron temperatures are 500 K or less. Unfortunately, researchers could not determine electron temperatures in this region from the available high-altitude data.

Orbiter's discoveries completely revolutionized thoughts about Venus' ionosphere. The region turned out to be much more complex and variable than expected. The ionosphere declined markedly at solar minimum. Its density also was much lower. Nightward ion flow was greatly reduced at solar minimum. This resulted from a greatly reduced electron density in the dayside upper ionosphere.

Solar activity varied greatly over Orbiter's lifetime. The variations affected the properties of the ionosphere on the planet's nightside. When solar ultraviolet radiation was most intense at solar maximum, the ionosphere extended to its highest level. Also, transport of ions from the dayside was the main source of the nightside ionosphere. By contrast, at solar minimum, nightward ion transport lessened and the main source of the nightside ionosphere appeared to be electron precipitation.

In the upper ionosphere and the magnetotail near Venus, the effects of solar extreme ultraviolet radiation are significant. This is especially true for the altitude profile of magnetic field, electron density, and temperature in the nightside ionosphere.

Researchers discovered that electron density decreases by about one order of magnitude from high to low flux of solar extreme ultraviolet radiation. Also, the electron temperature changes by a factor of at least 2. The induced magnetic field also increases by 2 to 3 nT. In the lower ionospheric regions from 200 to 600 km (124 to 373 miles), the effects differ. At lower extreme ultraviolet fluxes, there is a slightly reduced electron density and a high temperature. These conclusions were based on analysis of the Orbiter's data from 1979 to 1987. The results are in accord with entry phase observations. Phase III measurements of the electron density above the ionospheric density peak were lower than measurements at solar maximum during the earlier parts of the Pioneer mission.

Although the Orbiter's measurements covered more than a solar cycle, changes in altitude of periapsis created problems in separating altitude structure from variations due to the solar cycle. The evolution of the spacecraft's orbit allowed study of the main nightside ionosphere only during solar maximum. Also, researchers had to confine measurements of the upper ionosphere to periods near solar minimum. The results still show that variations in solar extreme ultraviolet radiation strongly control the structure of the nightside ionosphere.

The Orbiter's electron temperature probe made important measurements during Phase III. The median electron density at the ionospheric peak at about 140 km (87 miles) altitude was unchanged from its value at solar maximum. However, the ionosphere was increasingly depleted at high altitudes. At 200 km (124 miles) altitude, the density was reduced by a factor of 7. The electron temperature, by contrast, was reduced by a factor of 2 at 140 km (87 miles) and greatly enhanced at higher altitudes. It even exceeded its value at solar maximum. It was enhanced over the solar maximum value by a factor of 1.3 at 200 km (124 miles) and by a factor of 2 at 500 km (311 miles).

These results generally supported earlier conclusions that a reduced nightward ion flow at low levels of solar activity depletes the nightside upper ionosphere. The lack of a variation in electron density near the peak of electron density over the period between solar maximum and the final entry of Orbiter led to another important conclusion. Nightward ion transport is not as important as local ion production by energetic particles in forming the peak density layer. The decrease in electron temperature at low altitudes suggests that low densities of the upper ionosphere during Phase III were unable to support heat conduction from the dayside ionosphere. Consequently, the lower nightside ionosphere was cooled by collisions with ions and neutrals. Some of its heat also was conducted downward to cooler regions.

Another important feature of the nightside ionosphere is a deep trough in electron density. This trough typically appears between the main peak of the ionosphere and the upper ionosphere. For example, this trough was observed on either side of periapsis at an altitude of about 180 km (112 miles) on two consecutive orbits.

Also during Phase III, the ion mass-spectrometer data showed that there were lower numbers of all ion species in the midnight dusk sector than at solar maximum. The most prominent change was the decrease in oxygen ions. It was more than one order of magnitude from solar maximum to solar minimum. The light hydrogen ion is produced in the hydrogen bulge region by charge exchange between oxygen ions and hydrogen transported from the dayside. Its concentration drops by a factor of 4.

Another interesting phenomenon discovered by Orbiter was the disappearing ionosphere. This occurs under solar maximum conditions when solar-wind pressure increases beyond normal. It also occurs during low solar activity when dayside ion production falls to a low value. A disappearing ionosphere is defined as the state when the ion density above the main ionosphere peak becomes greatly reduced. For both cases, there is a similar reduction in the number density of oxygen ions. Scientists concluded that both reductions result from decrease in the transport of ionization from the dayside ionosphere. More than 25 of

these conditions were recorded during the Pioneer mission.

Several wave phenomena were detected during the entry phase of the mission. The following were observed: neutral density waves of several hundred kilometers wavelength, plasma density fluctuations with wavelengths of about one kilometer, and plasma waves with even shorter wavelengths.

The kilometer-sized waves were prominent during Phase III of the mission. They were often quasi-sinusoidal and occurred in a relatively narrow altitude layer of 145 to 155 km (90 to 96 miles). This area was just above the layer where electron density peaked. Investigators did not observe these waves above the sharp gradient, at the peak layer, or below it. They suggested that the waves are generated by the steep density gradient between the main nightside ionosphere from a rapidly flowing plasma above. There was a tendency for the waves to rise slightly to higher altitudes as dawn approached.

Plasma waves were measured by the electric field detector throughout the low altitude ionosphere during the entry phase. The waves fell into two classes. A wideband signal in regions of low magnetic field was restricted to the 5.4 kHz channel and lower. The waves had a roughly constant burst rate above an altitude of 160 km (100 miles) and were attributed to acoustic mode waves generated by precipitating electrons from the solar wind. Whistler mode waves in the 100 Hz channel were attributed to lightning. However, these waves might also result from gradient drift instabilities in a horizontal magnetic field. Unfortunately, the spacecraft could measure only the horizontal component of the field. Without a measurement of the radial field, it was not possible for scientists to resolve this question.

Solar-Wind Interaction

The Sun's upper atmosphere, or solar corona, is so hot that it is almost completely ionized. Even heavy atoms, such as iron, have lost many of their electrons. This ion-electron gas expands rapidly from the Sun, reaching speeds of over 400 km/sec (about 1 million mph), and forms the solar wind. At such speeds, the solar wind requires three days to reach Venus and four days to reach Earth. When Venus was between the Sun and Earth, meteorologists used solar-wind data from Pioneer Venus to warn of impending solar-wind disturbances on their way to Earth.

Interaction of the solar wind with a planet is similar to the interaction of the atmosphere with a supersonic aircraft. As an aircraft travels through air at subsonic speeds, pressure waves propagate ahead of the plane at the speed of sound. They warn of the plane's approach and divert air molecules out of its path. However, when an aircraft travels at supersonic speeds, the warning cannot be transmitted ahead, and a shock wave forms in front of the plane. This shock diverts the air around it. The solar wind travels faster than the speed of pressure waves that could divert solar-wind flow around a planet. Consequently, a shock wave, or bow shock, forms in the solar wind in front of each planet.

The bow shock of Venus is in many respects similar to the bow shock of Earth. This might be expected because the properties of the solar wind are similar at Earth and at Venus. However, there are differences. At Venus, the ionosphere, which extends only a few hundred kilometers above the surface, deflects the solar wind. On Earth, by contrast, the strong terrestrial magnetic field deflects the solar wind at a distance of over 10 Earth radii, tens of thousands of kilometers above the planet's surface. This results in a much larger bow shock at Earth than at Venus. According to present models, the bow shock's distance from the planet could affect the energies of particles the shock reflects back into the solar wind. However, the wave phenomena at Venus, in association with the reflected beams, seem equal to the terrestrial wave phenomena in amplitude, in frequency of occurrence, and in other properties.

Another way in which Venus could differ from Earth in its solar-wind interaction is that the solar wind can reach Venus' neutral atmosphere. As a result, processes that are thought to be important for comets could occur at Venus. In comets, the neutral atmosphere becomes ionized either by solar ultraviolet radiation or by the exchange of an electron between a heavy neutral cometary ion and a light solar-wind ion, usually a proton. This process adds mass ("mass loads") to the solar wind and slows it down. Since the solar wind has a magnetic field that connects the slowed down solar-wind plasma to the freely flowing plasma far from the comet, a long magnetic tail is formed behind a comet, joining the slow and fast ionized gas.

Venus' neutral atmosphere is bound to the planet by gravity far in excess of a comet's. While Venus' gravity can hold an atmosphere, the comet's cannot. However, some of the neutral atoms of Venus' atmosphere do reach the solar wind and can be lost through photoionization and charge-exchange processes. There is both direct and indirect evidence that Venus acts very much like a comet in its interaction with the solar wind. First, Venus' bow shock is slightly weaker than Earth's shock. This would occur if charge exchange behind the shock led to absorption by Venus' atmosphere. Second, Venus has a comet-like magnetic tail. This would occur if the magnetic field, draped across the dayside of the planet, became mass-loaded. Third, direct observations have been made of ions from Venus flowing beside and behind the planet with a velocity almost equal to that of the solar wind.

The location of the bow shock as observed by Pioneer was somewhat surprising. Before the Pioneer mission, a common belief was that any planetary magnetic field of Venus would be too weak to hold off the solar wind. Hence, the size of the bow shock would be determined by the size of the planet itself and would be relatively unchanging. However, Pioneer Venus observed a shock that is 35% larger than the shock observed by the Soviet Veneras 9 and 10 spacecraft. Why should the size of the shock change? Soviet measurements occurred at solar minimum, whereas Pioneer Venus' were initially at solar maximum. Scientists speculated that the change in the solar cycle, in particular in the flux of ultraviolet radiation, caused changes in Venus' upper atmosphere. These altered the rate of processes such as photoionization and hence the solar-wind interaction. Scientists investigated this speculation further during the extended mission of Pioneer Venus when solar activity began to decline. They confirmed that the bow shock distance does change with the solar cycle (Figure 6-44).

An electric field detector was carried to Venus for the first time on Pioneer Orbiter. The instrument measured the electric field associated with oscillations of ions and electrons. It provided evidence for a plasma-wave mechanism that couples the magnetosheath's energy to the ionospheric plasma by whistler waves. It also provided the basis for some interesting and important comparisons among planetary bow shocks.

When scientists compare the plasma emissions at Venus, Earth, Jupiter, and Saturn, they see an evolution in properties. The waves at Saturn are quite unlike those at Venus. The ratio of solar-wind velocity to the pressure-wave velocity, or Mach number, determines the strength of the bow shock. The major change in solar wind with distance is that Mach number increases with distance from the Sun. This provides experimental verification that the processes in the shock change with the shock strength. The electric field detector also provided evidence for lightning on the planet, which confirms similar Soviet observations below the cloud tops. Pioneer Venus was not equipped with instruments to search visually for lightning, yet it detected the electromagnetic waves that lightning created. On almost every low altitude nightside pass of Orbiter, it received signals typical of those generated by lightning discharges.

As periapsis began to rise during Phase II, researchers discovered long ionospheric tail rays that extended more than 3000 km (1870 miles) downstream. This is called the ionotail. The tail rays are thought to be ionospheric hydrogen and oxygen ions accelerated to velocities high enough for them to escape the planet. Their discovery is important to studies of how the water of ancient Venus' oceans might have been scavenged from the atmosphere over geological time.

The plasma analyzer made measurements of conditions in the ionosheath downstream of the planet during Phase III. Researchers found a depletion of energetic ionosheath electrons downstream from the terminator, similar to that in the Mariner 10 data. There are several explanations for this condition. If the depletion is due to atmospheric scattering, there would be electrons traveling along draped magnetic flux tubes threading through Venus' neutral atmosphere. These electrons would lose energy from impact ionization with oxygen. Atmospheric loss could provide a natural process for electrons at energies of about 100 eV to be selectively removed. Energetic electron depletion might alternatively be a strong draping that connects the depletion region magnetically to the weak downstream bow shock. This connection could reduce the electron source strength. It is not clear from the data whether the energetic electron depletions observed by Mariner 10 and Pioneer Venus Orbiter result from depletion by atmospheric scattering or from a reduced source strength.

The Exosphere

The exosphere forms the outermost fringe of the atmosphere. In this region, atoms move in ballistic trajectories and rarely collide with each other. Orbiter's ion mass spectrometer discovered that the number of hydrogen atoms increased steadily through the night, then decreased quickly through the day. The atoms were effectively "trapped" by the very low temperature of the nightside exosphere. Hydrogen atoms were so scarce in the dayside exosphere that oxygen replaced them in dominance. The oxygen atoms are unusual in that they are very hot. They are produced by decomposition of ionized oxygen molecules. This process at lower altitudes is the mechanism by which ultraviolet sunlight heats the atmosphere.

Hydrogen in the exosphere, as identified from Lyman alpha glows, showed two components. At lower altitudes, there was a component of the exosphere at 275 K. However, this component was negligible above 3000 km (1860 miles) and allowed Orbiter to detect a nonthermal component. There is now general agreement that various reactions drawing on the energy of the ionosphere produce this nonthermal component of the exosphere.

In the exosphere, collisions with ambient gases do not slow the quickly moving atoms of oxygen and hydrogen. They rise thousands of miles into space and form the first obstacle to the solar wind approaching Venus. Some of these atoms attain velocities high enough to escape into space.

Hydrogen and oxygen atoms in the exosphere escape by different processes. On the dayside, extreme ultraviolet sunlight ionizes oxygen atoms, then the solar wind carries them away. On the nightside, many more hydrogen atoms have charge-exchange collisions with hot protons. The hot protons capture electrons and become fast moving hydrogen atoms that possess enough velocity to escape Venus' gravity. Cool hydrogen atoms lose an electron and become cool protons. These move too slowly to escape. They remain in the exosphere.

The Intrinsic Magnetic Field

Except for Venus and the Moon, and possibly Mars, every planet visited by spacecraft has a magnetic field that is thought to be internally driven. Some scientists speculated before Pioneer reached Venus that perhaps the planet had an internal magnetic field too weak for previous missions to detect. However, Orbiter probed thoroughly for a field with highly sensitive instruments and found none.

During Phase III, the magnetometer made repeated measurements from midnight to about 4:30 a.m. at altitudes below 185 km (115 miles). Data from this phase were from the bottom of the nightside ionosphere. They were important because researchers wanted to obtain information about the possibility of an internal planetary field contributing to the observed magnetic field. In this region, explored by Orbiter during Phase III, it was found that the magnetic field was generally stronger at comparable altitudes than it was at times of high solar activity. Also, at solar minimum, this increase, coupled with a decrease in electron density, caused the ratio of the magnetic pressure to the thermal pressure to approach unity at this altitude. At solar maximum, however, the ratio was much less than unity.

Researchers observed another major difference between conditions at the start of the mission and at the entry phase. From 160 to 200 km (100 to 125 miles), the magnetic field pressure exceeded that of the ionospheric plasma. However, below 150 km (93 miles), the induced field was weaker, diminishing sharply. This permitted researchers to search for an intrinsic field during Phase III. Pioneer measurements clearly show that Venus' intrinsic magnetic field is extremely weak. The data showed no evidence of a planetary field. Venus has the lowest magnetic moment of any planet visited by spacecraft so far. This field is so weak that it can play no role in the interaction of Venus with the solar wind.

One of the principal unsolved problems of geophysics is the nature of the source of the terrestrial dynamo that generates the magnetic fields of Earth and the other planets. Scientists hoped that a measurement of a magnetic field of Venus, a planet which appears in many respects to be Earth's twin, would help clarify the effect of spin rate on the dynamo process. Venus spins on its axis much more slowly than does Earth, once in 243 Earth days. Dynamo theories predict that a planetary dynamo, such as that generating Earth's field, should depend on spin rate. If Venus' dynamo were identical to Earth's, but weaker in proportion to the spin rate, the planet would have a magnetic field that could easily be detected. However, it does not, so other explanations are needed.

A planetary magnetic dynamo requires a highly electrical conducting liquid core. The absence of a conducting core may explain why Earth's satellite, the Moon, does not have a magnetic field. Unfortunately, it does not explain the absence of a field of Venus. Under the temperatures and pressures in the core of Venus, there should be a highly conducting fluid. However, the composition and electrical conductivity of the fluid may be different from those of Earth. Although Venus appears to be Earth's twin in size, it may not be a twin in chemical composition since it formed at a different place in the solar nebula and probably at a different temperature.

Another possible difference is the weakness of any energy source which would drive Venus' dynamo. Present thinking about our planet's dynamo is that a solid inner core is growing at the center of the Earth. As this core grows, it releases its latent heat of fusion into the surrounding fluid. Scientists calculate that this energy source is stronger than the once popular radioactive heating mechanism. Pressure and temperature at the core of Venus are only slightly less than at Earth's core. However, this difference may be sufficient to prevent solidification of Venus' inner core. This could be true even if the internal composition of the two planets are the same.

Lack of a magnetic dynamo on Venus today has implications for Earth. Suppose the reason for lack of a dynamo is that Venus has a totally liquid core. Then Earth may not have had a magnetic dynamo and an intrinsic magnetic field until its solid inner core began to form. Today, Earth is protected by a strong magnetic field that isolates its atmosphere from the solar wind. As a result, there is very little loss of Earth's atmosphere to the solar wind. If the Earth did not always have a strong magnetic field, there would have been times when it was not protected. Known magnetic reversals of Earth's field also would have led to periods when our planet was not protected from the solar wind. These effects would have to be considered in determining how our planet evolved so that life could originate and develop on it.

If we gain a better understanding of the terrestrial dynamo process, scientists may be able to infer some of the internal properties of the planet. On the other hand, if some of these internal properties become known through other means, they may be able to use the absence of a magnetic field of Venus to help understand the dynamo process. In short, all that can be unambiguously stated is that Venus at present does not have a magnetic dynamo. The nature of the source of planetary magnetic fields still remains one of the major unsolved problems of geophysics.

Interplanetary Magnetic Field

The interplanetary magnetic field originates from the solar dynamo, which generates a magnetic field on the Sun. The field is borne outward by the solar wind and varies with conditions on the solar surface that produce the solar wind. Researchers combined magnetometer data from Orbiter at 0.7 AU (astronomical unit) from the Sun with similar data from the IMP-8 spacecraft at 1.0 AU. They compared the long-term behavior of the interplanetary magnetic field over a solar cycle at these two locations. They discovered that at Venus there was an enhancement of the typical field magnitude during declining solar activity compared with the field at maximum or minimum solar activity. This is different from fields in the vicinity of Earth. Here, we observe high fields most frequently during solar maximum conditions. This suggests that the intensity of fields from transient solar disturbances, such as coronal mass ejections, depends upon

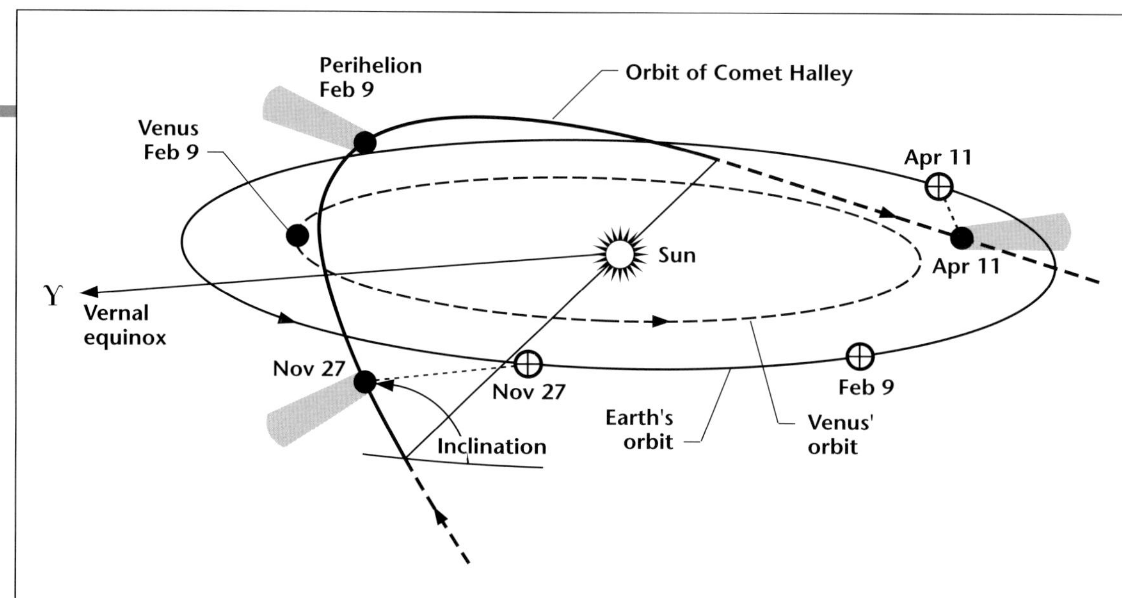

Figure 6-46. The orbits of Earth, Venus, and Comet Halley. The comet passed perihelion above Venus in February 1986. Orbiter then had a unique opportunity to observe the cometary activity at the important period of perihelion passage when the comet was closest to the Sun and most active.

position within the heliosphere. However, these results apply only to the solar cycle that Orbiter observed. We need more data to extend the results into a general rule of interplanetary field intensity. Scientists do not yet have a clear explanation of why the strong transients they detected at Venus peaked in the declining phase of solar activity.

The observations from Pioneer were significant because data on the strength, orientation, and variability of the interplanetary magnetic field are required for studies of how the solar wind interacts with planets and comets. The magnitude of the interplanetary field affects the structure and shape of bow shocks at the planets. Orientation of the magnetic field determines the efficiency of solar wind coupling with strongly magnetized planets and the mode of ion pickup from the ionospheres of weakly magnetized planets and from comets.

Comets Observed by Orbiter

Comet Halley passed within 40 million km (25 million miles) of Venus only 5 days before the comet's perihelion on February 9, 1986. At perihelion, the comet was 87.9 million km (54.6 million miles) from the Sun. Orbiter, at the time, was close to 40.2 million km (25 million miles) from the comet (Figure 6-46).

The Science Steering Group agreed to forego normal Orbiter observations of Venus for 70 days from late December 1985 to early March 1986. They devoted the spacecraft's resources instead to observations of the comet with the ultraviolet spectrometer. Each day, mission controllers maneuvered the spacecraft so that the spectrometer could observe the region near the comet's nucleus. At the same time, the solar panels had to gather sunlight and the antenna had to point toward Earth. More than 40 maneuvers, though complicated, were highly successful. The only losses of data were during superior conjunction in January 1986. At this time, Venus and Orbiter were on the far side of the Sun from Earth. A solar flare also interrupted radio communications on February 3.

Researchers obtained data in near real time for ultraviolet emissions from hydrogen, oxygen, carbon, and hydroxyl radicals in the comet's coma. From these data, they calculated the production rates of water and carbon-bearing ices from the nucleus. The production rate of water rose from 10 tonnes (approximately 1.1 U.S. tons) per second at 1 AU inbound to 50 tonnes per second shortly after perihelion. Then it fell slowly to 40 tonnes per second at the time when the Soviet Vega 1 spacecraft

Figure 6-47. Orbiter obtained valuable data about the release rate of materials from the comet's nucleus during its perihelion passage.

encountered the comet on March 6. (See Chapter 7 for results of Vega's encounter.)

After perihelion, the water production rate varied with a complex 7.4-day periodicity. These results provided a unique description of the comet's behavior during the otherwise poorly observed perihelion passage. Other spacecraft could observe the comet on its inbound and outbound paths only. The results from Orbiter coupled with those from other sources (Interplanetary Ultraviolet Explorer (IUE)) showed that Halley lost about 270 million tonnes of water during its perihelion passage. If the comet's density is 0.3, the loss of water would amount to about 10 meters of material from active areas of the nucleus (Figure 6-47).

From February 2 to 6, 1986, a special series of operations allowed Orbiter's ultraviolet spectrometer to acquire a spin-scan image of Halley's entire hydrogen coma (Figure 6-48). At that time, the coma was about 25 million km (15 million miles) across. The image clearly showed the effects of solar radiation pressure on the trajectory of cometary hydrogen. It also showed the signature of the different hydrogen atom velocities associated with the two main production processes: photodissociation of water molecules and of hydroxyl radicals.

In addition to Comet Halley, Orbiter observed six other comets. Among these, it observed Comet Encke at 0.58 AU outbound. Investigators deduced Encke's water production rates from measurements of atomic hydrogen. The comet

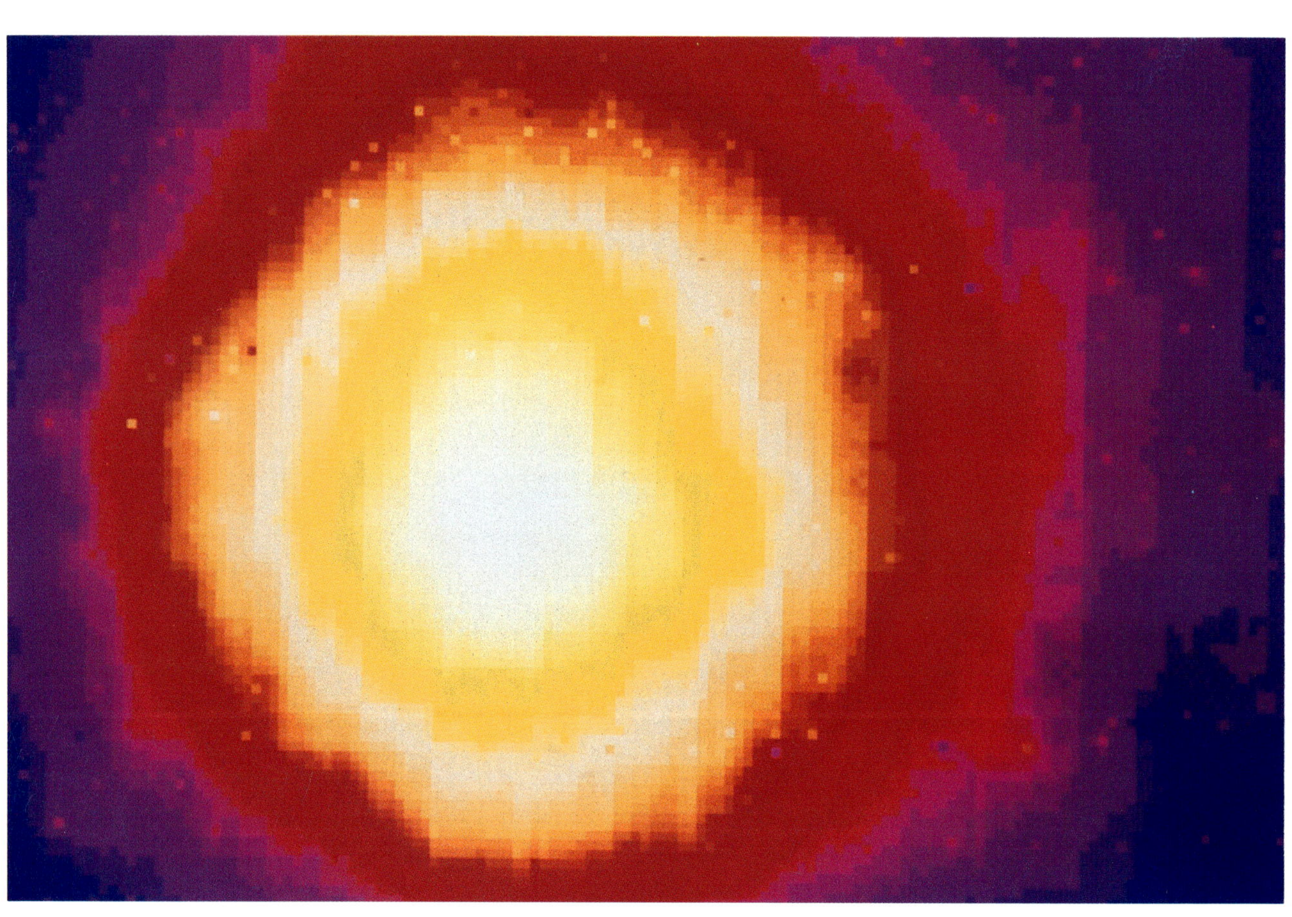

Figure 6-48. The spacecraft imaged the coma and hydrogen halo of the comet. Scientists determined the comet's gas composition, rate of water loss from the nucleus, and ratio of dust to gas in the coma and in the nucleus.

was losing water faster than anyone expected. It appears from the data that water ice and dust are distributed unevenly in the nucleus' cometary material.

When combined with data from IUE, the Orbiter data demonstrated a profound and unexpected difference between the comets' visual and ultraviolet light curves. Activities of other comets observed from Orbiter (Giacobini-Zinner in September 1985, Wilson in March-April 1987, Nishikawa-Takamizawa-Tago in April 1987, McNaught in November 1987, and Machholz in September 1988) fell within a factor of 2 of those scientists expected. Comparison of carbon/hydrogen ratios for Nishikawa-Takamizawa-Tago with those for Halley and Wilson led to a prediction that Nishikawa-Takamizawa-Tago, like Halley, but unlike Wilson, is a periodic comet. Scientists subsequently confirmed this prediction when they calculated the comet's orbit more precisely as an ellipse.

Observations of Giacobini-Zinner from Pioneer Orbiter coincided with the passage of the International Cometary Explorer (ICE) spacecraft through the comet's tail. The comet had just passed perihelion and was about 160 million km (100 million miles) from the spacecraft. Also, the Pioneer observations showed that Giacobini-Zinner is much more active than Encke, but less active than Halley. Lyman alpha ultraviolet emissions from the

hydrogen corona of Giacobini-Zinner were detected on either side of the nucleus as far as 5 million km (3 million miles) from it.

Oceans on Venus?

There have long been speculations that early in its history Venus had a temperate climate and possessed oceans like Earth's. These oceans vaporized as Venus grew hotter. A runaway greenhouse effect some three billion years ago resulted as a cool early Sun increased its luminosity. The oceans evaporated, and solar ultraviolet radiation split the water molecules into oxygen and hydrogen. The lightweight hydrogen atoms easily attained escape velocity from the planet and sped off into space. The discovery by Pioneer that heavy hydrogen (deuterium) is 150 times more plentiful on Venus than on Earth has been taken as evidence that Venus once had 150 times as much water in its atmosphere as today. The heavier deuterium could not reach escape velocity as readily as ordinary hydrogen. This suggests there was enough water on Venus to cover its surface to a depth of several feet.

When Orbiter made its final descent into Venus' atmosphere, it found evidence of 3.5 times as much water as that from the hydrogen/ deuterium ratio. Investigators discovered an unexpected escape mechanism capable of accelerating both hydrogen and deuterium from the planet. A lot more hydrogen must have escaped than previously thought. In turn, this means that there must have been more water on early Venus. Other theorists suggested that conditions on an early Venus might have developed an almost explosive pouring of hydrogen into space. That process would have carried along many deuterium atoms, too. If such a process did occur, deep oceans like those on Earth could have been lost to Venus in only a few hundred million years.

There are other reasons cited for why Venus should have possessed early oceans. All the terrestrial planets are thought to have formed from a mix of planetesimals moving around the Sun in fairly eccentric orbits. As they grew from these planetesimals, all the planets would have received similar proportions of volatiles. Also, the planets should have received similar amounts of volatiles from cometary impacts. Since Venus has about the same abundances of at least two other volatiles, nitrogen and carbon, it should have had the same abundance of water.

However tantalizing speculative theories may be that Venus once had terrestrial-type oceans, we need much more intensive analysis of available data and new missions to Venus to resolve the uncertainty.

Summary of Major Results from Pioneer Venus

The following text highlights Pioneer's findings about Venus or confirms earlier observations. During the mission, Pioneer scientists

- Obtained radar altimetry for nearly all the surface of the planet and many radar images; discovered volcanic and tectonic features such as rift valleys, mountains, continents, and volcanoes. Found that there is a unimodal distribution of topography (quite unlike the bimodal distribution on Earth) and a dearth of elevated regions of continental size. Confirmed the existence of great troughs (rift valleys); however, researchers found no evidence for continuous ridge systems that are typical of the terrestrial plate tectonics system.

- Obtained measurements of the gravity field. When combined with radar altimetry results, this measurement showed that the interior behavior of Venus is more like that of Earth than Mars or the Moon. However, there is a great difference between Venus and Earth. On Venus, there is a strong positive correlation of gravity with topography at all wavelengths.

- Determined the structure of the clouds globally and vertically—their layers, distribution of different sized particles, composition, and optical properties—confirming results from earlier Soviet probes.

- Made refined measurements of composition and abundances of major, minor, and noble gas species in the lower, mixed atmosphere and in the upper, diffusively separated atmosphere.

- Discovered much structure in the polar regions of the atmosphere, thereby clarifying our understanding of the circulation pattern in those regions.

- Discovered that sulfur dioxide is an important absorber of ultraviolet radiation at wavelengths below 3200 angstroms, but that another absorber must be present to account for absorption at longer wavelengths.

- Detected radio signals that some researchers believe originate from lightning discharges in the clouds of Venus, thereby confirming some observations the Venera probes made.

- Obtained much new data about atmospheric state properties (temperature, pressure, density) globally and vertically from the surface through the clouds and into the upper atmosphere.

- Obtained measurements of vertical profiles of wind velocities at four probe locations and global wind measurements at the cloud tops.

- Determined the sinks for solar radiation and the sources and sinks of infrared radiation in the lower atmosphere and clouds at four locations characterizing daytime, nighttime, low latitude, and high latitude conditions.

- Discovered that the high atmosphere well above the cloud tops is much colder at night than in the daytime.

- Combined these observations into a conceptual general meteorological model for comparative meteorological studies.

- Mapped the airglow on the dark side of Venus.

- Provided strong support for a greenhouse effect that, coupled with global dynamics, explains the high surface temperature.

- Determined the global properties of the ionosphere—its ion composition, temperature, flows, electron concentration and temperature, modification of ionospheric properties by input from the solar wind, and the production and maintenance of a nightside ionosphere.

- Determined the nature of the solar-wind interaction with the planet. This included temporal and spatial studies of the location of the bow shock and ionopause and of particle and energy input to the atmosphere over a complete solar cycle.

- Confirmed that Venus has little if any intrinsic magnetic field, and set a very low upper limit on a magnetic moment of the planet.

- Determined how the ionosphere varies over the 11-year cycle of solar activity.

- Determined how the solar activity cycle affects the atmosphere in general and the interaction of the planet with the solar wind.

- Made important discoveries about the rate of evolution of materials from cometary nuclei.

- Worked in cooperation with other spacecraft missions to map the positions of over 30 gamma-ray burst events.

CHAPTER 7

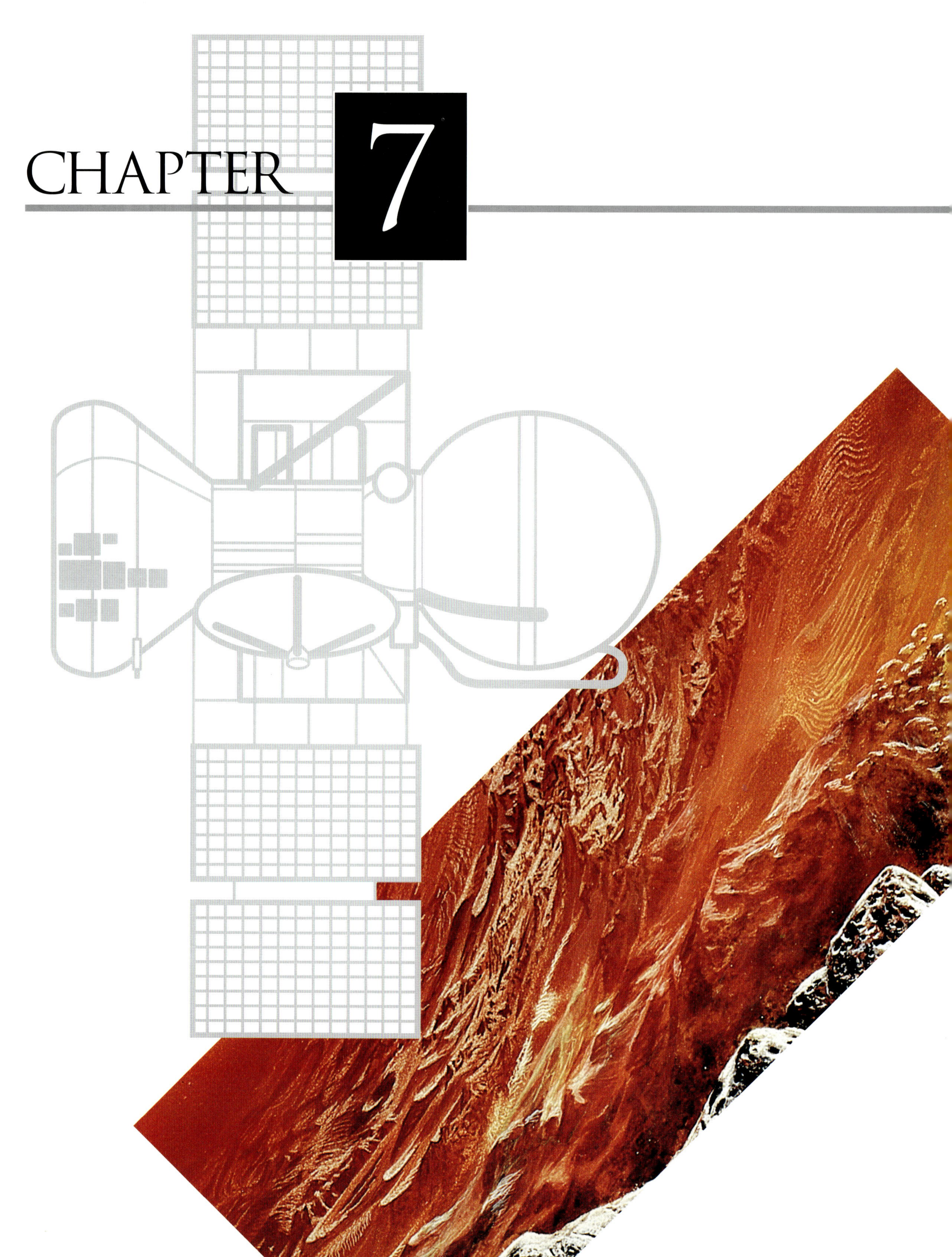

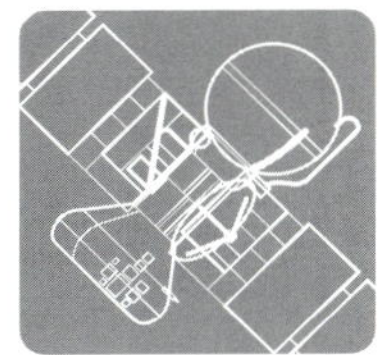

SOVIET STUDY RESULTS

R. Z. Sagdeev, V. I. Moroz, and T. Breus

Some years before NASA published the first edition of this book in 1983, Soviet space scientists graciously contributed this chapter. In it, they detailed their Venera missions 4 through 12 (1967-1978). They also mentioned the "upcoming" (1984) Vega project at the end of their text. To bring events up to date, our American authors have returned and added their own text (1994) at the chapter's end. They describe the flights of the Soviet Veneras 13 through 16 (1981-1982). They also give results of Vegas 1 and 2, including the successful Comet Halley flyby in 1986.

Venus, the planet nearest Earth, has always been of interest to the Soviet Space Program—it has sent the largest number of unmanned space probes there. The planet's many features that are similar to our own Earth has prompted this keen interest in Venus. The two planets' mass and geometry are indeed similar, and they receive roughly equal energy from the Sun.

Some 20 years ago, scientists thought that Earth's "sister planet" was its exact replica. They envisioned it with a slightly warmer surface, hydrosphere, and, possibly, biosphere. Yet, as the first studies revealed, there are drastic differences in climate. The temperature on the Venusian surface averages 735 K (about 462°C, or 864°F). However, the average temperature of Earth's surface is 15°C (59°F). Furthermore, Venus' entire surface, regardless of latitude or time of day, seems to be uniformly heated. This situation is distinctly different from conditions on Earth.

All these unique features of the Venusian atmosphere, however, have been established only in the era of space exploration.

Soviet Spacecraft

In the second half of the 1950s, radio telescopes yielded data about the high temperature of Venus' surface. So unexpected was this information, not all scientists believed it. To settle the issue, the first Soviet interplanetary automatic stations to Venus had "surface phase state" sensors onboard. These sensors could determine whether the vehicle had landed on a solid surface or if ocean waves were rocking it.

On October 18, 1967, Venera 4, the first spacecraft to descend into Venus' atmosphere with a parachute, had no such sensor onboard. However, for this mission, the spacecraft had protection against the extremely high temperatures it encountered. This protection allowed it to take actual measurements of the conditions it faced. Subsequent Venera spacecraft—Venera 5/Venera 6 (1969) and Venera 7/Venera 8 (1972)—added to the information (see Table 7-1). These probes yielded detailed information about variations in temperature, pressure, and density of the Venusian atmosphere with altitude. Venera 7 and Venera 8 made soft landings and transmitted signals directly from the planet's hot surface. Instruments aboard Venera 8 took the first scattered solar radiation measurements. They also furnished information about soil composition, including uranium, potassium, and thorium.

Table 7-1. Soviet Space Vehicles That Studied Venus, 1967 to 1978

Space vehicle		Date		Landing site			Measurements
Name	Type	Launch	Approach	Latitude	Longitude	Solar angle, deg	
Venera 4	Descent module + flyby vehicle	6/12/67	10/18/67	19	38	–20[a]	Descent module: Temperature, pressure, density, wind velocity; CO_2, N_2, H_2O content at altitudes of 55 to 25 km; ion number density in the ionosphere, magnetic field Flyby vehicle: 11_{α} — and 0I/1300Å — radiation in upper atmosphere; ion flux in region of solar-wind flow around planet; magnetic field
Venera 5	Same as above	1/5/69	5/16/69	–3	18	–27	Temperature, pressure, wind velocity, CO_2, N_2, H_2O content at altitudes of 55 to 20 km
Venera 6	Same as above	1/10/69	5/17/69	–5	23	–25	Same plasma measurements as on Venera 4
Venera 7	Descent module (soft landing)	7/17/70	12/15/70	–5	351	–27	Temperature
Venera 8	Same as above	3/26/72	7/22/72	–10	335	+5	Temperature, pressure, solar scattered radiation (from 55 km to surface), wind velocity
Venera 9	Descent module (soft landing) + artificial satellite	6/8/75	10/22/75	32	291	+54	Descent module: Temperature, pressure, wind velocity; CO_2, N_2, H_2O content, solar scattered radiation (several filters), clouds (nephelometer), panoramic survey of surfaces Satellite: TV survey of clouds, IR radiometry, spectroscopy of the day- and nightside; photopolarimetry; energy spectra of ions and electrons, electron and ion number densities and temperatures, magnetic field in region of solar-wind interaction with planet; radio occultations
Venera 10	Same as above	6/14/75	10/25/75	16	291	+62	Same as above
Venera 11	Descent module (soft landing) + flyby vehicle	9/9/78	12/25/78	–14	299	+73	Descent module: Temperature, pressure, wind velocity, composition (mass spectrometer); solar scattered radiation spectrum; nephelometer, thunderstorm activity Flyby vehicle: Upper-atmosphere UV spectrum
Venera 12	Same as above	9/12/78	12/21/78	–7	294	+70	Same as above; gas chromatograph and measurements of particle-composition of cloud layer

[a]Minus sign denotes night landing (the Sun below the horizon). First generation vehicles landed at night (except Venera 8, which landed near the terminator). It was necessary since information was transmitted directly to Earth. Since Venera 9 information was relayed via the artificial satellite from the lander and the landing was made during the day, this was widely used to study solar radiation propagation in the atmosphere (to check the greenhouse hypothesis).

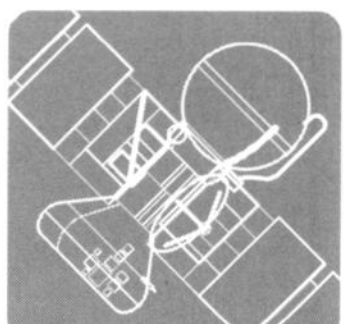

Veneras 4 and 6 also obtained unexpected results in plasma and magnetic measurements. They discovered a shock wave in the solar wind near Venus like the one near Earth. The shock front of Venus, however, was much closer to its surface. Before the spaceflight to Venus, scientists hypothesized that the number density of charged particles in Venus' ionosphere could exceed by three orders of magnitude the number density of charged particles in the main peak in the terrestrial ionosphere. Ion number densities that Venera 4 measured during its descent on Venus' nightside did not confirm that suggestion, nor did Mariner 5's radio-occultation observations about electron number densities on the ionosphere's nightside and dayside. In Venus' ionosphere, the maximum number density of charged particles was about the same as on Earth. Mariner 5 observed a distinct upper boundary of the dayside ionosphere at an altitude of 500 km (310 miles). Within the boundary, the electron number density decreased by two orders of magnitude within an altitude range of only 50 to 100 km (31 to 62 miles). The boundary was similar to the plasmapause—the upper bound of Earth's thermal plasma envelope. Because of this similarity, scientists gave the name *ionopause* to the Venus phenomenon. However, Earth's plasmapause is much farther from the planet's surface, roughly 20,000 km (12,428 miles).

Although large-scale features typical of solar-wind flow around both Venus and Earth are similar, the magnetic field Venera 4 first measured near the planet seemed insignificant—only about 10 gamma (10^{-4} gauss) at an altitude of 200 km (124 miles). The surface magnetic field in Earth's equatorial region is about 50,000 gamma. Until recently, it had been thought that Venus' intrinsic magnetic field might play a significant role in forming the pattern of solar-wind flow around the planet, as it does in the case of Earth.

Operating an automatic interplanetary probe in Venus' hot and dense atmosphere was technically difficult. Nevertheless, in the 1960s, a team of scientists designed spacecraft for Venus research. The academician S. P. Korolev and then G. N. Babakin, Corresponding Member, U.S.S.R. Academy of Sciences, headed the team. NASA lauched Pioneer Venus 11 years after Venera 4, almost at the same time as Veneras 11 and 12 were launched.

It often happens in science that the solution to one problem leads to new, more complicated problems. Spaceflights to Venus were no exception. They showed that climatic and atmospheric conditions, so similar to Earth for some physical parameters, are generally quite different from those on Earth. What are the reasons for these differences? Can the climate and composition of Earth's atmosphere experience the same changes in the foreseeable future? If so, what would cause such changes: altered external conditions, environmental pollution, or something else? Such questions prompt many scientists throughout the world to consider exploration of Venus a top-priority task.

Venus can be a natural "cosmic laboratory" for studies in comparative planetology. The value of such research becomes more apparent because it is impossible to realize experiments on such a scale under Earth conditions.

Any planet's atmosphere is a complex system with many interactions and feedbacks. Its composition, for instance, is determined by how and under what conditions the planet formed, and by outgassing processes from its solid body. Other factors include reactions among atmospheric gases, the upper atmosphere's structure (from which light gases escape into the interplanetary space), and so on. The character and rate of many atmospheric processes depend on temperature, which in turn depends on the atmosphere's composition. The latter consideration is most essential for Venus. The gaseous and aerosol composition of the Venusian atmosphere allows some solar radiation to penetrate down to the surface. The opacity of the atmosphere is high, however, for infrared radiation. As a result, the surface temperature remains high. The phenomenon, which we call the greenhouse effect, is much more conspicuous on Venus than on Earth. On Earth, the greenhouse effect adds about 35°C (63°F) to the surface temperature.

A fuller understanding of what is taking place on Venus required sophisticated chemical analyses of the atmosphere and an exact knowledge of the altitudes and spectral regions where solar radiation is absorbed. Scientists also needed to study the nature of the clouds that prevent astronomers from seeing the lower layers of the atmosphere.

After the first-generation Venera probes made plasma and magnetic measurements, scientists were faced with many new problems. With theories and concepts existing at the time, it might have been possible to find solutions to some of the problems. In particular, scientists wanted to explain the weak intrinsic magnetic field near Venus. For example, they could use theories of how magnetic fields originate and maintain themselves near planets on the basis of planetary dynamos. These theories predict that a planet, if it has an intrinsic magnetic field, must rotate rapidly and have a liquid, conducting core. Scientists had used close values of mean densities of terrestrial planets to build similar models of their inner structures. Consequently, planetologists could attribute Venus' absence of an intrinsic magnetic field to its slow rotation (about 243 terrestrial days).

Scientists observed shock waves near both Venus and Earth. But Venus, they knew, had a much weaker intrinsic magnetic field than Earth. What is the obstacle—different from Earth's—that retards the solar wind and forms a shock wave near Venus?

Indeed, a strong intrinsic magnetic field protects Earth, its atmosphere, and ionosphere against the solar wind's direct effect. However, for Venus, the solar wind could interact directly with its atmosphere and ionosphere, causing ionization, compression, and heating of the ionosphere and atmosphere. The solar wind, flowing around the planet's conducting ionosphere, together with the interplanetary magnetic field, could induce electric currents in the ionosphere and thus produce induced magnetic fields. If these induced fields are strong enough, they could brake the solar wind and form an induced magnetosphere, rather than an intrinsic one, near the planet.

All these assumptions rested on the observed similarities and differences in the solar wind's pattern flowing around Venus and Earth, and they had to be verified. Much more complex and accurate measurements were needed.

To conduct more detailed experiments in the deep layers of Venus' atmosphere, interplanetary probes needed heavier and more sophisticated instruments. More importantly, the vast amount of data gathered by the instruments had to be transmitted back to Earth. Accordingly, the first-generation probes, which had not been intended to deal with such problems, were succeeded by Veneras 9 through 12 (see Figure 7-1). Whereas the earlier probes had entered the Venusian atmosphere in their entirety, the new Venera probes separated into an orbiter and a lander some time before landing. Depending on mission profile and ballistics, the orbiter either became

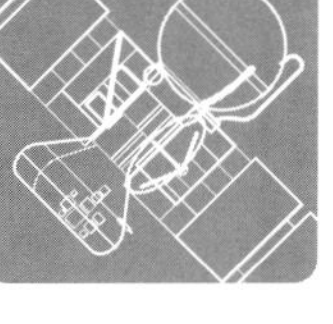

an artificial satellite of Venus (Veneras 9 and 10) or it flew past the planet and entered an orbit around the Sun (Veneras 11 and 12) (Figure 7-2). The orbiters carried instruments to study the planet's radiation at various wavelengths, the interplanetary plasma and magnetic fields, and to conduct astronomical observations.

In 1975, Veneras 9 and 10 splendidly demonstrated the capabilities of a new generation of spacecraft. For the first time, a panoramic view of another planet was transmitted from its surface to Earth (Figure 7-3). A series of investigations looked at the atmosphere's optical properties. They determined the general features of the cloud structure. The clouds are in a layer about 20 km (12 miles) thick, with a lower boundary at an altitude of 50 km (31 miles). Radiation fluxes were measured in several spectral regions and the water vapor content was derived from the intensity of the absorption band. Scientific equipment onboard the orbital vehicles—Venus' first artificial satellites—produced important results.

A series of plasma and magnetic radio-occultation observations (Veneras 9 and 10 orbiters) made it possible to study in detail the solar-wind flow pattern around the planet, and discover a plasma-magnetic tail of the planet. The observations also allowed scientists to investigate the character of the magnetic field and the properties of the dayside and nightside ionosphere, and to identify atmospheric ionization sources in the planet's deep optical umbra.

Analyses of Veneras 9 and 10 experimental data indicated new problems. But expertise in designing sophisticated scientific equipment that could operate under very difficult conditions (enormous decelerations, high temperatures and pressures) solved most of them in the Veneras 11 and 12 probes that reached Venus late in 1978. The construction of a huge, 70-m (230-ft) diameter parabolic reflector at the Deep Space Communication Center also greatly improved data reception from the landers.

Recent scientific results from the new generation of Soviet Venera probes are discussed in the sections that follow. Table 7-1 summarizes launch dates, descent module landing coordinates, and other data.

Chemical Composition of the Venusian Atmosphere

Until 1967, scientists assumed, because of the planet's similarity to Earth, that the main chemical in Venus' atmosphere was nitrogen. Besides nitrogen, scientists expected to find a small amount (1% to 10%) of carbon dioxide, whose absorption bands they had observed as far back as the 1930s. But even simple chemical sensors on the first Venera probes proved the very opposite to be the case. The most abundant gas in the atmosphere is carbon dioxide (96.5% according to estimates), whereas nitrogen makes up just over 3%. At the time, it was impossible to get reliable information about the content of the atmosphere's many small constituents: water vapor, oxygen, carbon monoxide, sulfur compounds, and noble gases. These constituents play a tremendous part in the life of the atmosphere. They absorb solar and thermal radiation (the greenhouse effect), participate in chemical reactions, condense to form cloud layer particles, and also contribute to other processes.

The abundance of noble gases and their isotopes is of particular interest. These isotopes fall into two groups: radiogenic isotopes and primordial isotopes. The radioactive decay of elements formed radiogenic isotopes. Primordial isotopes have survived since the formation of the Solar System's planets some 4.5 billion years ago. From the absolute and relative

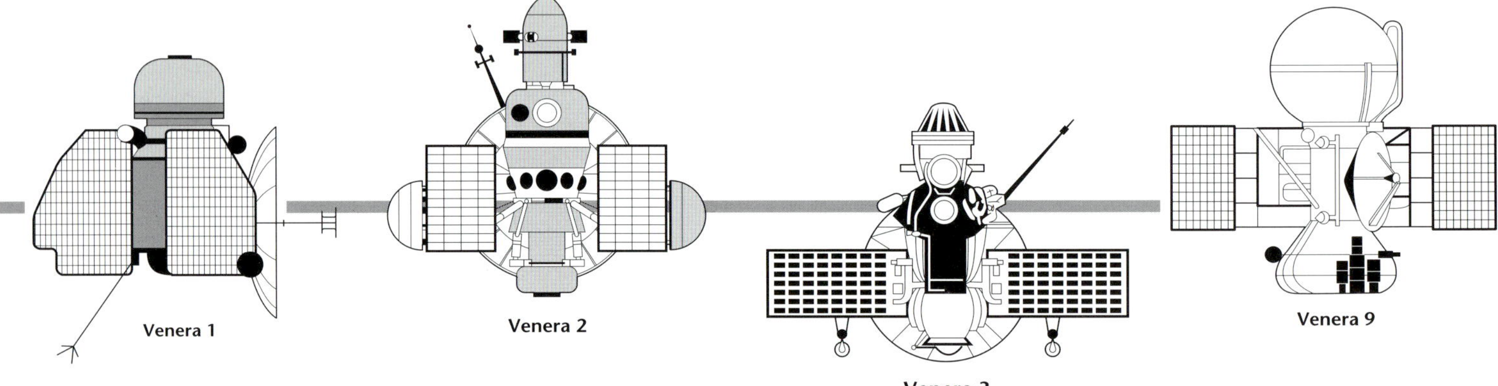

Figure 7-1. In the two decades that the United States sent four spacecraft to Venus, the Soviets attempted 29 missions (15 were successful). Although some of the failures were never officially admitted, U.S. or European sources detected them. These seven illustrations show the evolution of the Soviet spacecraft to explore Venus. It came from many sources and was not a part of the Soviet authors' contribution to this chapter. We have included it to place the U.S. and Soviet missions in perspective.

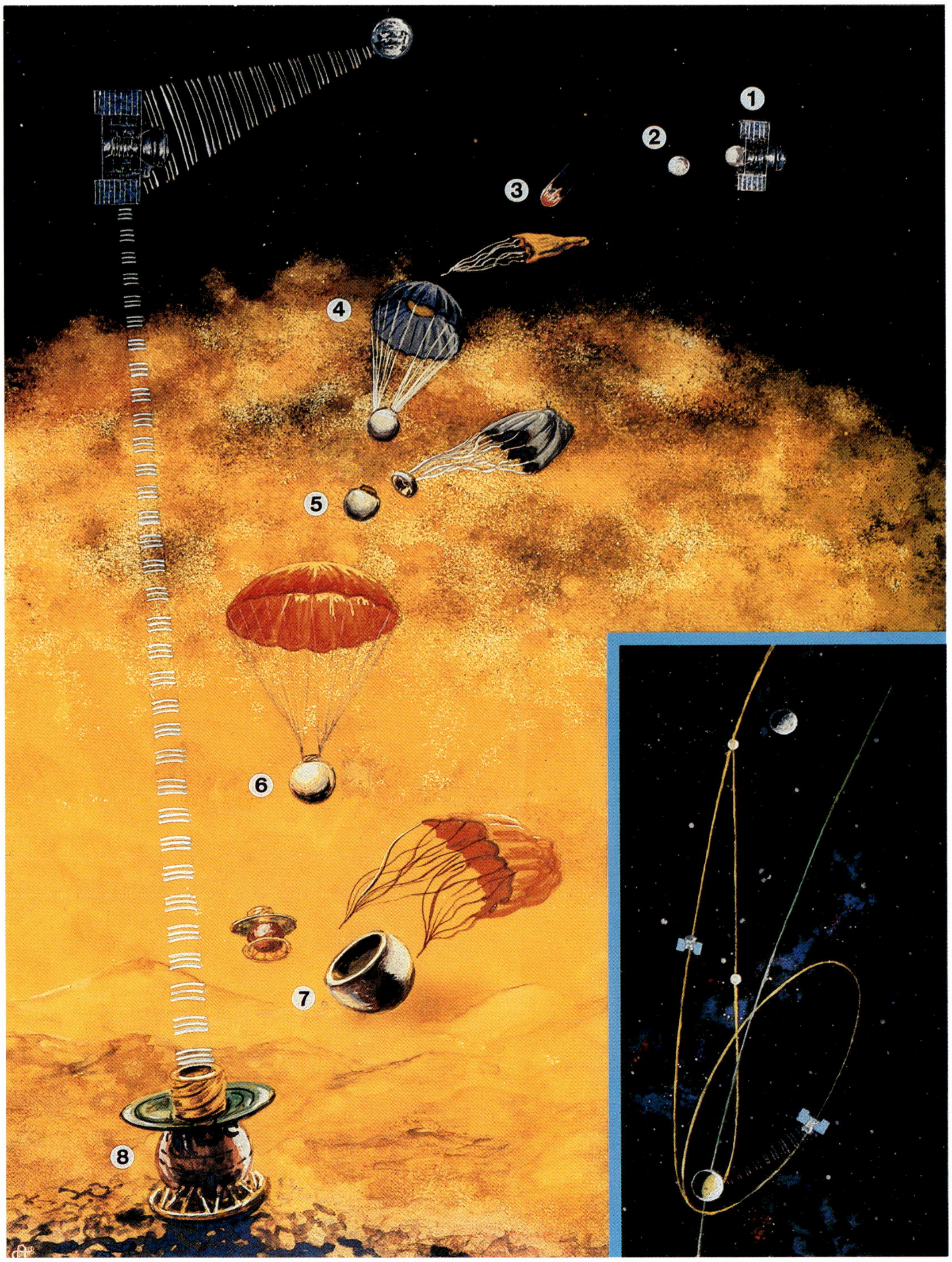

Figure 7-2. Landing scheme of the Soviet second generation automatic spacecraft (Veneras 9, 10, 11, 12).

1) Interplanetary spacecraft on Venusian orbit.

2) Separation of descender and orbiter two days before the landing.

3) Entry into the Venusian atmosphere.

4) Deployment of auxiliary and displacement parachutes.

5) Jettisoning of hatch.

6) Deployment of decelerating parachute at 66 to 62 km (41 to 38.5 miles) and beginning of telemetry data transmission.

7) Jettisoning of lower sector of thermal protection shell and jettisoning of decelerating parachute at about 48 km (30 miles) altitude.

8) Landing and data transmission to Earth via the flyby bus.

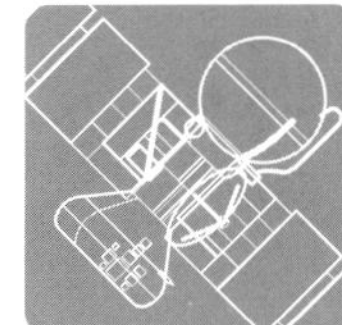

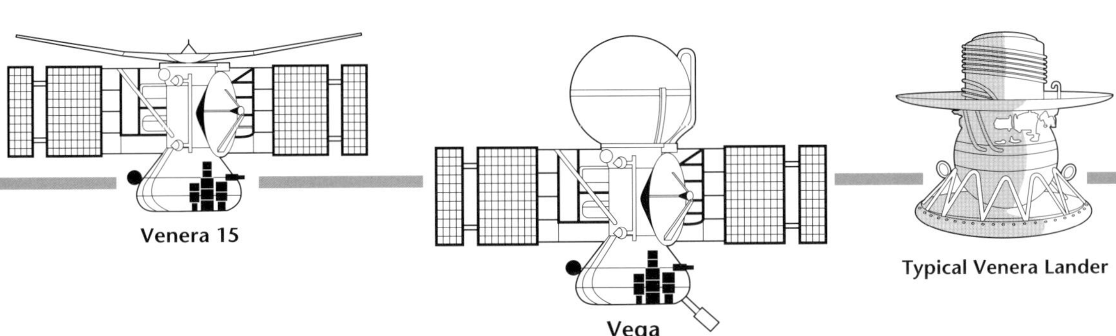

Figure 7-3. Panoramic view of the Venusian surface at 291 east longitude, 32 north latitude, relayed by Venera 9 descent module. Numerous stone blocks with sharp edges are around the spacecraft, a fact testifying to their comparatively young age. On the planet's surface at the landing site, much small-grained substance resembling dust or sand is visible. After the lander's impact, a dust cloud rose that registered on a photometer for a few minutes.

content of primordial isotopes, we can gain some insight into the Solar System's history, in particular, about conditions in which the protoplanetary nebula gave rise to the planets, and about their formation process. Argon isotopes will be discussed as an example.

For fine chemical analysis of atmospheric gases, Soviet investigators used a mass spectrometer, a gas chromatograph, and an optical spectrometer. (The mass spectrometer takes microscopically small gas samples, ionizes them, and sorts them according to their mass with a high frequency electric field.) A group of scientists headed by Vadim Istomin (Institute of Space Research, U.S.S.R. Academy of Sciences) conducted the mass spectrometer experiment. The instruments (Figure 7-4) on both vehicles switched on at an altitude of about 24 km (15 miles) and operated until touchdown. These instruments scanned the mass range from 10 to 105 atomic units in 7 seconds. The gas sampling time was under 5×10^{-3} seconds, and the sampling rate was once every 3 minutes. The instruments took a total of 22 samples and transmitted about 200 mass spectra to Earth. The mass spectrum in Figure 7-5 is an average over 7 of 200 mass spectra.

The mass spectra show several peaks. These peaks correspond to the molecules carbon dioxide and nitrogen, and the atoms carbon-12, carbon-13, oxygen-16, oxygen-18, and nitrogen-14 (from decomposition of carbon dioxide and nitrogen molecules inside the instrument). Also corresponding to peaks are three noble gases: neon, argon, and krypton. Quantitative data appear in Table 7-2. The presence of krypton (about 6.5×10^{-5}%) is noteworthy. Instruments on the Pioneer Venus probe detected no krypton.

In Istomin's experiment, every single record of the mass spectrum shows krypton. Estimates averaged over tens of records showed that the relative abundances of the main krypton isotopes with atomic weights 84, 86, 83, and 82 are comparable to those on Earth. The argon results were extremely surprising. The radiogenic isotope argon-40 and the primordial argon-36 are present in Venus' atmosphere in equal amounts. On Earth, argon-40 is 300 times more abundant than argon-36.

A full explanation of this anomaly is a matter for the future, but M. Izakov (Institute of Space Research) has proposed an elegant hypothesis.

He assumes that Venus derived the greater part of its atmosphere from the protoplanetary nebula. Earth (and Mars) captured relatively little gaseous material from it, and most of their atmospheres were outgassed from their interiors. According to this hypothesis, the meteorite and asteroid accumulation process, which gave rise to all the planets 4.5 billion years ago, proceeded more rapidly for Venus. This happened because the planet is closer to the Sun, and the meteorite bodies were denser there. The capture of gas also was more rapid. Before the new data, scientists believed the atmospheres of the Earth group of planets (Venus, Earth, and Mars) were of secondary origin, formed by degassing from their interiors. The argon-36 anomaly for Venus, however, casts doubt on this.

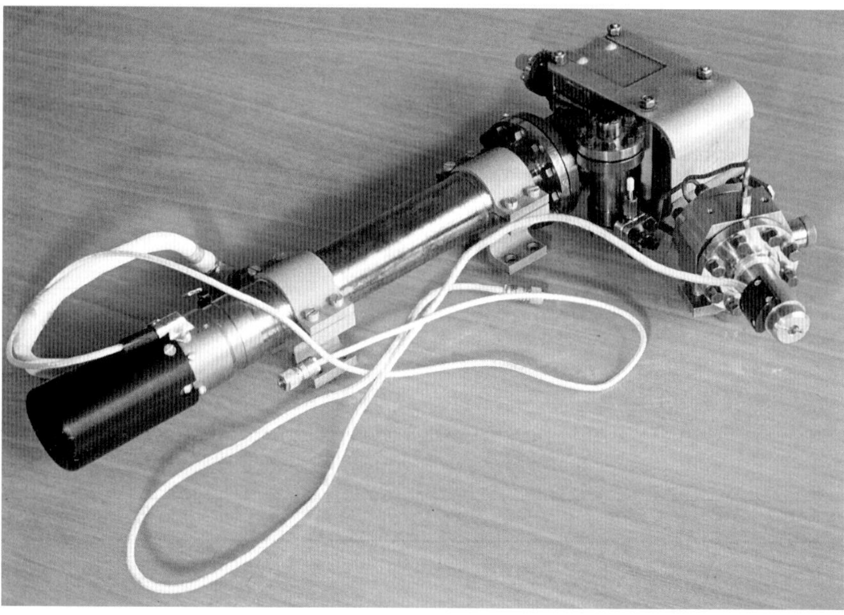

Figure 7-4. A general view of the mass spectrometer carried by the Venera spacecraft.

The atmosphere of Venus was also chemically analyzed by the Sigma gas chromatograph (Figure 7-6). Lev Mukhin of the Institute of Space Research supervised this experiment. (Gas chromatographic analysis is based on different degrees of adsorption of various gases by porous substances. The heart of the gas chromatograph is a column filled with a specific sorbent. The instrument pumps an atmospheric gas sample through the column. There the mixture separates into individual components. Various constituents of the mixture leave the column one by one, and a special ionization detector records them.)

A chromatograph was also installed onboard the Pioneer Venus Large Probe (V. Oyama at Ames Research Center supervised this experiment). Oyama (1979) reported that no carbon monoxide was found, but Venus' atmosphere contained a large amount of molecular oxygen (exceeding the upper limit from the Soviet experiment). Oyama later reported (1980) that he had misidentified the relevant chromatographic peaks, and the missing carbon monoxide was found. Oyama's data revealed another aspect that has not been explained: the presence of relatively large amounts of water vapor—approximately 0.5% at an altitude of 44 km (27 miles) and 0.1% at 24 km (15 miles).

Water absorbs light in several spectral bands, some of which (7200, 8200, and 9500 angstroms) are quite distinct in the spectra from the optical spectrophotometer (Figure 7-7) onboard the Veneras 11 and 12 descenders. (V. Moroz supervised this experiment.) From the bands' intensity, scientists could determine water content in the Venusian atmosphere at different altitudes. This quantity proved very small (2×10^{-3}% near the surface and 2×10^{-2}% at 50 km, or

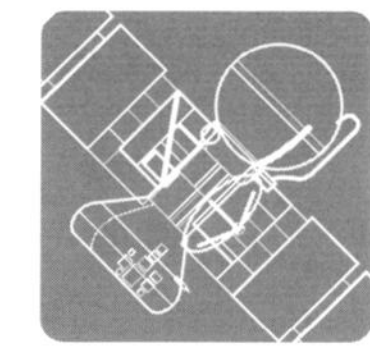

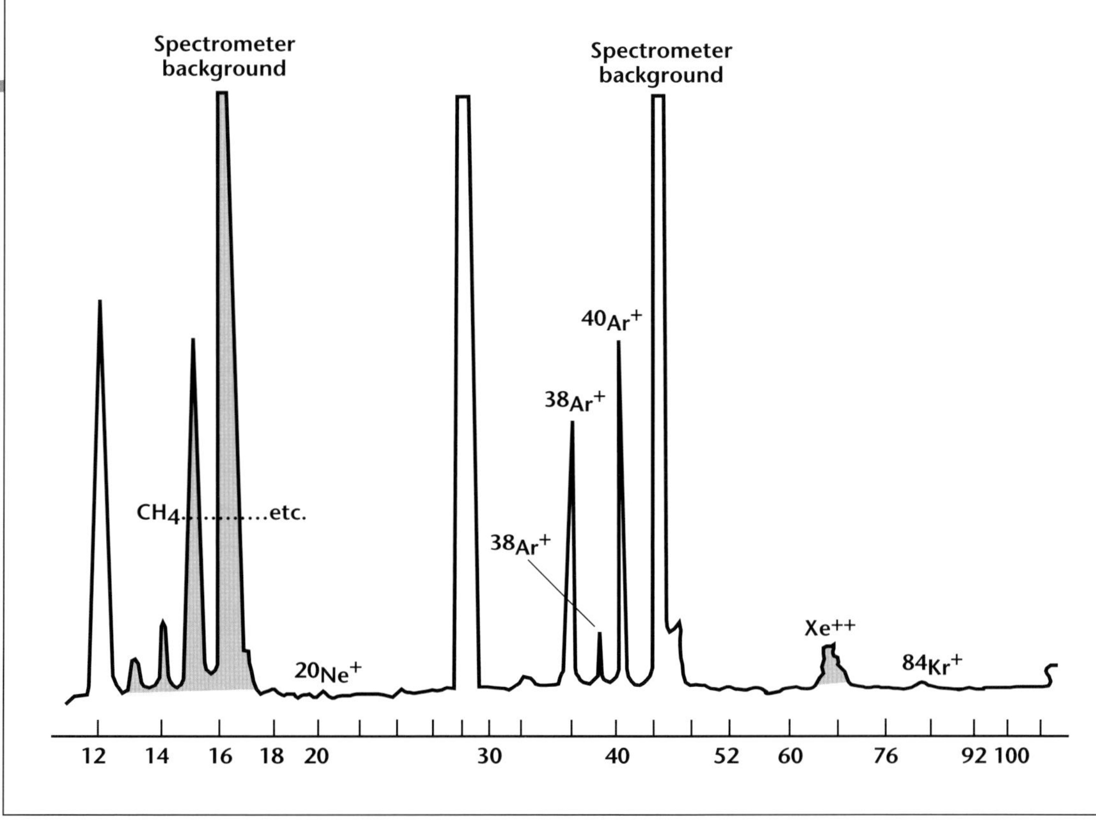

Figure 7-5. Averaged mass spectrum (the sum of seven separate spectra) obtained in the regime of noble gas analysis.

31 miles). Oyama's experiments had yielded a quantity several orders of magnitude greater.

Parallel measurements with a chromatograph and a mass spectrometer provided independent control of the results. The Venera 12 chromatograph did not detect water vapor. From this fact, it follows that, at an altitude below 24 km (15 miles), water vapor content is below 0.01%. The Veneras 11 and 12 mass spectrometers registered a slight excess in the oxygen-16 mass peak as compared with oxygen-18 (if the oxygen-18/oxygen-16 ratio is assumed to be exactly equal to Earth's). Note that oxygen-18 and oxygen-16 are formed in the instrument from carbon dioxide. If this excess is due to the water contribution (the molecular weight of water also is 18), the water vapor abundance correlates reasonably well with the optical measurements.

There is a simple way to verify whether the quantity of water vapor varies from site to site. The height dependence of temperature that Pioneer Venus' Large Probe obtained can be compared with infrared radiation fluxes measured by the same vehicle. This comparison makes it possible to calculate the mean absorption coefficient for thermal planetary radiation (into which the diffuse solar light penetrating deep in the atmosphere is

Figure 7-6. Gas chromatograph carried by the Veneras 11 and 12 spacecraft.

Table 7-2. **Chemical Composition of the Atmospheres of Venus and Earth**

Gas	Content by volume, %	
	Venus	Earth
Carbon dioxide	96.5	3.2×10^{-2}
Nitrogen	3.5	78.1
Water vapor	2×10^{-3}[a]	0.1
Oxygen	10^{-3}	21.0
Carbon monoxide	3×10^{-3}	10^{-4}
Sulfur dioxide	1.5×10^{-2}[b]	10^{-4}
Hydrogen chloride	4×10^{-5}[c]	—
Hydrogen fluoride	5×10^{-7}[c]	—
Methane	10^{-4}	1.8×10^{-4}
Ammonia	2×10^{-4}	—
Sulfur	2×10^{-6}[d]	—
Noble gases:		
Helium	2×10^{-3}	5×10^{-4}
Neon	1.3×10	1.8×10^{-3}
Argon	1.5×10^{-2}	0.9
Krypton	6.5×10^{-5}	1.1×10^{-4}
Mean molecular weight	43.5	28.97

[a]Mixing ratio near surface. At an altitude of 50 km, it is an order of magnitude higher; at 70 km, an order of magnitude less.
[b]Mixing ratio below 20 km; at 70 km, it is four orders of magnitude less.
[c]Mixing ratio above 60 km (only the data for ground-based spectroscopy available).
[d]Gaseous sulfur is meant (molecules S_2, S_3, S_4, S_5, S_6, S_7, and S_8); estimate refers to altitudes below 40 km.

transformed). This coefficient depends strongly on atmospheric water vapor concentration. The calculated water vapor concentration was found to correspond closely with optical measurements.

The total amount of water vapor in Venus' atmosphere appears to be disastrously small. If the planet's entire water vapor (2×10^{-3}%) condensed, it would form a liquid layer no more than 1 cm thick. Obviously, there can be no seas, oceans, and liquid water on Venus' surface—the temperature is too great for that. All of Venus' water is either in its crust or in its atmosphere. This is yet another anomaly, no less odd than the argon-36/argon-40 ratio.

There is nothing extraordinary about the atmosphere's high carbon dioxide concentration. Almost all of Earth's carbon dioxide is bound up in carbonates. On Venus, all carbon dioxide—because of the high temperature and absence of liquid water—is in the atmosphere. Total amounts of carbon dioxide on both planets are roughly equal. But the concentration of water on Venus presents a problem. Three explanations are possible: (1) Venus formed with less water; (2) at the early stages of evolution, water vapor dissociated, hydrogen escaped into the interplanetary space, and oxygen vanished through chemical reactions; and (3) water is bound up in minerals (where there are rocks that retain water very well at high temperatures).

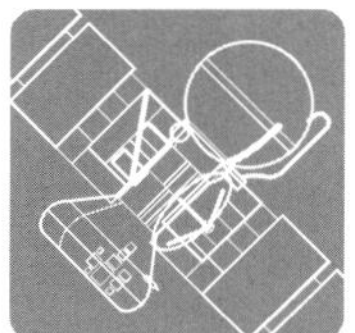

Solar Radiation and Clouds in Venus' Atmosphere

Both the Veneras 11 and 12 landers carried spectrophotometers. From an altitude of 65 km (40 miles) until touchdown on Venus, they registered, for the first time, the daylight sky spectrum and the angular distribution of brightness at 10-second intervals. These measurements showed that a large amount of solar radiation reaches the planet's surface. Significantly, this is scattered rather than direct sunlight. Since the cloud cover at 60 to 70 km (37 to 43 miles) scatters solar radiation, an observer could not see the Sun from Venus' surface nor from an altitude of 55 km (34 miles). In terms of energy, it is unimportant what sort of radiation penetrates Venus' atmosphere—direct or scattered. An evaluation of solar energy reaching the surface (3%) and Venus' thermal radiation confirmed a pronounced greenhouse effect. This effect results in high temperatures in the atmosphere's deep layers and at the Venusian surface. The observation confirms the hypothesis that Carl Sagan put forward as far back as 1962.

According to Veneras 11 and 12 data, the energy distribution in the scattered sunlight spectrum changes as the probe penetrates deeper into the atmosphere. Just as on Earth, the effect results from two types of scattering. The first is aerosol scattering of light by cloud particles. The second is Rayleigh scattering by carbon dioxide and nitrogen molecules. The probes also detected light absorption in ultraviolet, which probably belongs to gaseous sulfur molecules.

There are several layers of clouds in Venus' atmosphere at altitudes from 50 to 70 km (31 to 43 miles). Their boundaries are distinct in the curves showing the decrease in scattered sunlight intensity with the probe's descent (Figure 7-8).

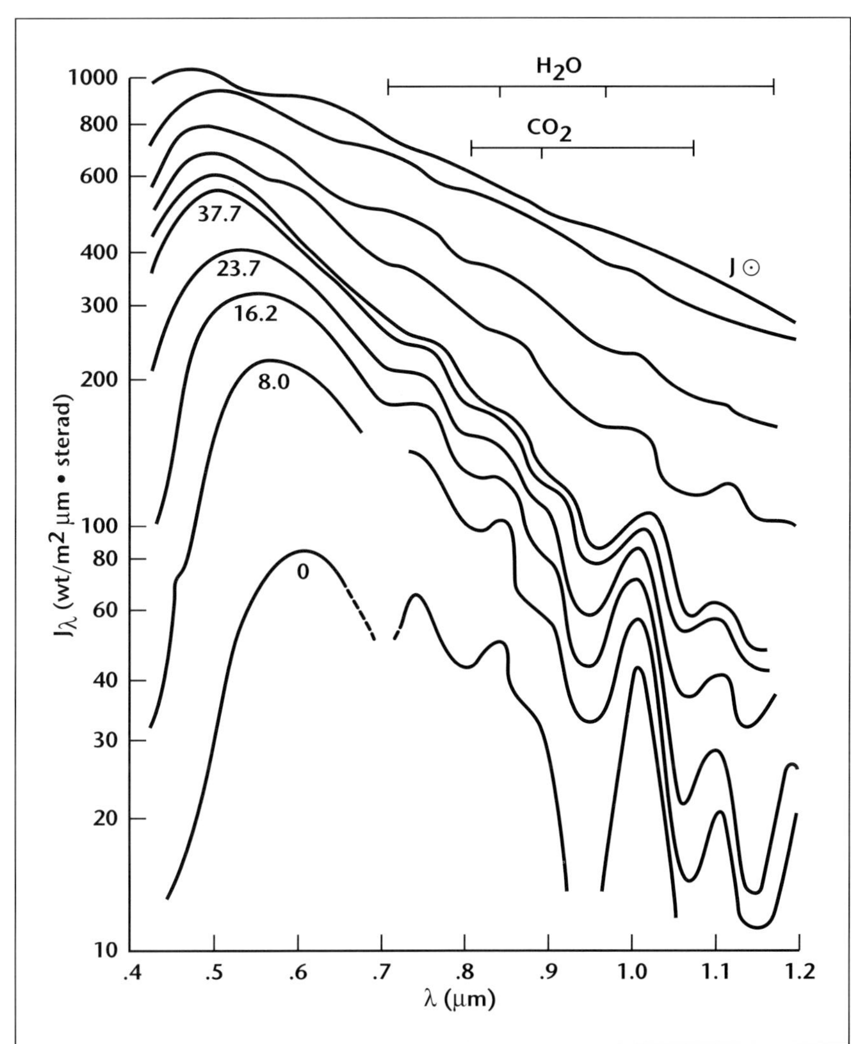

Ground-based observations fixed the approximate position of the cloud cover's upper boundary. Veneras 9 and 10 nephelometers and photometers, however, first observed the lower boundary.

Veneras 9 and 10 nephelometer experiments (M. Marov, Institute of Applied Mathematics, U.S.S.R. Academy of Sciences) made it possible not only to determine the cloud cover's lower boundary, but also to estimate cloud particle concentration, size, and the atmosphere's refractive index. To a limited extent, the

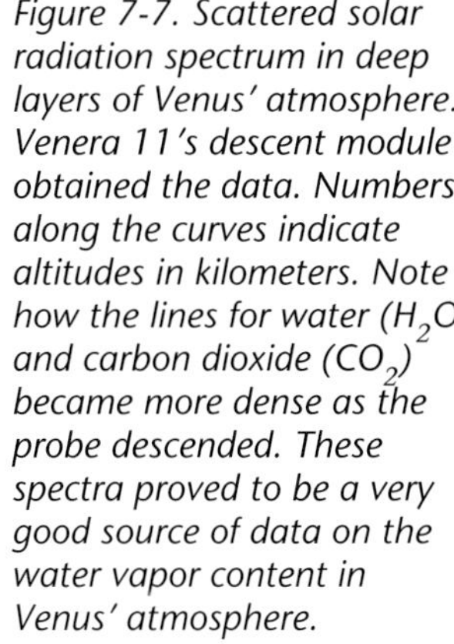

Figure 7-7. Scattered solar radiation spectrum in deep layers of Venus' atmosphere. Venera 11's descent module obtained the data. Numbers along the curves indicate altitudes in kilometers. Note how the lines for water (H_2O) and carbon dioxide (CO_2) became more dense as the probe descended. These spectra proved to be a very good source of data on the water vapor content in Venus' atmosphere.

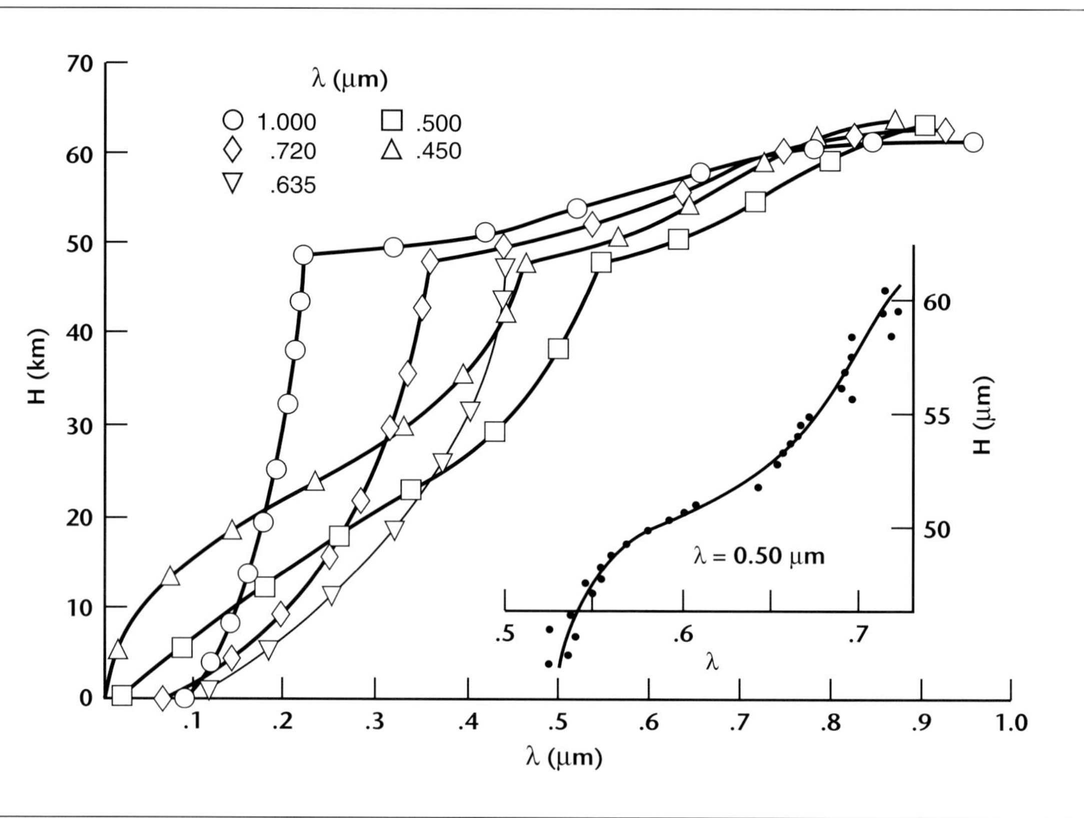

Figure 7-8. Radiation intensity from the zenith as a function of altitude for some wavelengths. Venera 11's descent module obtained the data. Symbols along the curves indicate wavelengths. The sharp change in the steepness of the curves at an altitude slightly less than 50 km (30 miles) is a result of crossing the lower cloud layer boundary.

Venera 11 mission repeated these observations. The Pioneer Venus Large Probe enabled R. Knollenberg and D. Hunten to study in great detail the particle-size distribution.

Venusian clouds are relatively transparent. The meteorological visibility inside the clouds is several kilometers. There are three layers. The upper layer is at 57 to 70 km (35 to 43 miles), the middle at 52 to 57 km (32 to 35 miles), and lower at 49 to 52 km (30 to 32 miles). Particles are of three types: large (7 microns in diameter), medium-sized (2 to 2.5 micron), and small (average diameter 0.4 micron). Only small and medium-sized particles are present in the upper layer. The other two layers have all three particle types. Large particles account for no less than 90% (in terms of mass) of the entire cloud cover.

The composition of Venusian clouds has long baffled scientists. The simpler hypotheses, based on Earth analogies (liquid or frozen water, mineral dust), were discarded when ground-based observations yielded data on the optical properties of the cloud particles. Since there is hydrochloric acid in Venus' atmosphere, scientists put forward yet another hypothesis. They speculated clouds consisted of hydrochloric acid droplets. But a number of considerations made it necessary to abandon this assumption, too. In terms of optical properties, a suitable candidate is sulfuric acid (H_2SO_4) which is present as tiny droplets in Earth's stratospheric clouds. Sulfur compounds reach the atmosphere all the time from Earths' interior, and chemical reactions produce particles that are in Earth's stratospheric clouds. An analogy appears quite reasonable here, since a sulfur compound (SO_2) and pure sulfur in the gaseous state occur on Venus.

Also in terms of refractive index and the infrared absorption coefficient, sulfuric acid is a suitable candidate for the main component of Venusian cloud particles. This, however, does not account for the planet's yellowish color. Scientists have suggested that the clouds

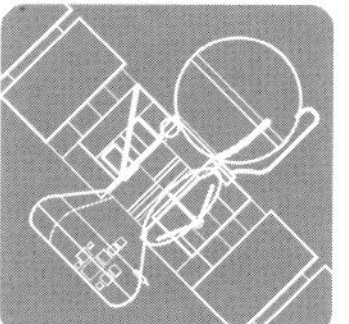

contain larger particles of solid sulfur, in addition to particles of concentrated sulfuric acid. Nephelometric experiments revealed that only small and medium-sized particles could consist of sulfuric acid. The large particles must have a different composition. It was originally assumed they did consist of sulfur.

The Venera 12 mission included, for the first time, an experiment on the direct chemical analysis of cloud particles. (Y. Surkov, Institute of Analytical Chemistry and Geochemistry, U.S.S.R. Academy of Sciences, conducted the experiment.) Cloud layer particles were collected on special filters and analyzed with an x-ray fluorescent spectrometer. The instrument subjected a sample to hard radiation from a radioactive source. As a result, the inner electron shells of atoms (K-shells) were excited, which generated characteristic x-rays whose spectrum was recorded and used to identify the sample's composition. In fact, the composition was determined only at the element level since molecules or any types of bonds could not be determined. At altitudes from about 61 km (38 miles) down to 49 km (30 miles), the most abundant element among cloud-cover particles is chlorine. Either sulfur is not present at all or there is only about 1/20 as much sulfur as chlorine. Thus, it appears that the cloud cover's large particles consist of chlorine compounds, although it is not apparent which specific compounds these are.

Winds, Storms, and Night-Sky Glow

Ground-based observations had already established that Venusian winds are unusual. Near the upper boundary of clouds, the speed of fairly regular atmospheric streams is nearly 100 m/sec (328 ft/sec). These swiftly flowing atmospheric masses form a single stream as they sweep above the slower atmospheric layers and solid body of the planet. The rotation period of the planet's body is very long—243 Earth days. Venus' rotation is retrograde, opposite to the rotation of Earth and the other planets in the Solar System. The clouds move, together with the upper part of the atmosphere, in the same retrograde direction, completing one rotation in 4 days at an altitude of 65 to 70 km (40 to 43 miles).

Measurements of the lander's descent velocity made it possible to determine the wind profile down to the surface. As the lander approached the planet's surface, the wind gradually subsided. Within the last 10-km (6-miles) thick layer of atmosphere, the wind speed was only about 1 m/sec (3 ft/sec). To measure wind velocity on the surface, the Veneras 9 and 10 landers carried conventional wind vanes.

The existence of clouds in the atmosphere and the highly intensive dynamic processes that occur there made it quite probable that storm phenomena might be present. The objective of experiments that L. Ksanfomaliti (Institute of Space Research) supervised was to find effects in Venus' atmosphere similar to terrestrial thunderstorms. Storm discharges generate low-frequency electromagnetic pulses. Ksanfomaliti used a low-frequency (8 to 100 kHz) spectrum analyzer with an external antenna in the experiment and did, in fact, observe pulse radiation similar to that typical in Earth's thunderstorms (Figure 7-9). After receiving Veneras 11 and 12 mission results, scientists analyzed the night-side observation data that Veneras 9 and 10 had earlier obtained. It turned out that Venera 9 had, indeed, registered a short-lived glow on Venus' nightside. The glow was possibly storm-generated. Estimates suggest that the number of storms on Venus could be even greater than on Earth.

For a long time, many ground-based observers have noted a weak nightglow (the ashen light of Venus). It seems possible that this effect arises during periods of particularly high storm activity. Besides, another effect—a constant night airglow undetectable from Earth—results from chemical reactions in the upper atmosphere. In the visible spectrum, this airglow only occurs when molecular oxygen bands are excited in a carbon dioxide rich atmosphere such as Venus'. Veneras 9 and 10 orbiters were the first to register the bands. (V. Krasnopolsky, Institute of Space Research, supervised the experiment.)

The Sun's ultraviolet radiation (in the hydrogen and helium lines) is scattered by corresponding atoms in the planets' upper atmosphere. The excited atoms re-emit ultraviolet quanta and produce line-scattered radiation. Measurements of its intensity can be converted to hydrogen and helium concentrations. These lightest of elements make up the outermost portions of the atmospheres of Earth, Mars, and Venus. Veneras 11 and 12 flyby probes each carried an instrument to measure radiation intensity in the upper atmosphere in 10 different ultraviolet intervals of the spectrum, which included hydrogen and helium lines and lines of several other elements. V. Kurt (Institute of Space Research) supervised the experiment, which also involved French physicists J. Blamont and J. L. Bertaux. An analysis of the high-quality spectra provided some estimates of the composition and structure of Venus' upper atmosphere.

Experiments conducted during the descent of Veneras 11 and 12 into Venus' atmosphere studied three basic problems: fine chemical analysis of atmospheric gases, nature of clouds, and thermal balance of the atmosphere. Of these, the chemical composition studies were considered the most essential. All the experiments were successful. The scientific instruments on the Pioneer Venus probe were similar to those on the Venera probes—a gas chromatograph, a mass spectrometer, and some optical instruments. A comparison of the results is of great interest.

In April 1979, Soviet and American scientists who had participated in both missions met at the Institute of Space Research, U.S.S.R. Academy of Sciences, Moscow. During that meeting, they compared data from the different probes and discussed the implications. The meeting's published results made it clear that the space science community had succeeded in studying the fine chemical composition of Venus' atmosphere. The investigations of both the Soviet and American probes had cleared a way for solving the mysteries about Venus.

Solar-Wind Interaction with Venus—Bow Shock and Intrinsic Field

The first experimental observations of Venus' bow shock were obtained from descending and flyby trajectories of Venera 4, Venera 6, Mariner 5, and Mariner 10. The properties of the plasma were measured by Venera 4 with charged-particle traps. K. Gringauz, Institute of Space Research, U.S.S.R. Academy of Sciences, headed the experiments. S. Dolginov and his colleagues (Institute of Earth Magnetism and Radiowave Propagation, U.S.S.R. Academy of Sciences) measured the magnetic field.

The various types of charged-particle traps, or wide-angle detectors, are actually a system of electrodes—a collector and several grids. Various voltages—direct current, gradually changing direct current, and alternating current—are usually applied to these grids, which makes it possible to analyze the trapped particles by their energies and charge signs. Scientists observed the shock wave as a sharp, simultaneous increase in the interplanetary plasma and amplitude of magnetic field fluctuations that occurred some distance from Venus.

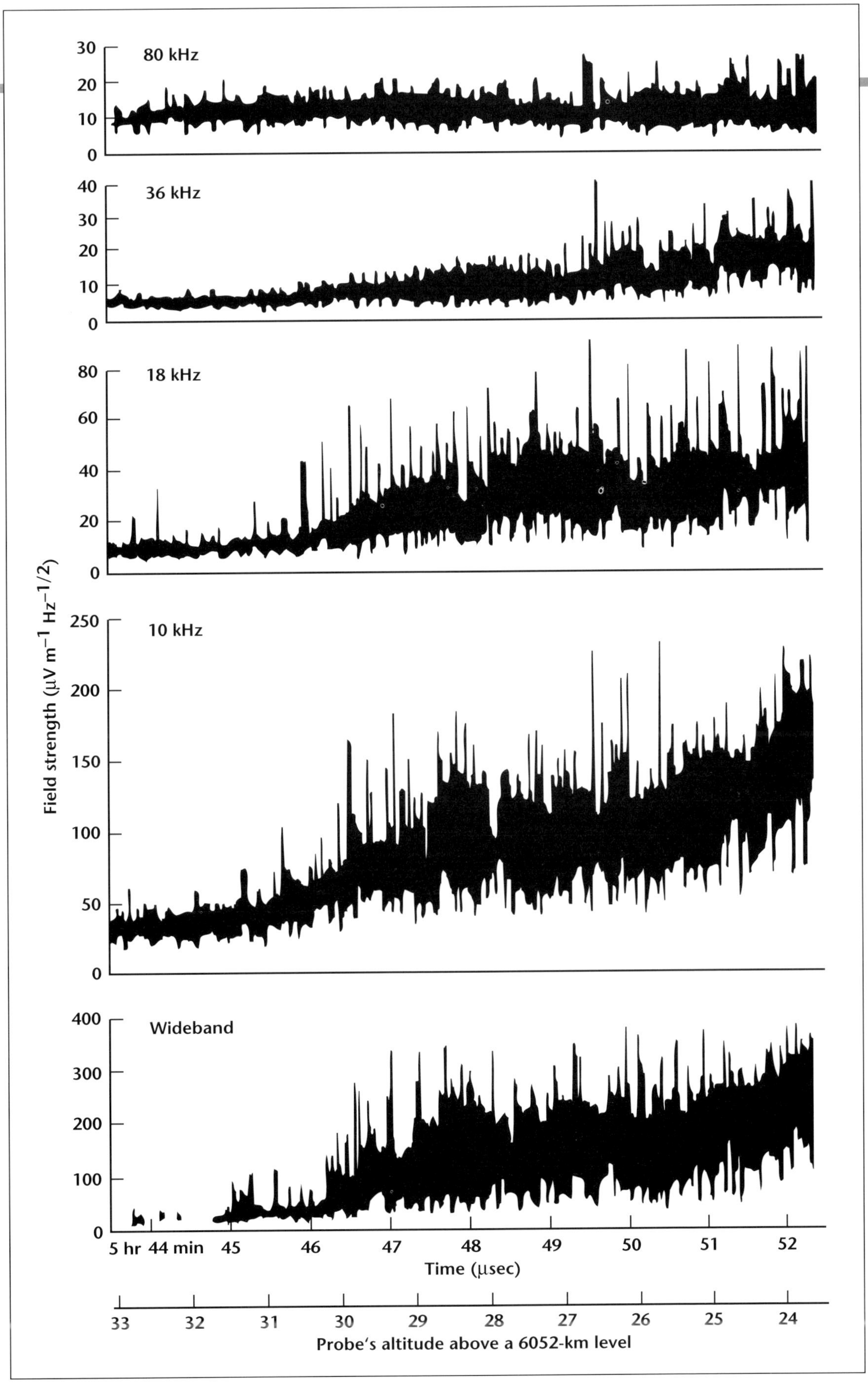

80 kHz
30
20
10
0

36 kHz
40
30
20
10
0

18 kHz
80
60
40
20
0

10 kHz
250
200
150
100
50
0

Wideband
400
300
200
100
0

Field strength (μV m^{-1} Hz$^{-1/2}$)

5 hr 44 min 45 46 47 48 49 50 51 52
Time (μsec)

33 32 31 30 29 28 27 26 25 24
Probe's altitude above a 6052-km level

Figure 7-9. The "GROZA" experiment of the Venera 11 descent module recorded these radio noise bursts. The bursts are plotted against altitude for various frequencies. Lightning strikes in the planet's atmosphere evidently caused the noise.

Systematic observations of the interactions of the solar wind with Venus were performed with plasma and magnetic instruments onboard the first Venus orbiters, Veneras 9 and 10. The plasma properties were measured with wide-angle analyzers, Faraday cups, and retarding potential analyzers (RPA) (K. Gringuaz, Space Research Institute) and with narrow-angle detectors and electrostatic analyzers (O. Vaisberg, Space Research Institute). The magnetic measurements were made by S. Dolginov, Institute of Earth Magnetism and Radiowave Propagation.

An electrostatic analyzer is, in its simplest form, two curved concentric plates separated by a small gap. A potential difference is applied to the plates. Particles entering the gap pass through it only if they have a certain energy/charge unit ratio. This energy corresponds to the applied potential difference. By applying different potentials to the plates, an energy spectrum of particles can be obtained.

Figure 7-10 shows 32 bow shock crossings by Veneras 9 and 10. These data are from the wide-angle analyzers and show the mean front position, based on data of 86 crossings by Pioneer Venus (Slavin *et al.*). The shock front position near Venus is close to the surface—about 0.3 Venus radius in the frontal subsolar area. Two circumstances explain the differences in the mean front positions of Soviet and American vehicles. These spacecraft crossed the front at different latitudes, and the measurements occurred during different phases of the solar activity cycle.

Veneras 9 and 10 also took measurements with electrostatic analyzers, which showed that the asymmetry of Venus' bow shock was linked to the solar wind's anisotropic nature. The bow shock's radial distance in the polar direction is approximately 2000 to 3000 km (1243 to 1864 miles) greater than in the equatorial direction.

After the experiments on Venera 4 by S. Dolginov and his colleagues, Venus' magnetic moment was initially estimated as 5 to 8×10^{-21} gauss cm^3 (10 gamma on the surface). After reviewing Veneras 9 and 10 data, this estimate was lowered and the intrinsic field on the planet's surface was assumed not to exceed 5 gamma.

Magnetic field measurements at altitudes from 140 to 200 km (87 to 124 miles) showed that most field values did not exceed the threshold sensitivity of the instrument, or 2 gamma. Thus, it was confirmed that Venus' intrinsic magnetic field is all but absent.

Plasma Magnetic Tail

All trajectories of Soviet vehicles that have landed on planets or put artificial satellites into orbit have approached planets from their nightside and have allowed observations of the planets' wake at altitudes greater than 1500 km (932 miles). Veneras 9 and 10 entered the dayside only to latitudes above 32°. These vehicles penetrated deep into the planet's optical umbra and allowed detailed measurements of the distribution of the plasma and magnetic field. Their measurements showed that a plasma-magnetic tail with typical features exists near Venus, some of the features being similar to the tail of Earth's magnetosphere. In particular, the oppositely directed bundles of magnetic field lines along the Sun-planet direction were present on Venus. In other words, the magnetic field component along the Sun-planet direction was essentially higher than the others.

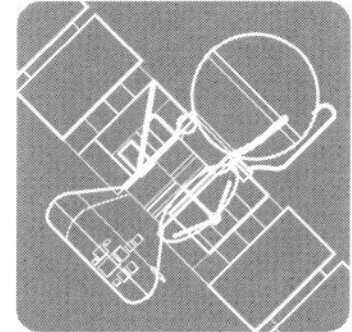

These field line bundles in the tails were separated in the layer where the magnetic energy density had a deep minimum. This layer is similar to the "neutral-sheet" of Earth's magnetosphere. The data from wide-angle analyzers showed that plasma properties and distribution in the tail also resemble Earth's magnetotail. At the tail boundary and in the transition region, a characteristic change in differential ion spectra was observed similar to that in Earth's boundary layer, or plasma mantle. The plasma features deep in the tail resemble those in Earth's plasma sheath.

Figure 7-11 shows regions of solar-wind interaction with Venus. These regions include the shock wave, the transition region (A) behind the shock front, and the plasma-magnetic tail. The B-region corresponds to the corpuscular penumbra, or boundary layer. Data from electrostatic analyzers also indicated a tail boundary that separated plasmas with different

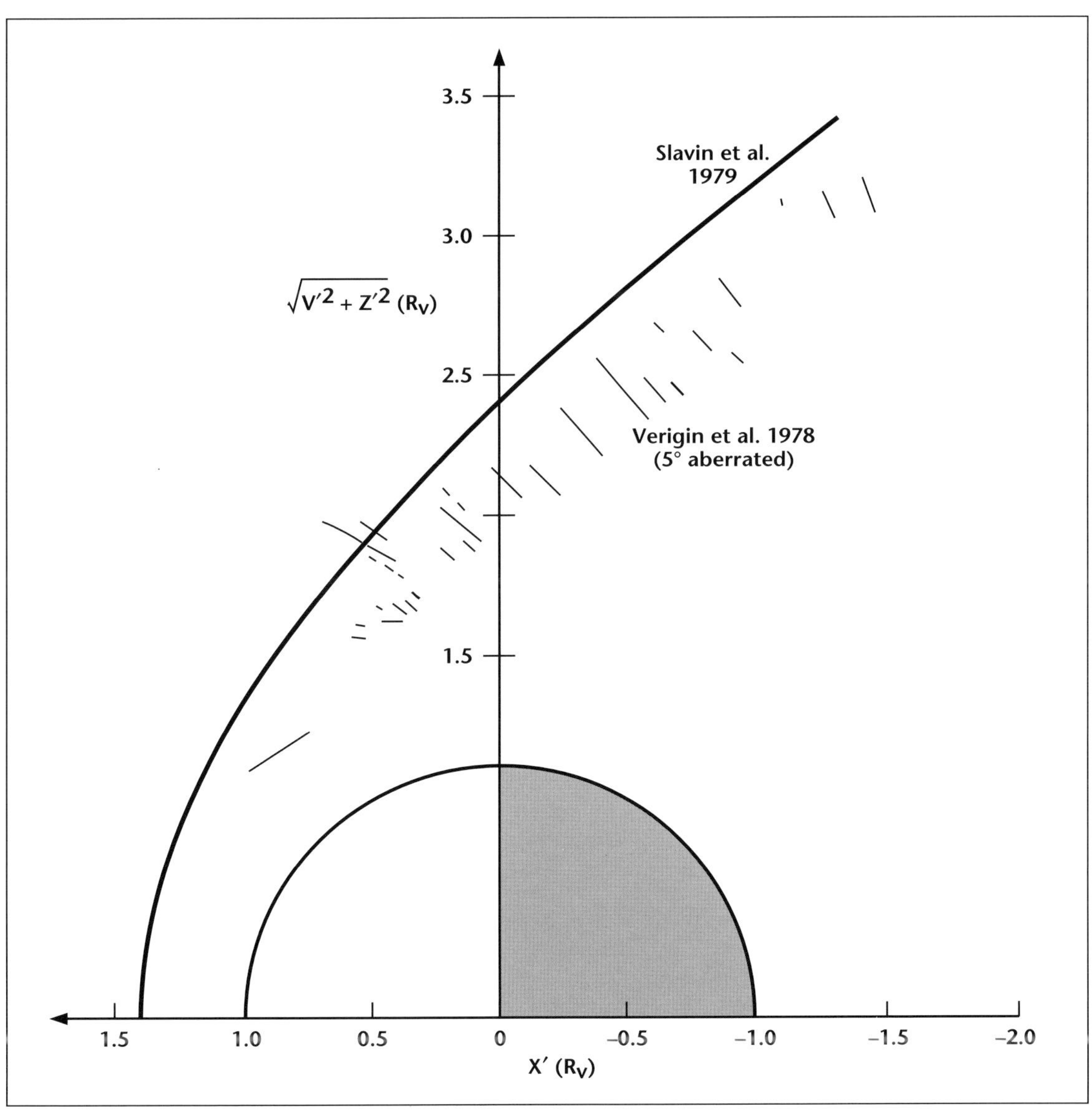

Figure 7-10. Position of the shock front near Venus (measured by Veneras 9 and 10). The lengths of the short curves and the points show the parts of the orbit where the spacecraft crossed the front. The solid curve shows the average position of the front (determined from Pioneer Venus data). The cylindrical system of coordinates is used where the X′-axis is oriented to the solar-wind direction.

properties. Outside this boundary, plasma was evidently of solar-wind origin but disturbed by its interaction with the obstacle. Inside the boundary, plasma was cooler and had a smaller bulk velocity, probably an accelerated or heated plasma of planetary origin. Such a boundary layer could appear, and its properties would resemble the boundary of two liquids. One boundary moves and, because of viscous interaction with the lower liquid, accelerates and heats it. When the solar-wind plasma with the frozen-in magnetic field moves relative to the ionospheric plasma, the boundary separating these liquids can be unstable. For instance, the boundary begins to move or fluctuate because of increasing solar-wind pressure. The bubbles of the solar-wind plasma flow are then pressed into the ionosphere, tearing away from the flow. This condition also could occur with ionospheric plasma rising up in the transition region. A variety of processes cause plasma instabilities, smear the boundary, and dissipate solar-wind energy and its subsequent transfer into the ionosphere.

In Figure 7-11, the region extending to 5 Venus radii (C-region), where the regular ion fluxes are absent, is positioned under the corpuscular penumbra, which is the corpuscular umbra region that does not coincide with the optical shadow of Venus. The behavior of the electron fluxes was quite different from the measured ion fluxes. The fluxes were everywhere, including the corpuscular umbra. Only their intensity decreased (Figure 7-12), and the character of the spectrum changed; that is, high-energy tails appeared in the spectrum. Apparently, electrons and ions inside the tail were subjected to some acceleration processes.

It was likely that in the far tail regions of Venus, the boundary layer gradually thickened and merged with the plasma sheath as it does for Earth. As in the plasma sheath of Earth's magnetosphere, accelerated ion fluxes with energy greater than 2 keV (C-region in Figure 7-11) were observed near the neutral-sheet plane. These fluxes occurred when the B_x component of the magnetic field reversed its sign (x-axis was along the Venus-Sun line—Figure 7-13). Thus, the large-scale pattern, magnetic field topology, and plasma distribution in the Venusian tail showed a striking resemblance to Earth's magnetosphere.

Nature of the Obstacle Forming a Shock Wave

An extended tail near Venus with properties similar to those in Earth's magnetosphere seems rather striking. Before the Pioneer Venus experiments, this tail led the American specialist C. T. Russell to revise the magnetic field estimates that Soviet specialists previously made. He increased the estimated value of Venus' intrinsic magnetic field.

More careful study and detailed revisions of the data for magnetic and plasma measurements near Venus have begun. An analysis of magnetic measurement data on Veneras 9 and 10 showed that the tail's magnetic field properties had one essential difference. This difference became apparent after comparing data the two spacecraft obtained simultaneously. One spacecraft was in undisturbed solar wind and the other in the planet's tail region.

During each measurement, the magnetic field topology—two field line bundles stretched along the tail—was preserved. However, in several instances, the plane of the neutral sheet separating these bundles changed its orientation. Sometimes this plane was located vertically, almost parallel to the meridian plane, but this is not typical, for example, of Earth's magnetotail. By comparing the magnetic data two spacecraft obtained at the same time, E. Eroshenko (Institute of Earth

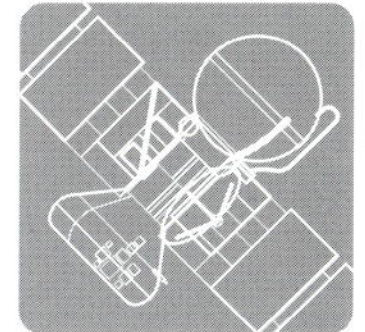

Magnetism and Radiowave Propagation) showed that the neutral-sheet plane in the tail always remained perpendicular to the transverse component of the interplanetary magnetic field. It rotated with the rotation of this transverse component.

The conclusion is that the measured magnetic field is not the planet's intrinsic field. Rather, it is the field of the "magnetic barrier" that currents flowing in Venus' conductive ionosphere induce. In other words, magnetic field tubes of the solar plasma flowing around the planet encounter an almost ideal conductor: the ionosphere. They cannot penetrate it and they deform, retarding especially strongly near the stagnation subsolar point of the ionosphere. The magnetic field accumulates at the subsolar region and forms a magnetic barrier. Still flowing around the planet, the solar wind carries the ends of the field tubes retarded at the frontal part of the planet. The tubes drape the planet and stretch tail-like on the nightside. Thus, the field line bundles elongate in opposite directions on the two sides of the planet. The orientation of the plane separating these bundles depends on the orientation of the magnetic field in the undisturbed solar wind. In the simplest case, if the interplanetary magnetic field vector lies in the ecliptic-horizontal plane, field lines of the tubes draping the planet are in opposite directions on the dawn and dusk sides. In this case the neutral-sheet plane is parallel to the meridian plane. If, however, the interplanetary-field vector is in the meridian plane or near it, the neutral-sheet plane will either partially or completely coincide with the ecliptic plane. It is very difficult to distinguish this case from

Figure 7-11. Schematic representation of the near-planet shock wave (dotted line) and Venus' magnetosphere from Veneras 9 and 10 data. Arrows show the direction of the solar-wind plasma flow. The A-region is the transition layer behind the shock front. The B-region is the boundary layer. The C-region is the corpuscular shadow. The D-region (solid line) is the magnetosphere boundary. The E-region is the plasma sheath which contains a neutral sheet separating magnetic field lines directed toward each other.

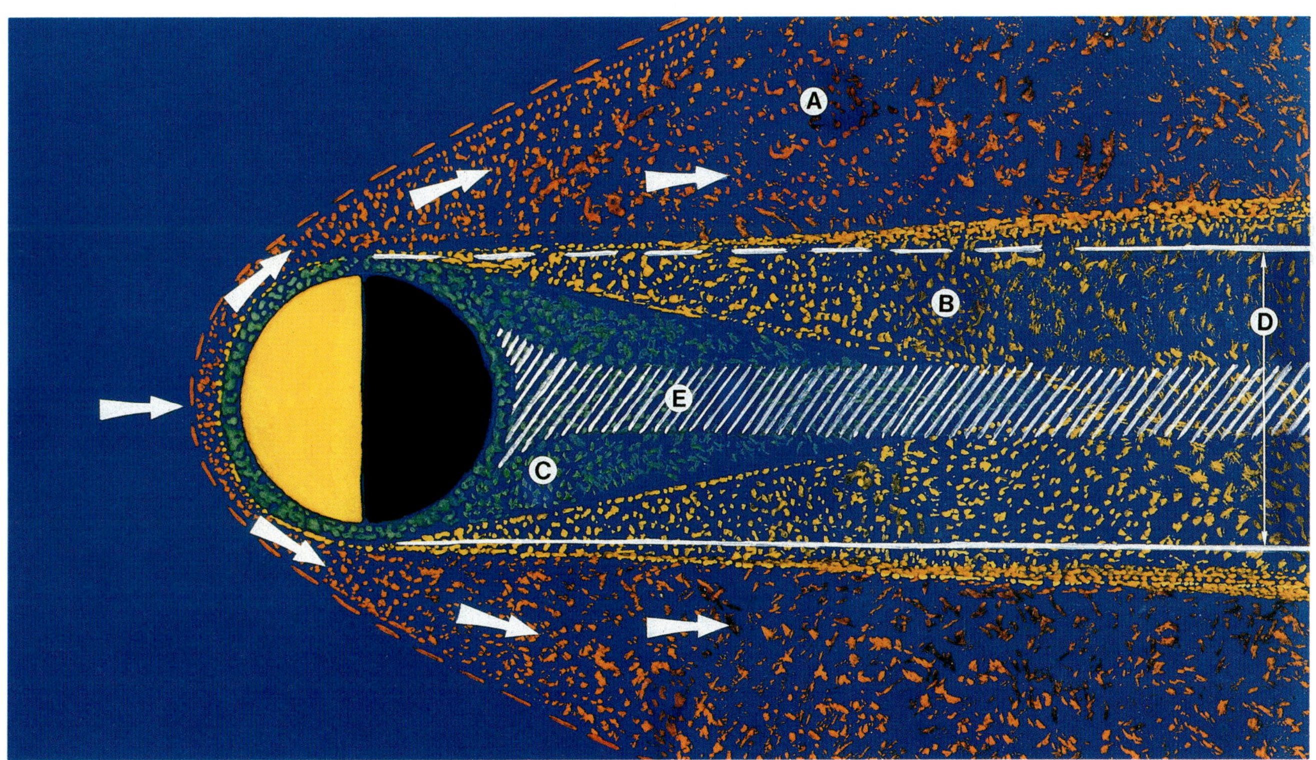

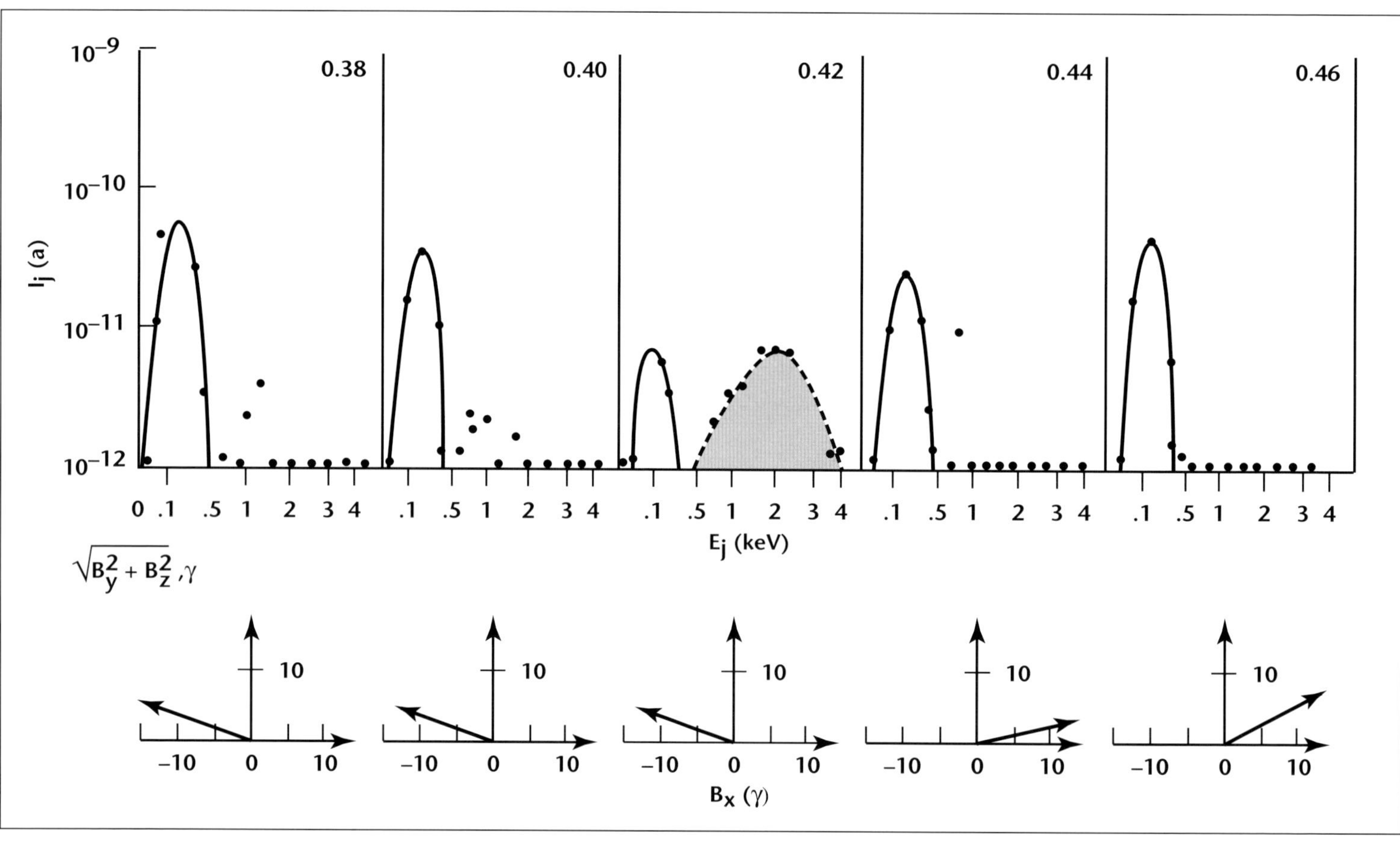

Figure 7-12. Ion energy spectra that Venera 10 obtained on April 19,1976. The spacecraft measured the intense flows of energetic ions (shaded part of the 0.42 spectrum) in the region of the planet tail where the magnetic field B_x-component changed its sign (B_x-component turn is shown underneath the spectra between 0.42 and 0.44). These flows are part of the plasma sheath of the Venusian tail.

the intrinsic magnetosphere tail, with the dipole axis near the polar axis, as for Earth.

The problem remained unsolved for currents that form the induced magnetosphere flow. Another unsolved problem was how an extended induced magnetic tail can form.

After Veneras 9 and 10 experiments and on the basis of research by American investigators (P. Cloutier and R. Danniel), E. Eroshenko assumed that currents are induced in the ionosphere itself and are mainly in its maximum. The region from the ionosphere maximum to its upper boundary is 200 to 300 km (124 to 186 miles) on the dayside.

Soviet laboratory simulation experiments (at the Space Research Institute, headed by I. Podgorny) were very important in understanding tail formation in the "induced" magnetosphere. In these experiments a Venusian artificial ionosphere was formed from vaporization products of a wax sphere placed in a hydrogen plasma flow with a frozen-in magnetic field. On the artificial ionosphere's dayside, a sharp boundary formed, over which the magnetic field increased with the "magnetic barrier." Field lines were parallel to the ionospheric boundary. Measurements on the wax sphere's nightside showed that a long tail forms (up to 10 radii of the sphere) with the field orientation in the tail being typical of the observed Venusian magnetosphere (Figure 7-14).

The experiments on Pioneer Venus finally confirmed that Venus has practically no intrinsic magnetic field and that a magnetic barrier forms on its dayside.

If the assumption that the induced current flow inside the ionosphere is correct, the upper ionosphere boundary should coincide with the magnetic barrier's upper boundary. However, it does not. From Pioneer Venus data, the

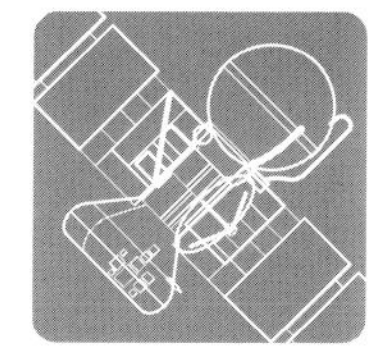

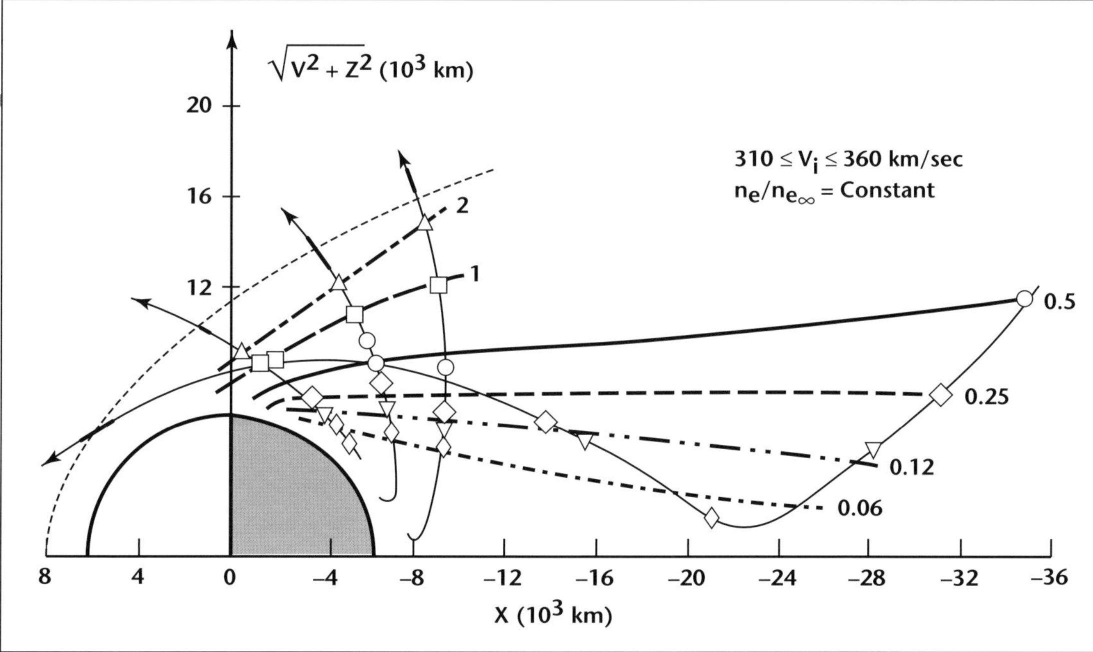

Figure 7-13. Distribution lines for constant number densities of the plasma's electron component near the region where the solar wind interacts with Venus (from Veneras 9 and 10 data). Electron measurements corresponding to velocities of solar wind, v_i, in the narrow interval 310 to 360 km/sec (193 to 224 miles/sec) were chosen for the analysis. Numbers along the lines designate the values of electron number density, h_e, relative to their values in the solar wind.

barrier's magnetic field usually decreases sharply on the upper ionosphere boundary, or ionopause, simultaneously with the growth of the thermal ionospheric plasma's concentration and temperature. That is, the field behaves as if there is a conductor carrying a current in the ionopause region at 50 to 100 km (31 to 62 miles). Sometimes Pioneer Venus detected high values of the magnetic field inside the ionosphere in the region of the main maximum.

It is evident that, in the ionosphere itself, strong currents could flow. C. T. Russell associated that phenomenon with the discovery of magnetic "flux ropes" in Venus' dayside ionosphere. American specialists (F. Johnson and W. Hansen) and Soviet specialists (T. Breus, E. Dubinin *et al.*, Space Research Institute) gave qualitative explanations and estimated flux-rope characteristics.

In the dayside ionosphere, a special set of magnetic field tubes from the magnetic barrier, which results from the instability of the ionopause as it fluctuates due to varying solar-wind pressure, apparently can press in the ionosphere, tear off the solar-wind flow, and submerge into the ionosphere. With these tubes moving in such a manner, the field-aligned current can twist them into spirals and make their cross sections more compressed as they submerge deeper into the ionosphere. Pioneer Venus data showed that the entire dayside ionosphere was often filled with these flux ropes or their pieces.

Dayside and Nightside Ionospheres of Venus

Scientists investigated properties of Venus' dayside and nightside ionospheres by observing radio occultations. This was during the flybys of Mariners 5 and 10, Veneras 9 and 10, and the long mission of Pioneer Venus Orbiter.

In 1967, ion traps on Venera 4 made the first direct measurements of the ion number density's upper limit in Venus' nightside ionosphere. In 1978-1979, Pioneer Venus, using various mass spectrometers and plasma analyzers, measured ion and electron number densities, temperatures, and ionosphere composition. The spacecraft made these direct measurements down to 140 km (87 miles) on both the dayside and nightside of Venus.

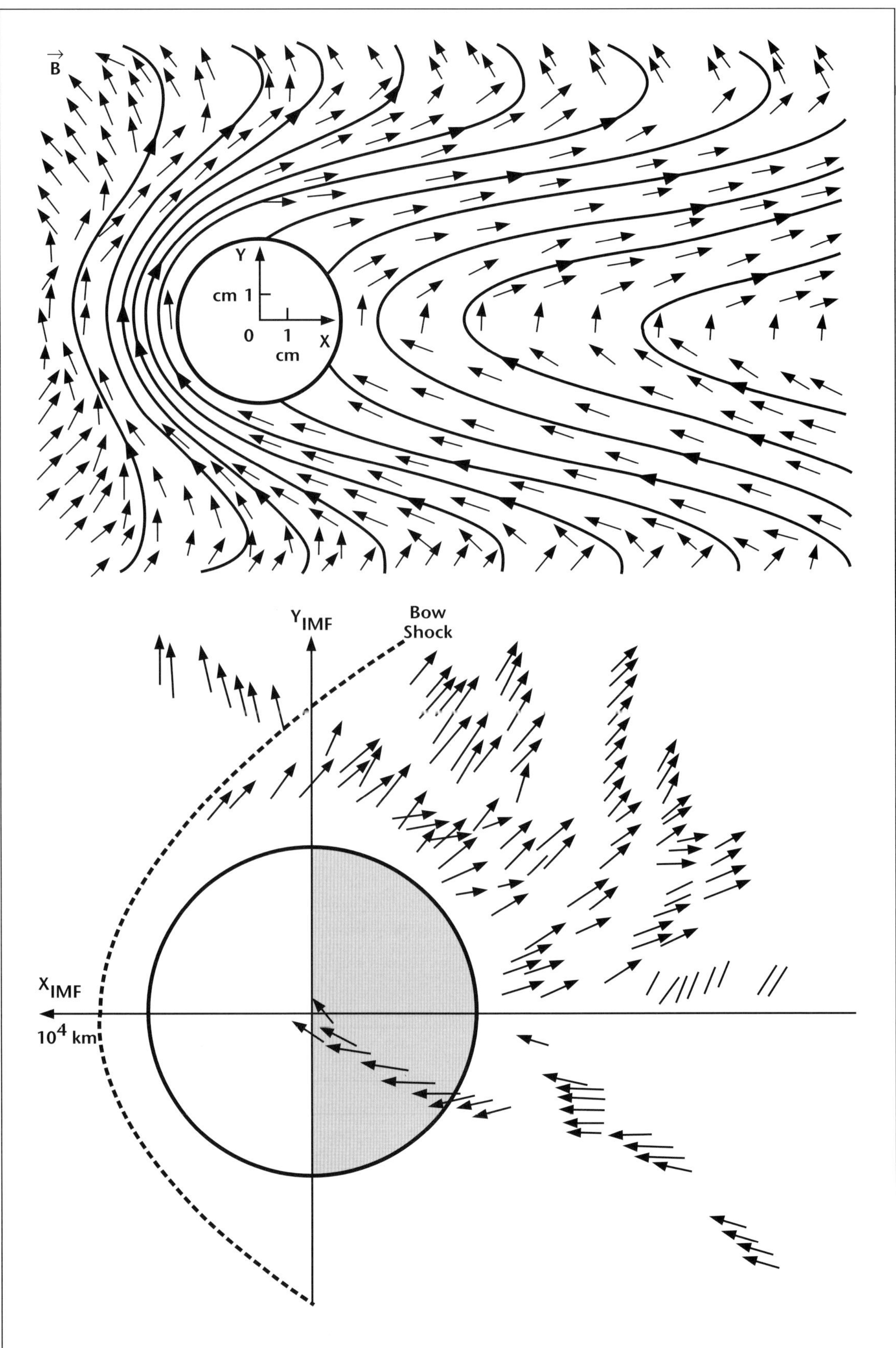

Figure 7-14. Comparison of laboratory model of induced magnetosphere (top of figure) with the field topology in the tail of Venus' magnetosphere measured during the Veneras 9 and 10 experiments. Projection of magnetic field vectors appears in the system of coordinates rotating together with the interplanetary magnetic field vector.

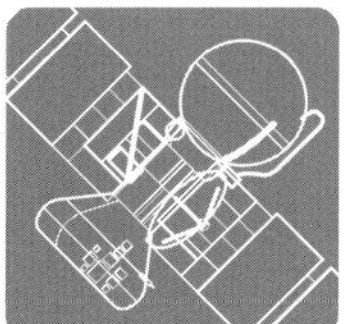

Venus' Dayside Ionosphere
Early experiments and radio-occultation observations during Mariner 5 and 10 flybys of Venus indicated that a sharp upper boundary —an ionopause—exists on electron number density profiles in the dayside ionosphere.

The ionopause heights of these profiles were very different: 500 km (310 miles) on Mariner 5 and 350 km (217 miles) on Mariner 10. The dynamic pressure of the undisturbed solar wind during Mariner 10's flyby was higher than during Mariner 5's. Based on this difference, American investigators suggested that the solar wind could compress the Venusian ionosphere (S. T. Bauer). As a result, the electron number density profile should be distorted, and the significant flow of the solar wind could then penetrate to the ionosphere. According to some estimates (C. T. Russell), the value of the incoming solar-wind flow could be 30% of the total solar flux. As a result, the shock wave might "settle down" on Venus' surface and become attached rather than detached (C. T. Russell). As the data from Veneras 9 and 10 showed (N. Savich, Radioelectronics Institute), the ionopause has a distinct dependence on solar zenith angle. Near the subsolar region, the ionopause was at 250 to 280 km (155 to 174 miles). With an increase in the Sun zenith angle x, the ionopause height increased. This dependence had the following form: $1/\cos^2 x$. In other words, it corresponded to variations with zenith angle of the solar wind's dynamic pressure $pv^2 \cos^2 x$ (p is density and v velocity of the solar wind).

In the stagnation region, where $\cos^2 x = 1$ and the dynamic pressure is maximum, the ionopause is much nearer the surface. At the flanks, with an increase in x, it moves farther away from the surface and experiences greater variations in height. Beginning with a zenith angle of approximately 58° to 60°, a region appeared above the main ionization maximum. This region had an almost constant electron number density on the order of 10^3 cm^{-3}. It also displayed an extension of roughly 300 km (186 miles) or more, the so-called "ionosheath." The Pioneer Venus data showed that heights of the upper ionospheric boundary vary considerably. The amplitude of its variations increased with zenith angle, but the character of the boundary behavior was generally the same as that shown by Veneras 9 and 10 data. The large range in ionopause heights that Pioneer Venus measured was due to differences in measurement techniques. Data that gave the positions were from various sensors that were subjected to the effect of the vehicle potential, especially near the terminator. During transfer from the illuminated to nonilluminated portion of an orbit, the photocurrent from the vehicle decreases in the shadow. Consequently, the potential of the free body in the plasma decreases, which affects the zero reference in measurements with traps.

Another reason might be that the very low position of its periapsis may have caused the Pioneer Venus trajectory in the ionosphere to give a horizontal rather than vertical cross section. The results then would depend on horizontal plasma variations, which perhaps were even greater than usually appear in radio-occultation data.

In any case, according to radio-occultation observations on Pioneer Venus and Veneras 9 and 10, these ionopause variations were less striking. However, this problem required further analysis and correlation.

With increasing distance from the subsolar point, the boundary between the solar wind and the ionosphere becomes unstable. The magnetohydrodynamic boundary layer

develops because of viscous interaction of two plasmas, instabilities, and dissipation of energy. Its thickness grows to the flanks. Possibly the ionosheath formation on the electron number density profile is associated, in a yet unknown way, with the formation of this boundary layer.

How much solar wind penetrates to Venus' ionosphere? Is it 30% of the flux coming toward the planet, or is it less?

Based on indirect data (T. Breus, Space Research Institute) and theoretical estimates (P. Cloutier and R. Danniel), the absorption should be negligibly small. Actually, it should not exceed 1% because the shock front position near Venus is sufficient to follow the law of magnetohydrodynamic flow around an impenetrable obstacle. Pioneer Venus results later confirmed this value.

Venus' Nightside Ionosphere

It became evident after radio-occultation experiments onboard Mariners 5 and 10 and Veneras 9 and 10 that Venus' nightside ionosphere was irregular. Electron density profiles in the nightside ionosphere sometimes had two narrow maxima of roughly the same order of magnitude. These maxima were 5 to 10 km (3 to 6 miles) apart. Sometimes the number density in the upper maximum exceeded that in the lower one. It was natural to associate irregular electron density variations in the nightside ionosphere with the influence of solar-wind flows. It was just such an assumption that Soviet and American specialists made after their respective Venera 4 (1967) and Mariner 5 experiments. But it was still obscure how the solar wind penetrated to such low heights in regions far from the terminator. (This was before Veneras 9 and 10 experiments and before discovery of the plasma magnetic tail near Venus.) The assumptions and estimates on how solar-wind electron fluxes ionized Venus' nightside atmosphere seemed inconclusive.

American researchers suggested another hypothesis. They assumed that hydrogen and oxygen ions forming in the dayside ionosphere were transported with the solar-wind flux to Venus' nightside. The ions then diffused down to the heights of the main maximum of the night ionosphere and exchanged charge with neutral molecules of CO_2 and O_2. As a result, ions O_2^+, O^+, and CO_2^+ formed, and the nightside ionosphere consisted of these ions.

Veneras 9 and 10 measured electron fluxes at an altitude of 1500 km (932 miles) in the region of Venus' optical umbra (see Figure 7-12). K. Gringauz and his colleagues Verigin, Breus, and Gomboshi suggested that these fluxes can ionize the atmosphere and form the upper maximum of the night ionization.

Calculations showed that, because of these electron fluxes, the maximum of the electron number density could really form, which corresponded to the radio-occultation measurements of Veneras 9 and 10 (Figure 7-15). The fact that electron density variations in the flux at altitudes of 1500 km (932 miles) correlated well with those in the ionosphere's upper maximum also argued in favor of the assumption. The calculated and experimental profiles, however, coincided only when the neutral atmosphere density in the calculations (that is, an initial ionizable material) was more than an order of magnitude less than in available models. The neutral temperature also might be lower than in these models. Veneras 9 and 10 radio-occultation measurements (N. Savich) also showed the neutral temperature to be much lower (about 100 K) than had been suggested before. Other observations need

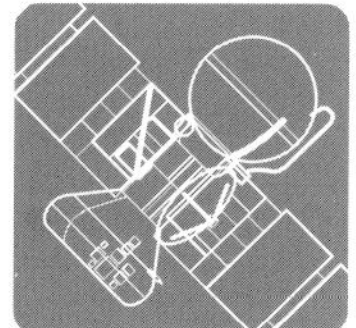

explanations, too. For example, scientists knew that electron fluxes coming into the atmosphere caused nighttime glows. Experiments, however, did not show these glows. Another question puzzled scientists: How were electrons at 1500 km (932 miles) able to reach 140 km (87 miles)?

An explanation is also needed for the ionization source that produces the second maximum in the nightside ionosphere, which frequently has the same order of magnitude as the upper one. Ionization sources such as ion transport from the dayside ionosphere and diffusion and charge-exchange of ions with atmospheric molecules can hardly account for one or two very narrow maxima that have been observed in experiments. Electrons with energies greater than 70 eV, which Soviet scientists had used in the calculations described earlier, could not reach the lower maximum because they "died" at higher altitudes.

American specialists (D. Butler and J. Chamberlain) and a Soviet specialist (V. Krasnopolsky) hypothesized that the lower maximum formed as a result of meteor ionization at an altitude level where the number density of neutrals was 10^{12} to 10^{13} cm^{-3}. This level was actually lower by about 20 km (12.5 miles) than that for 2×10^9 cm^{-3}, at which the upper ionization maximum that K. Gringauz and his colleagues had estimated is formed. Meteor ionization could produce a rather narrow maximum. Despite criticism and correction of the available neutral atmosphere models, Soviet investigators followed this hypothesis based on their own data.

Eventually, Pioneer Venus data verified the results of calculations that, in turn, confirmed this hypothesis. These data indicated that the number density of neutral components and plasma temperature at the height of the ionization upper maximum was several orders of magnitude less than in available models (Figure 7-16). The neutral temperature in Venus' nightside atmosphere was about 100 to 140 K.

Pioneer Venus detected fluxes of electrons with energies less than or equal to 250 eV (the upper threshold of the instruments) at an altitude of 140 km (87 miles). The intensity of the flux was sufficient to produce ionization equal to that measured experimentally. This information was conclusive evidence that the Soviet hypothesis for an electron source of ionization in Venus' upper ionosphere was correct.

Pioneer Venus measured velocities of the O^+ ion transport from the dayside to the nightside ionosphere. These velocities were sufficient to sustain the nightside ionosphere. However, the maximum of the ionization so formed gradually decreased with increasing height in the region above the maximum. Soviet data showed that the thickness of the ionization layer at the maximum half-width level exceeded by about two times the thickness of the experimental profile layer.

From these observations, it became clear that electron fluxes help form the narrow upper maximum of ionization in the planet's nightside ionosphere. It is even possible that double-component electron flux (consisting of electrons with energy less than 70 eV and greater than 350 eV) forms double maxima of very irregular ionization. It also is possible that accelerated fluxes of ions that Veneras 9 and 10 detected in the tail form the lower maximum (T. Breus, A. Volacitin, and H. Mishin). The transport of O^+ ions from the dayside ionosphere contributes mainly to the formation of the ionosphere's upper region.

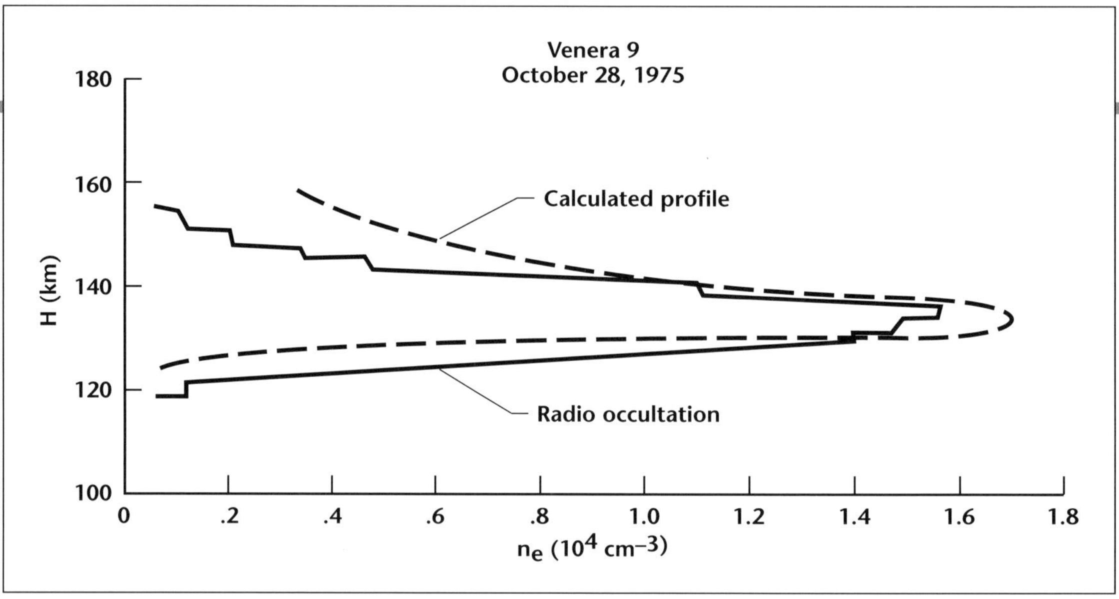

Venera 9
October 28, 1975

180
160
140
120
100
H (km)

Calculated profile
Radio occultation

0 .2 .4 .6 .8 1.0 1.2 1.4 1.6 1.8
n_e (10^4 cm^{-3})

Figure 7-15. Comparison of the electron-number density profile in Venus' nightside ionosphere (from Venera 10 data that measured electrons ionizing the atmosphere) with the profile obtained by radio-occultation measurements.

Where do electron fluxes appearing in the planet's optical umbra form? How do they enter the atmosphere at altitudes of 100 to 140 km (62 to 87 miles)?

Veneras 9 and 10 detected a plasma-magnetic tail near Venus. This discovery provides at least a partial answer to these questions. For the present, it allows appropriate assumptions to be made.

Indeed, in the plasma sheath, acceleration of solar-wind particles was observed, the latter flowing into the tail from its flanks. Also, acceleration of ions and electrons in the day-side ionosphere could occur and these could be transported to the tail and picked up by the solar-wind flux.

Different mechanisms in the tail can accelerate electron fluxes. These fluxes can precipitate and then be injected into the atmosphere at low altitudes to produce an irregular source of ionization. Such a source essentially depends on the properties of the solar-wind and the situation in interplanetary space.

The plasma and magnetic experiments the Soviets conducted near Venus for over a decade were very useful. At the XVII General Assembly of the International Association of Geomagnetism and Aeronomy in Canberra, Australia (December 1979), results of magnetic and plasma measurements near Venus were summarized. Here is a list of basic results obtained by Soviet (Veneras 9 and 10) and American (Pioneer Venus) investigators. The list also includes theoretical work and models that contributed much to the interpretation of the results:

- Discovery of the plasma-magnetic tail (Venera vehicles)
- Identification of the induced nature of the magnetic field measured near Venus (Venera vehicles and Pioneer Venus)
- Determination of the shock front position (Venera vehicles)
- Detection of the shock front asymmetry (Venera vehicles)
- Hypothesis of an electron source of night-side ionosphere ionization (Venera results and calculations)
- Confirmation of the Venus "induced" tail in laboratory simulation experiments (Soviet data)
- Evidence for the pressure balance at the iono-pause, sustained by the "magnetic barrier" and the ionosphere thermal plasma pressure

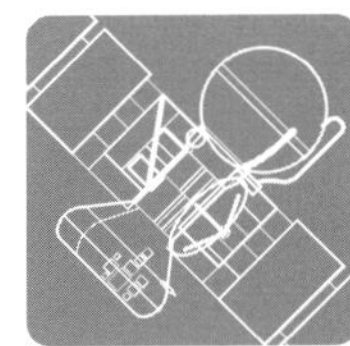

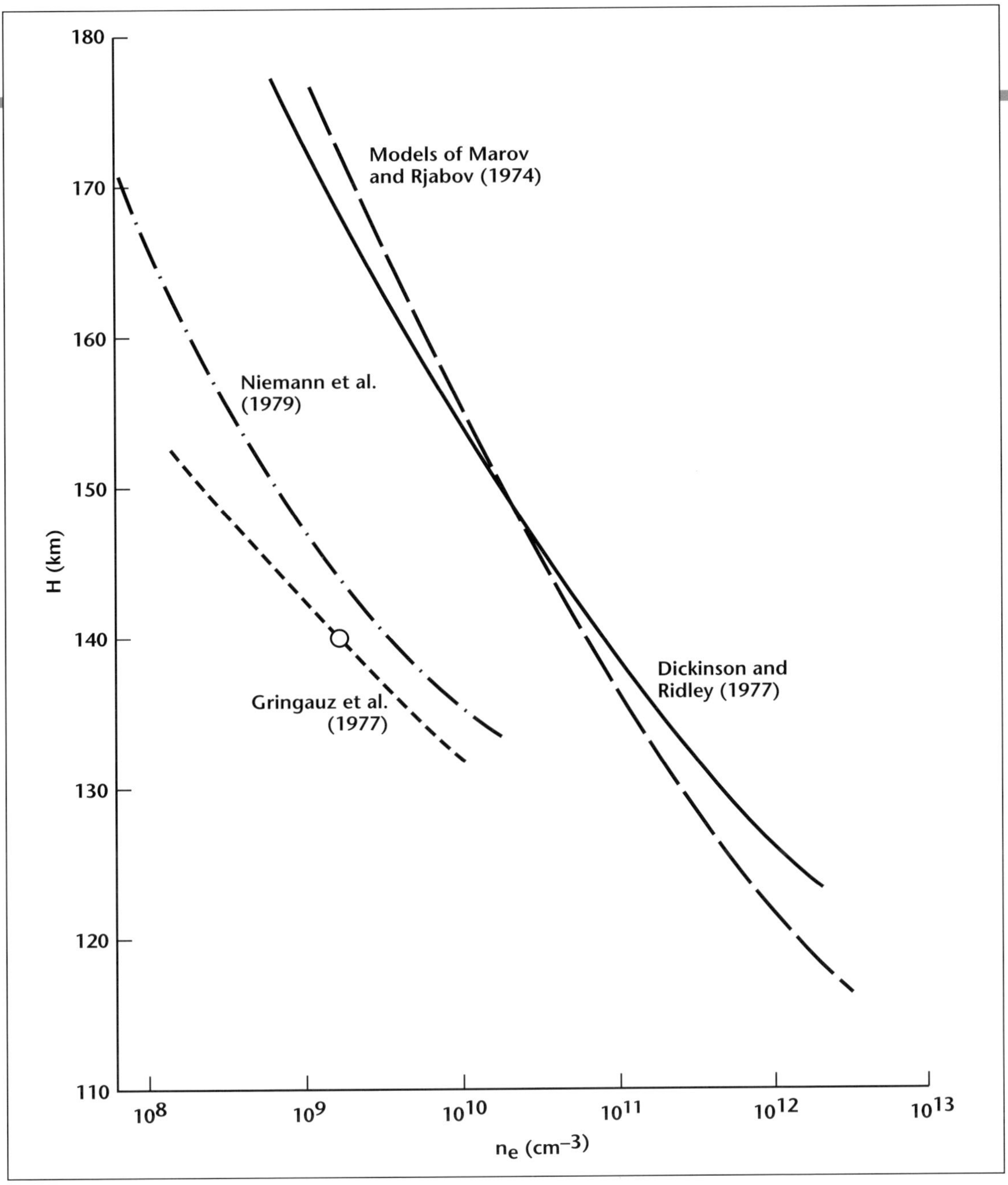

Figure 7-16. Dependence on a height, h, of number density of neutral particles, n_n, according to the models by M. Marov and O. Rjabov (Institute of Applied Mathematics, U.S.S.R. Academy of Sciences), R. Dickinson, and E. Ridley. The dependence $n_n(h)$, suggested by the group headed by K. Gringauz (Space Research Institute, U.S.S.R. Academy of Sciences), agrees with the results of H. Niemann et al. *obtained from Pioneer Venus.*

on the one hand and by solar-wind streaming pressure on the other (Pioneer Venus)

- Discovery of magnetic "flux ropes" in the ionosphere (Pioneer Venus)
- Explanation of the nature of the magnetic flux ropes (Soviet and American interpretation of results)
- Detection of the magnetic field increase before the ionopause in laboratory and numerical experiments, confirming the existence of the magnetic barrier (Soviet results).

Prospects for Further Research

Not everything we have learned about Venus appears here. Our knowledge of the planet has been enriched considerably. But has Venus ceased to be a mystery planet? Unfortunately (or fortunately), the answer is no. Venus still has many mysteries. While earlier puzzles were unraveled and many problems were solved, new mysteries arose which are much more difficult to understand.

Some of the problems yet to be solved are:

- We still have no true explanation for the higher content of primordial inert gases on Venus.
- It is entirely unclear why there is so little water in the Venusian atmosphere. Has Venus formed without water? Is water hidden in the crust, or was it lost during the planet's evolution? Why is the vertical profile of water vapor concentration so extraordinary?
- We have not yet determined the chemical composition of the cloud cover particles.
- We do not understand the mechanism responsible for the motion of the atmosphere at altitudes of 40 to 70 km (25 to 43 miles), the four-day rotation.
- How active is the planet's interior? Is there volcanic or seismic activity?
- Finally, we do not know when the present temperature conditions of Venus' atmosphere and surface set in. Did these conditions exist when Venus formed? Or was Venus' climate more moderate during a sufficiently long initial epoch?

How should the exploration of Venus continue? Evidently, only spacecraft of different types can solve such diverse problems. To study atmosphere dynamics, balloons are indispensable. We also could use them to investigate the cloud cover's physical and chemical properties.

Descenders, or probes, are needed to study the chemistry of the minor constituents of Venus' atmosphere and its thermal budget. These spacecraft would operate along the usual descent trajectory from parachute deployment to touchdown. For best results, they should begin to function at the highest altitude possible, at no less than 70 km (43 miles). Finally, seismic observations require that instruments remain on the planet's surface for many months. Engineers must design this special equipment to operate at high temperatures. The technical problems are numerous, but we are hopeful that we can solve them. We also expect that new and more sophisticated instruments will appear.

Another interesting program was the Soviet-French Vega project. This program included two new spacecraft that were improvements on Veneras 11 and 12. These spacecraft would fly by the planet and jettison two landers for a soft landing on the planet. Each flyby also would inject two balloons to study atmospheric dynamics.

The remaining Russian contribution to this book (below) refers to the Vega mission. The new spacecraft's mission to Venus and to Halley's comet was highly successful. Its results appear in the next section of this chapter.

The Vega landers are designed to study chemical composition of inert gases, aerosol particles, thunderstorms, and other properties during their descent. These landers are equipped to measure pressure, temperature, chemical composition of Venus' soil, and possibly seismic activity.

A particularly fascinating Vega mission involved one of the brightest and most interesting comets in the Solar System. The comet Halley approaches the Sun once every 76 years. Such an event occurred in 1986, and Soviet scientists prepared a Vega mission to record the event.

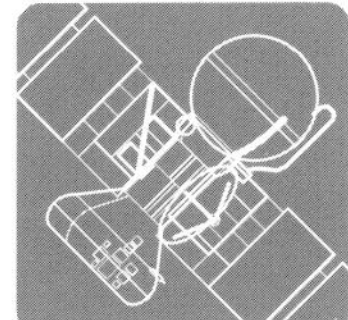

Comets can help us understand the Universe's origin and evolution. There is an assumption that comet nuclei are the material from which the planetary system formed. Until the Vega mission, astronomers could only study comets with ground-based instruments. We knew practically nothing about the structure of comets' nuclei, ionization sources, mechanisms for formation of plasma structures in comets' tails, and the reasons for the comets' various shapes.

Conditions for observing the comet from Earth were relatively unfavorable in 1986. So studying Halley's comet from space was particularly important. To investigate Halley's comet, the European Space Agency launched the comet flyby spacecraft Giotto. Japan launched two spacecraft, Sakigake and Suisei.

The Soviets had not planned a special mission to the comet. However, Vega flyby vehicles to Venus were able to use a gravitational maneuver near the planet to travel on to the comet (Figure 7-17). These vehicles approached within several thousand kilometers of the comet and were able to photograph its nucleus. Among many other phenomena, they studied components of the dust and gas that evaporated from the nucleus, and ion concentrations. These three projects—European, Japanese, and Soviet—complemented each other, in terms of both scientific goals and equipment.

THE FINAL VENERAS AND THE NEW VEGA SPACECRAFT

R. O. Fimmel, L. Colin, E. Burgess

We have added the following material to this chapter to give you information that goes beyond the period covered by the Soviet authors. This material documents other events in the exploration of Venus. It covers the period from the extended Pioneer Venus mission to the beginning of NASA's Magellan mission to Venus.

After missing the 1976-1978 launch opportunity, the Soviets sent their next mission to Venus in September 1978. Veneras 11 and 12 were each a combination flyby and lander spacecraft. They arrived in December 1978. The flyby spacecraft gathered data on the ultraviolet spectrum of the upper atmosphere as they sped by Venus. They successfully telemetered these data back to Earth.

Both landers provided atmospheric data as they penetrated the atmosphere before landing safely on the surface. During the descent through the atmosphere, an instrument designed to search for "thunderstorm" activity recorded radio bursts that might be attributed to lightning. These data reached Earth about five days before Fred Scarf detected "whistlers" with the Pioneer Venus instruments (see Chapter 6). Pioneer's orbital configuration did not allow an earlier search for such whistlers.

The Venera landers found the ratio of argon-40 to argon-36 was several hundred times less than in Earth's atmosphere. Why did Venus have so little argon-40, a decay product of potassium-40? The amount of this potassium isotope in Venusian rocks is about the same as in terrestrial rocks. One possibility is that Venus may not have experienced as much volcanic activity as Earth. Arguing against this, however, are the images returned by Magellan showing that Venus has experienced a great deal of volcanic activity.

The issue might be resolved if atmospheres were the result of comet impacts rather than mainly the result of internal activity and evolution of volatiles from within the planets. In such a scenario, incoming material, not planetary material, would govern isotopic ratios. The atmospheric ratios would bear no relationship to ratios in the material of the planet itself.

Figure 7-17. Vega-Halley mission—the red line shows the flight trajectory of Halley's comet; the yellow line shows the trajectory of the flyby vehicles. The green and blue lines show the orbit of Venus and Earth, respectively. Top left shows the descent module separating from the flyby spacecraft at Venus. At the top right is a general view of a prototype of the Vega spacecraft: 1) the flyby spacecraft, and 2) the descent module. The final version carried a balloon probe and a lander in the descent module. In bottom right, the flyby spacecraft is passing by Comet Halley, almost nine months after its encounter with Venus.

Venera 12's lander settled on the surface near Phoebe Regio, where it measured a surface temperature of 480°C (896°F). It also recorded an atmospheric pressure 88 times greater than Earth's sea level pressure. The flyby spacecraft relayed telemetry from the lander. After roughly 110 minutes of data relaying, the flyby spacecraft went below the horizon as viewed from the landing site. Communication ended. The other lander, Venera 11, measured close to the same atmospheric pressure, but it recorded a temperature some 34°C (93°F) less than at the Venera 12 site about 725 km (450 miles) farther north in Phoebe Regio. This lander lost

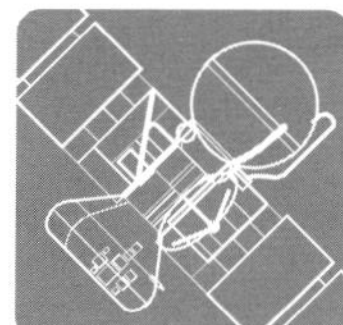

contact with the flyby spacecraft 95 minutes after landing.

No Russian spacecraft were sent to Venus at the 1980 opportunity. The next missions were Veneras 13 and 14, two spacecraft launched on October 30 and November 4, 1981 (see Table 7-3). Again the buses that transported the landers to Venus were flybys. This arrangement allowed more weight to be allocated to the landers for a given launch weight. Soviet scientists again targeted the landers for the Phoebe Regio area, where they landed on March 1 and March 5, 1982, respectively.

These landers accomplished the high technology task of gathering small soil samples from the planet's hot surface and examining them without exposing the interior of the landers to the high surface pressure and temperature. The samples were analyzed within closed chambers. A series of airlocks moved the samples by airstream to chambers of decreasing pressures. The last chamber had a temperature of only 30°C (86°F) and a pressure one-tenth of Earth's at sea level. Basaltic sand and dust was identified in the samples at both landing sites. Of the two varieties in the samples, one is scarce on Earth. No granite was found in the sample material. The landers had improved photo-imaging systems and returned excellent images to Earth for color reproduction. These showed the typical flat Venusian landscape strewn with flattened rocks and weathered lava flows. Fine rubble covered much of the surface at the Venera 13 site, but there was much less of this material at the Venera 14 site, which was 725 km (450 miles) farther south and 740 km (460 miles) to the west in a lower area. Fluid lavas from the Phoebe volcanoes may have covered this area. Plate-like rocks visible in the Venera 14 and other surface images so resemble sedimentary rocks that some Russian experimenters have suggested that Venus' high atmospheric pressures may have led to a sedimentation process in the atmosphere similar to that in bodies of water on Earth.

Dust suspended in the atmosphere colored the sky orange, somewhat like the skies of Mars. The light yellowish-orange color of soil and rocks suggests that the surface is heavily oxidized, again like Mars. This oxidation might imply that the planet once possessed large bodies of water, the oxygen from which became trapped in surface rocks while hydrogen escaped into space.

Venera 14's instruments also detected what scientists believed were slight seismic disturbances. However, the other lander did not detect any such disturbances.

Veneras 13 and 14 temperature and atmospheric pressure readings were very similar to earlier lander measurements. These landers operated successfully for much longer than their design lifetimes. Venera 13 survived for 127 minutes. Venera 14 lasted for 63 minutes at a lower elevation and at a pressure of nearly 100 atmospheres.

Soviet scientists launched Veneras 15 and 16 on June 2 and June 6, 1982. These spacecraft went into near polar orbit around Venus in October 1982. Like Pioneer Venus Orbiter, they had an orbit with a period of 24 hours. Periapsis was a few thousand kilometers over the northern hemisphere and apoapsis about 65,000 km (40,391 miles) above the southern hemisphere. Science data began flowing to Earth in late October. The spacecraft carried advanced side-looking radar, which produced surface radar maps at higher resolution than those from Pioneer Venus Orbiter. The spacecraft also were able to map the surface into higher northern and southern latitudes, supplementing the Pioneer data. A radar map

Table 7-3. Soviet Space Vehicles That Studied Venus 1979 to 1986

Space vehicle		Date Launch	Arrival	Activity
Name	Type			
Venera 13	Lander Flyby	10/10/81	3/1/82	Landed in region of Phoebe Regio; analyzed soil samples; returned colored photo images from surface
Venera 14	Lander Flyby	11/4/81	3/5/82	Same as Venera 14 but from another site
Venera 15	Orbiter	6/2/82	10/10/82	Radar maps of the planet to high northern and southern latitudes
Venera 16	Orbiter	6/6/82	10/14/82	Same as Venera 14
Vega 1	Lander	12/15/84	6/11/85	Lander continued work of earlier landers, but in Aphrodite Terra region.
	Probe			Probe carried by balloon around planet sampling atmosphere.
	Flyby			Flyby spacecraft continued to a flyby of Halley's comet on 3/6/86
Vega 2	Lander Probe Flyby Halley	12/21/84	1/15/85	Same as for Vega 1 Encounter 3/9/86

atlas was published confirming the earlier Pioneer Venus Orbiter maps, which showed Venus as a very complex planet. The Soviet maps revealed many shield volcanoes, volcanic domes, lava flows, chaotic terrain, long ridges, great depressions, parallel mountains and valleys, regions of intersecting ridges and valleys, and many impact craters. Later, all these features appeared with greater detail in NASA Magellan images. Planetologists interpreted some features as evidence of plate tectonics and earthquake-type displacements. A major surprise was that erosion has degraded many craters. The spacecraft also used infrared spectrometers to map Venus. These images showed warmer temperatures at the poles than at the equator and confirmed that there is little difference between day and night temperatures.

The New Vega Spacecraft

After the Venera series, the Soviet Union continued Venus exploration with a new type of spacecraft, Vega. These advanced spacecraft consisted of a flyby bus, an atmospheric balloon probe, and a soft lander. The balloon probes each carried four scientific experiments, the landers carried nine. The multipurpose bus carried the probe and lander to Venus and then continued to a rendezvous flyby with Comet Halley.

The mission started in December 1984 with launches of two identical spacecraft, Vega 1 and Vega 2. Both arrived at Venus in June 1985. Explosive bolts fired, and the descent capsules separated from the flyby bus. The Vega 1 descent capsule entered the atmosphere on June 11, 1985; Vega 2 entered on June 15.

At about 125 km (78 miles) above the surface, each descender separated into a lander and a balloon-sonde. The landers were targeted for the area north of Aphrodite Terra, where they soft-landed successfully. The balloon-sondes used a French-designed balloon and drifted through the atmosphere for about two Earth days at an altitude of about 50 km (31 miles). Venusian winds carried them along at an average speed of about 65 m/sec (about

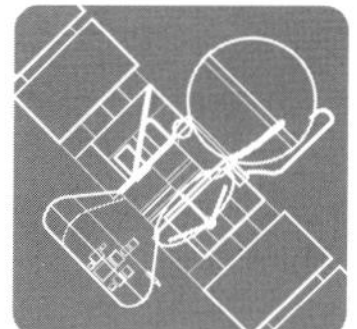

146 mph). As each balloon traveled some 10,000 km (6214 miles) around Venus, an international team of scientists evaluated their paths. Instruments recorded changes in light intensity but produced no conclusive evidence of lightning flashes.

The landers continued the Veneras' earlier work with more advanced instrumentation. Soil analysis by the Vegas discovered anorthosite-troctolite, a rock that is quite rare on Earth but common on the primitive crusts of the Moon and Mars. As they descended through the atmosphere, the landers also used instruments to sample the clouds to determine their sulfuric acid content. Both landers provided gas chromatograph and chemical reactor data that showed that clouds between 48 and 63 km (30 and 39 miles) contained one milligram of sulfuric acid per cubic meter. A mass spectrometer and an aerosol collector confirmed these concentrations.

Small samples of clouds were excited by x-ray to reveal the presence of other components. These included elemental sulfur, chlorine, and phosphorous. Other instruments determined the way in which light is diffused by the clouds and discovered that the particles have a size of about a tenth of a micron. (One micron is one thousandth of a millimeter.) The particle size determinations differed from the trimodal distribution that Pioneer Venus probes (1978) and Venera probes measured. The spacecraft identified two cloud layers at 50 and 58 km (31 and 36 miles) above the mean surface and each was about 5 km (3 miles) thick. The cloud layer results differed from earlier Soviet probes, suggesting that major changes occur in the clouds over large regions, since the two Vega probes entered the atmosphere some 1500 km (932 miles) from each other.

Comet Halley

In 1986, the flyby Vega spacecraft hurtled past Comet Halley on March 6 and March 9. Each spacecraft carried 15 scientific instruments for an international group of experimenters from nine countries. The first objective was to obtain a good look at the nucleus, which appears only as a star-like body from Earth.

The spacecraft observed the nucleus from distances of 8000 to 9000 km (4971 to 5593 miles). It was an elongated body some 14 km (8.7 miles) long and 7.5 km (4.7 miles) across, somewhat curved and irregular but definitely not two bodies. Images from the two spacecraft suggested a 53-hour rotation period for the nucleus. Even though dust clouds obscured the surface, the spacecraft were able to measure its reflectivity as being somewhat like that of the lunar surface.

Infrared measurements of the region near the nucleus suggested a surface temperature higher than predicted by the icy nucleus hypothesis. Scientists thought this high temperature might result from dust clouds close to the surface rather than from the surface itself. Ices must be present in the nucleus to give rise to the gas in the comet's coma, and evaporation of these ices would be expected to cool the nucleus. One explanation Russian experimenters suggested is that the nucleus' surface is covered with a thin refractory porous material. Solar radiation heats the top surface while the bottom is insulated and in contact with the icy material. Heat can be conducted internally to evaporate the ice while the resulting gases can escape through the porous material.

The spacecraft also made major contributions to studies of the comet's dust cloud. Scientists obtained hundreds of spectra from which they determined the dust's composition. Interaction of the comet with the solar wind was

investigated using another group of instruments, which also recorded crossings of the comet's bow shock. The information these instruments gathered is important to continuing studies of how Venus and other planetary bodies interact with the solar wind.

Studies of Venus, the other planets, and the comets in our Solar System will provide the key to a better understanding of Earth's evolution. Answering these questions is vitally important to our future, and efforts invested in such projects are certain to bear fruit.

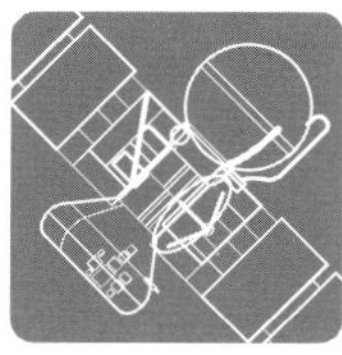

CHAPTER 8

VENUS AND EARTH

A COMPARATIVE PLANETOLOGY

Contributed by Thomas M. Donahue (Professor of Planetary Science, University of Michigan, and Interdisciplinary Scientist for Pioneer Venus Program)

In the early years of the space program, NASA and the planetary science community were enthusiastically selling missions like Pioneer Venus to members of Congress and the media. At that time, it was fashionable to argue that studying the other planets would help us understand Earth. There is certainly a sense in which this is true—and results from the Pioneer Venus mission demonstrate it well. However, there also is a sense in which it can be misleading. The best way, by far, to understand Earth's climate, weather, or any of the ways in which the planet works is to study Earth. But to understand Earth as one member of the collection of planets in the Solar System and, thus, to understand how that system formed and evolved, we must study more than Earth. It is necessary to study all the planets in detail. It would never have done to study Earth close at hand and the other planets only with telescopes. The results that Pioneer Venus achieved provide excellent examples of why this is true.

Perhaps the outstanding example was the enlightenment the mission brought to why Venus and Earth evolved into such different planets. Venus and Earth are among the four terrestrial planets. Mercury and Mars are the other two. There is good reason to believe that the four terrestrial planets have the same mixture of rocks, minerals, and volatile substances such as water and nitrogen. We base these beliefs on numerous observations of young solar systems elsewhere in the galaxy and sophisticated computer simulations. These terrestrial planets are the final products of the coalescence of a great swarm of objects called planetesimals. These grew by colliding with each other until, after 100 million years, only four planets (plus the asteroids) were left. At the end of this accretional epoch, a rain of comet-like objects from the outer Solar System may have coated these planets with volatile substances. But all terrestrial planets should have shared more or less the same endowment of original planetary material. Why, then, are Venus and Earth, which are such close neighbors and so similar in size and mass, so different?

Although they emerged from the same primordial nebula of gas and dust, Venus and Earth today are strikingly different. Why? Scientists speculate that, at one time, Venus may have had an abundance of water. What caused it to disappear? Can the processes that shaped Venus help us interpret Earth's continuing evolution? For example, can we learn about the greenhouse effect on Earth by studying Venus' past? This chapter provides insights into these and other questions.

A close look at the volatile inventory from Pioneer Venus and the Venera spacecraft data provides some clues. These data show that, whereas on Earth carbon in the form of carbon dioxide reacted with silicate rocks to form limestone, roughly the same amount of carbon dioxide exists as a gas dominating Venus' atmosphere. The reason for the difference is water. Earth has copious quantities of water necessary for the weathering of silicates leading to carbonate formation. The oceans also play a crucial role in limestone formation. Venus, as Pioneer measurements show, has less water by a factor of 250,000 than Earth. So what happened to the water that should have been as abundant on the early Venus as it was on the early Earth?

The Pioneer probe and orbiter measurements supplied an answer. When scientists studied deuterium—the heavy form of hydrogen—they found that deuterium is 150 times as abundant relative to ordinary hydrogen on Venus than it is on Earth. Deuterium's abundance strongly indicates Venus once had much more hydrogen than it now does. (Lighter gases escape more easily into space from a planet's gravitational field.) In fact, analysis showed that it must have had at least 300 times as much. That is, it had enough to form a planet-wide sea. This sea would have been between 4 and 25 meters deep if the hydrogen was in the form of liquid water.

Theoretical studies of the way hydrogen can escape from planets, which these observations inspired, showed that the original amount of water may well have been much greater. Some scientists estimate it as much as the equivalent of a full terrestrial ocean. The water might even have been liquid because the infant Sun was only about 75 percent as bright as today's Sun. With such a faint Sun shining, Venus' surface temperature would have been low enough for a while, perhaps, to have allowed an ocean to exist. But later, when the Sun grew brighter, the water would have evaporated, then converted into hydrogen and oxygen in the upper atmosphere by ultraviolet radiation and the hydrogen would have rapidly escaped. The process is called a runaway greenhouse.

In the runaway greenhouse, limestone would have released carbon dioxide gas, which would have entered the atmosphere. This would have resulted in the high surface temperatures on the planet today. Thus, it is possible to explain these high temperatures in terms of a greenhouse effect involving carbon dioxide. This explanation lends credibility to the argument that increasing carbon dioxide in the Earth's atmosphere also will increase its temperature. On the other hand, the Venus story inspired scientists to test the possibility that we might induce a runaway greenhouse on Earth. They found that Earth orbits too far from the Sun for that to happen.

The lack of an ocean may be relevant to understanding another great difference between Earth and Venus. Earth gets rid of its internal heat by means of convective motions in its fluid interior. These convective motions drive the Earth's crustal plates. Water may help provide the lubricant that allows some plates to dive beneath others in the planet's subduction zones. On the other hand, Pioneer Venus and Magellan mission radar found no evidence of plate motions on Venus, certainly not on the scale prevailing on Earth. Instead, these radar images clearly showed that at one time Venus shed its internal energy by volcanic activity. But even Venus' volcanism seems to have stopped a good while ago. Recent

Magellan measurements indicate that Venus now has a very thick, strong, and, therefore, dry crust. The same seems to be true of Mars. Even the interior of Venus may now be dry.

There is another excellent example of a class of similarities and differences among Solar System objects. This example concerns the ionospheres of terrestrial planets and the interaction of the solar wind with them. This particular observation only became clear when spacecraft carried out *in situ* observations. In the development of this understanding, Pioneer Venus played a central role. The ionospheres of Venus and Earth are very different and the solar wind interaction with them even more so. Apart from the difference in basic atmospheric constituents, the major reasons for the dichotomy are related to Venus' slow retrograde rotation and the absence of an intrinsic magnetic field. These two traits may well be related. In the case of Earth, the field seems to be created by an internal dynamo in the electrically conducting core. However, we do not completely understand the dynamo mechanism for Earth's much-studied field. Therefore, it is too much to expect that we would understand the reason for the absence of a magnetic field on Venus. One observation is certain, though, as Pioneer Venus and the Venera missions demonstrated in exquisite detail. The solar wind pushes close to Venus' ionosphere and interacts directly with it. On Earth, it stands far off because of the shielding that Earth's field provides.

After Pioneer Venus began to orbit Venus, experts formed another hypothesis involving the solar wind. They believed the interaction of the solar wind with Venus' atmosphere would have a direct analog in its interaction with comets, which are also intrinsically unmagnetized. But, when space missions encountered comets, such as Giacobini-Zinner and Halley, these experts learned again the need for direct measurements. The ease with which the solar wind picks up ions from the extended atmosphere of comets completely changed the physics of the interaction. In many ways, the analogy failed to hold. But only after intensive *in situ* study of the plasma environments of these three different objects were scientists able to clearly understand their natures.

There are other examples. Thus, several factors profoundly influence atmospheric circulation on Earth—in other words, Earth's weather and climate. These factors include the Earth's rapid rotation on its axis, the inclination of that axis (which is responsible for the seasons), and atmospheric pressure and composition. In particular, water, the crucial volatile present on Earth, scarce on Venus, is essential. In the case of Venus, tracking of the probes provided enough information about atmospheric circulation to form the basis for an eventual understanding. But clearly, we need more data on circulation in the deep atmosphere and the planet's radiation budget. Only with these data will we more completely understand the differences in the climates of the two largest terrestrial planets.

The Pioneer Venus Probe and Orbiter missions allowed us to make enormous strides in space sciences. They advanced our understanding of the way the objects of the inner Solar System work, how they originated, and how they evolved. But only more visits to Venus and the other members of the Solar System's family of planets, satellites, comets, and asteroids can make that understanding adequately complete and satisfying.

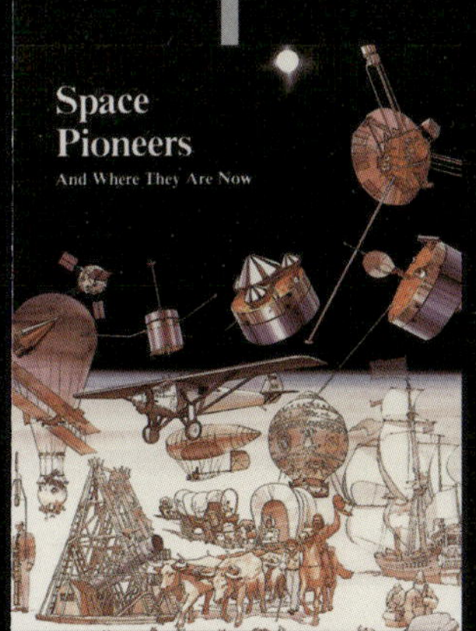

THE PIONEERS booklet published

PVO observes yet another comet in between its observations of Venus

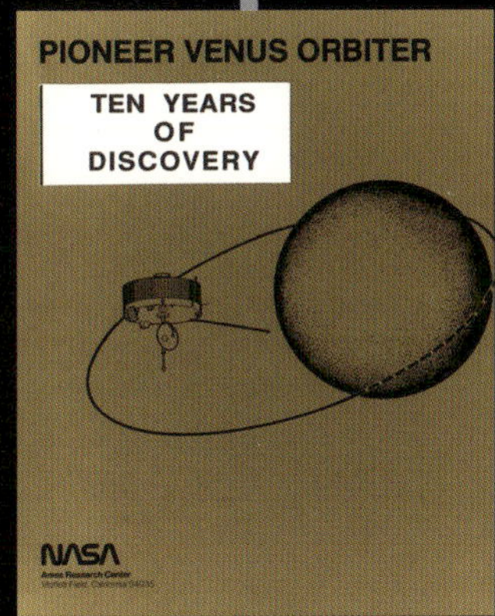

TEN YEARS OF DISCOVERY published; results of PVO science

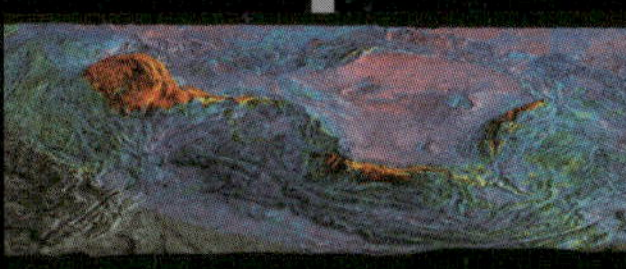

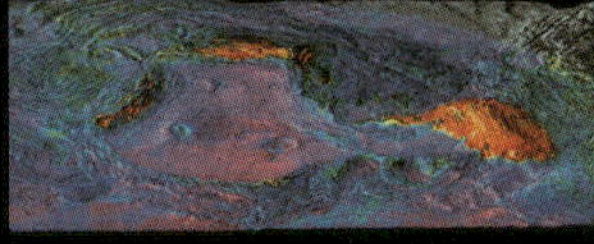

PVO ENTRY SCIENCE PLAN published

First Magellan detailed radar images of Venus

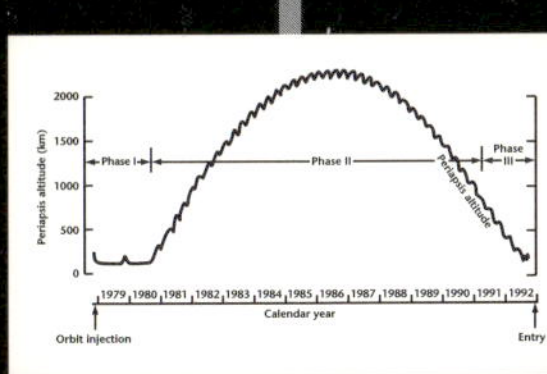

Periapsis falls closer to Venus; reaches lower thermosphere

Periapsis again controlled

PVO gathers details about lower ionosphere and atmosphere

In October PVO ends its 14-year mission; plunges into Venus' atmosphere

...WHILE ON EARTH

George Bush elected President

Soviets withdraw from Afghanistan

Iraq-Iran declare truce

Iraqis use poison gas on Kurdish civilians

1988

Hurricane Hugo

San Francisco, Loma Prieta earthquake

Valdez oil spill

Berlin Wall falls

US troops invade Panama

Tiananmen Square massacre

1989

Iraq invades Kuwait

Germany re-unites

John Major replaces Margaret Thatcher as British Prime Minister

Oil prices soar

Nelson Mandela released from prison in South Africa

1990

Soviet Union is dissolved

Operation Desert Storm frees Kuwait

Mount Pinatubo erupts

1991

US troops sent to Somalia

Serious destructive riots in Los Angeles

Hurricanes Andrew (Florida) and Iniki (Hawaii)

1992

APPENDIX A

Chronology of Exploration of Venus from Earth before the Pioneer Venus Mission

Date	Event
684 BC	Ninevah (Babylon) tablets record observations of Venus made as early as 3000 BC.
361 AD	Chinese annals record occultation of Venus by the Moon.
845	Chinese annals record an observation of Venus passing through the Pleiades.
1587	Tycho Brahe records an occultation of Venus by the Moon.
1610	Using the newly invented telescope Galileo discovers that Venus exhibits phases like the Moon.
1639	Horrox and Crabtree are first to observe a transit of Venus across the face of the Sun.
1643	Fontana claims that irregularities along the terminator of Venus are mountains.
1666	Cassini observes bright and dusky spots on Venus and claims Venus rotates in a little more than 24 hr.
1716	Halley records seeing Venus in daylight.
1726	Bianchini claims that Venus rotates in 24 hr.
1761	Lomonosov interprets optical effects observed during transit of Venus across the Sun as being due to an atmosphere on the planet.
1769	Captain Cook visits Tahiti to observe transit of Venus. Solar parallax determined to within a few tenths of an arcsecond.
1788	Schroter claims that his observations of Venus show that the planet rotates on its axis in 23 hr 28 min.
1792	Schroter concludes that Venus has an atmosphere because the cusps of the crescent phase extend beyond the geometrical crescent.
1807	Wurm determines the diameter of the visible disc of Venus to be 12,293 km (7639 mi.).
1841	De Vico claims, on the basis of his observations, that Venus rotates in a period of 23 hr 21 min on an axis included 53° to the planet's orbit.
1887	Stroobant explains that all the claims by astronomers of discovering a satellite of Venus were merely observations of faint stars.
1890	Schiaparelli concludes from his observations that Venus rotates in 225 days.

1907	Lowell produces drawings of Venus with broad dark lines that are hazy, ill-defined, and nonuniform. He concludes from his observations that Venus rotates in the same time that it revolves around the Sun, namely, 225 days.
1920	St. John and St. Nicholson, unable to detect any water vapor in its atmosphere, suggest that Venus is a dry, dusty world.
1922	Lyot measures the polarization of sunlight reflected from the clouds of Venus and introduces a new method of investigating the size and nature of particles in its clouds.
1927	Wright and Ross photograph Venus through ultraviolet filter.
1932	Adams and Dunham detect carbon dioxide in the atmosphere of Venus with a high-dispersion spectrograph on the Mount Wilson 100-in. telescope.
1942	Wildt shows that the high surface temperature of Venus could arise from a greenhouse effect in an atmosphere possessing a high proportion of carbon dioxide.
1945	Kuiper begins a long series of experiments with low- to high-resolution spectrographs to study rotational temperature of carbon dioxide at the cloud tops using infrared wavelengths.
1955	Hoyle suggests that the Venus clouds are a photochemical hydrocarbon smog.
1956	Mayer, McCullough, and Slonaker detect radio waves from Venus at 3-cm wavelengths, indicating that the surface temperature must be very high, about 330°C (626°F).
1956	Price makes the first radar sounding of Venus.
1957	Boyer discovers a 4-day rotation period of ultraviolet markings in Venus' clouds.
1960	Sinton and Strong establish temperature of the cloud tops as –39°C (–38.2°F), by infrared bolometry.
1960	Dollfus, using polarimetry, determines pressure at the cloud tops as 90 mbar.
1961	Opik proposes that clouds are thick dust consisting of calcium and magnesium carbonates.
1961	Sagan suggests that the high temperature of Venus' surface results from a greenhouse effect.
1961	Pettengill makes further radar observations of Venus and determines the astronomical unit with high precision.
1962	Kuz'min and Clarke show that the low radar reflectivity of Venus rules out any possibility of large bodies of water being on the surface.
1962	Carpenter and Goldstein, by radar observations of Venus, establish its rotation as being retrograde with a period of approximately 240 days.

1964	Deirmendjian proposes that the clouds are composed of water.
1966	Ash, Shapiro, and Smith analyze radar data and conclude that the diameter of Venus is 12,112 km (7526 mi.).
1966	Boyer and Guerin determine a cloud circulation of about 4 days from a study of ultraviolet photographs.
1966	Connes measures traces of HCl and HF in the atmosphere.
1967	Kuiper makes the first airborne observations of Venus.
1968	Eshleman and colleagues estimate surface temperature and pressure from radio, radar, and Venus probe data as 427°C (800°F) and 100 atm.
1970	Singer suggests that Venus lost its initial spin and obtained its present slow retrograde spin by impact of a satellite in a retrograde orbit.
1973	Young and Sill propose that the clouds of Venus consist of drops of sulfuric acid.
1973	Pollack observes Venus from a high-flying aircraft observatory and concludes that clouds are deep hazes of sulfuric-acid drops.
1973	Young describes observations of carbon-dioxide absorptions in the Venus atmosphere that show a 20% fluctuation of a 4-day period which represents upward and downward motions of the cloud deck on a planetwide scale.
1973	Goldstein's radar scans of Venus reveal huge, shallow craters on its surface.
1974	Goldstein produces high-resolution radar images of small areas of the planet's surface showing many topographic features.
1976	Carbon monoxide is detected in the upper atmosphere of Venus by Kitt Peak National Observatory. This gas had been detected earlier at lower altitudes through infrared spectroscopy.
1977	Radar images obtained at Arecibo indicate large volcanoes and craters on the planet.
1978	Barker identifies carbonyl sulfide in the Venus atmosphere.
1979	Sulfur dioxide is discovered in the atmosphere by observations from an ultraviolet satellite.

APPENDIX B

VENUS NOMENCLATURE AND MYTHOLOGY

M. E. Strobell and Harold Masursky
U.S. Geological Survey
Flagstaff, Arizona

During the last five years, committees of the International Astronomical Union (IAU) (1980) have chosen and approved names of Venusian surface features that appear on recently published maps (Masursky et al., 1980; Pettengill et al., 1980; U.S. Geological Survey, 1981) and a globe (U.S. Geological Survey and Massachusetts Institute of Technology, 1981). IAU committees developed this nomenclature, or naming system, to encourage scientists to discuss Venus' features. These included surface features, physical, chemical, and mechanical surface processes, and conditions within the planet's interior. All these features have led to the planet's present surface configuration.

Clouds and a dense atmosphere hide Venus' surface from visual observations. This fact prevented scientists from developing a naming system like the other terrestrial planets' before the mid-1960s. Early in the decade, monostatic and pulsed Earth-based radar systems were able to detect echoes from Venus' surface. With these systems, researchers were able to determine its spin-axis orientation and period of rotation.

At the same time, investigators recognized certain areas of anomalous, or unusual, reflectivity and brightness. Goldstein (1965) named the two brightest areas, which appear in images that California's Jet Propulsion Laboratory took in 1964, "Alpha" and "Beta." During the mid- and late-1960s, workers at other facilities (Carpenter, 1966; Dyce, Pettengill, and Shapiro, 1967; Rogers and Ingalls, 1969) confirmed these and other anomalous areas. At that time, each radar facility had its own informal naming system (Carpenter, 1966). In 1967, astronomers at the Arecibo facility, Puerto Rico, informally named features with high-delay Doppler frequencies for renowned physicists. They named one such feature, which they had not recognized previously, "Maxwell" (Jurgens, 1970). By 1969, scientists had recognized circular areas of very low reflectivity (Rogers, Ingalls, and Pettengill, 1974). In the early 1970s, they discriminated other irregular and elongate features on higher-resolution images.

When NASA completed plans for the Pioneer Venus mission, a Task Group for Venus nomenclature formed. It was established under the direction of the Working Group for Planetary System Nomenclature of the IAU. This task group's goal was to create a systematic plan for naming the features clarified by Pioneer Venus' altimetric and imaging systems. The group's goal also included naming those features appearing in a growing number of high-resolution Earth-based images.

The Task Group chose a theme in keeping with Venus' age-old feminine mystique (Table B-1, compiled by L. Colin). They named features for females, both mythological and real, who appear in the mythologies and histories of all world cultures. Circular, crater like features would be named for notable historical women while other features would bear the names of goddesses and heroines from myth and legend (IAU, 1977). The exceptions were "Alpha,"

Table B-1. Venus Mythology

Venus — Roman goddess of love and beauty, grace, fertility
Vesper — Latin, ancient Roman, evening star
Lucifer — Latin, ancient Roman, morning star

Aphrodite — Greek goddess of love, beauty, fruitfulness
Hesperos — Ancient Greek, evening star
Phosphoros — Ancient Greek, morning star

Quaiti — Egypt, evening star
Tioumoutiri — Egypt, morning star

Ruda — Arab, evening star

Helel — Hebrew, morning star

Ishtar (Istar) — Babylonia, Assyria, Mylitta, Chaldea, Sumeria

Astarte (Ashtarte) — Caanan, Phoenicia, Aramean, Southern Arabs, Egyptians

Athtar (Allat) — Arab

Ashtoreth — Biblical Israelite pagans

Anahita — Persia
Five names above are pagan semitic goddesses of love, fertility, maternity, sexual activity, and war.

Tai-pe — China, beautiful white one

Freya (Freyja) — Teutonic goddess of love, beauty, fertility
Frija — Old German
Frig (Friga) — Anglo-Saxon; Friday – 6th day of week
Frigg (Freia) — Old Norse

Chasca — Inca, goddess of love

Tlazolteotl — Mexico, goddess of love

Quetzalcoatl-Kukulcan — Post-classic Maya, lord of dawn

Noh Ek (Great Star), Chac Ek (Red Star), Sastal Ek (Bright Star), Ah Sahcab (Companion of the Aurora), Xux Ek (Wasp Star) — Mayan Venus

Cythera — Island birthplace of Venus

"Beta," and "Maxwell." These three names remained because they were, by then, firmly established in radar literature. The Task Group compiled an extensive list of female names that they could use. As scientists reduced Pioneer Venus data to map format, they applied names from the list to obvious features on those maps.

Two distinctive features stand out on Venus' topographic, reflectivity, and image maps. They are large radar-bright areas of highland terrain the size of terrestrial continents. These areas—Ishtar (Babylonian) and Aphrodite (Greek) Terrae—were named for goddesses of love. Ishtar Terra also is eye-catching on Earth-based images and on a mosaic that the astronomers at Arecibo compiled (Campbell and Burns, 1980).

Linear highland regions, which usually also are radar bright, take their names from other goddesses. Examples are Akna and Freyja Montes (mountains). Akna was the goddess of birth worshipped in Yucatan. Freyja was the principal Norse goddess and mother of Odin (Maxwell Montes is an exception). A high, relatively flat, and radar-dark area has the name Lakshmi Planum (plateau) to honor the Indian goddess of prosperity and fortune.

Low quasi-circular, or elongate, lowland plains that are generally radar dark have names from mythological heroines. For example, Helen Planitia (plain) is the name of the lady whose face "launched a thousand ships," and Sedna Planitia honors a beautiful Eskimo girl. Linear clefts, or canyons (chasmata), in Venus' surface are named for goddesses of the hunt or Moon. A single personage often had both attributes: Artemis was the Greek goddess of the hunt and of the Moon. Diana was her Roman counterpart.

Radar-bright linear features that coincide with an abrupt topographic change, such as a cliff (rupes), are named for hearth goddesses. For example, Vesta Rupes was named for the Roman goddess.

All circular features received names of notable deceased women. Irregular craters at or near mountain summits assume the names of classical women. For example, Sappho Patera has the name of the Greek poetess. Craters in plains areas received their names from modern women, such as the physicist Lise Meitner.

Scientists have traditionally applied the term "Regio" to any feature on a planetary surface that they cannot clearly define or understand, usually because image resolutions are not clear enough. They applied the term first to the albedo features on Mars and, more recently, to dark regions appearing on Voyager images of Ganymede. On Venus, astronomers originally used the term to describe the radar-bright features Alpha and Beta that Earth-based radar systems identified. The term now includes regions of somewhat elevated terrain that are smaller than continents but do not necessarily appear as discrete features on other data sets. These features take their names from titanesses and giantesses.

Other features (the radar-bright linear regions we know as lineae, or lines), have very low topographic expression at Pioneer Venus resolutions. Because of their low resolution, they are well known only in reflectivity images. These features are named for goddesses and heroines of war. These include women such as Hippolyta, the Greek leader of the Amazons, and Vehansa, the Teutonic war goddess. Features that are now designated as a linea or regio (region) may receive other generic feature designations at a later date. This can happen if future radar missions obtain higher resolution data that clarify these regions' true geomorphic expression.

Names of Venusian features appear in Table B-2.

Table B-2. Venus Nomenclature Assigned

Name	Latitude	Longitude East	Attribute
Chasmata (goddess of the hunt; Moon goddess); canyons			
Artemis	30–40S	125–145	Goddess of the hunt/Moon
Dali	21S	165	Goddess of the hunt
Devana	0-12N	289	Goddess of the hunt
Diana	15S	150	Goddess of the hunt/Moon
Heng-o	5N	348-358	Chinese Moon goddess
Craters (modern notable women)			
Eve	32S	0	Symbolizes the first biblical woman
Colette	65N	322	French novelist and writer
Lise Meitner	55S	321	German-Swedish physicist (1878–1968)
Sacajawea	63N	335	Shoshone Indian guide to the Lewis and Clark expedition to the Pacific Northwest (1786–1812)
Patera (classical notable woman); irregular, possibly volcanic craters			
Cleopatra	65N	10	Famous Egyptian queen; notable for her love affairs with Julius Caesar and Mark Anthony (69 B.C.–30 B.C.)
Sappho	15N	16	Greek lyric poetess of great power (580–610 B.C.)
Theodra	23N	280	Wife of Justinian; most famous and powerful woman in Byzantine history. Influential in passing laws that first recognized rights of woman (508–548 B.C.)
Linea (goddess and heroine of war); lines			
Antiope	40S	350	Amazon
Guor	20N	0	Valkyrie; Norse female warrior; means "battle"
Hariasa	19N	15	Germanic war goddess
Hippolyta	42S	345	Amazon
Kara	44S	306	Valkyrie maiden who, in Icelandic legend, sang so sweetly that the enemy could not defend themselves
Molpadia	48S	359	Amazon
Vihansa	54N	20	War goddess
Montes; mountains			
Akna	68N	318	Yucatan; goddess of birth
Freyja	73N	335	Mother of Odin in Teutonic mythology
Hathor	38S	323	Ancient Egyptian goddess of the sky
Maxwell	65N	45	James C. Maxwell; British physicist (1831–1879)
Rhea	32N	283	Titaness; Earth goddess
Theia	25N	281	Titaness in Greek mythology

Table B-2. Venus Nomenclature Assigned (Cont.)

Name	Latitude	Longitude East	Attribute
			Planum; plateaux
Lakshmi	67N	330	Indian goddess of fortune and prosperity
			Plainitiae (heroines)
Aino	45S	90	Finnish heroine; Vainamoinen, one of the Kalevala heroes, wished to marry her; she became a water divinity and thus escaped him
Atalanta	54N	162	Atalanta swore she would only marry the man who could beat her at a footrace. Melanion dropped three golden apples during the race and was able to win the race when Atalanta stopped to pick them up
Guinevere	40N	310	Wife of King Arthur and beloved of Lancelot
Helen	55S	255	Wife of Menelaus; Paris, son of Priam of Troy, fell in love with her and carried her off to Troy, thus precipitating the Trojan War
Lavinia	45S	350	Wife of Aeneas
Leda	45N	65	Wife of Tyndareus; Zeus, enamored of her charms, disguised himself as a beautiful swan and seduced her. She gave birth to Pollux and Helen (by Zeus) and Castor Clytemnestra (by Tyndareus)
Niobe	38N	120	Wife of Amphion of Thebes. She gave birth to 12 children, who were all killed by Artemus and Apollo
Sedna	40N	335	A beautiful Eskimo girl, who was wooed and won by a phantom bird who carried her off to a far shore. Sedna's father followed them, stole Sedna back, and started home with her. The phantom bird made a great storm come up, and the father, in fear, threw Sedna into the ocean. When she tried to climb back in the kayak, her father cut off parts of her fingers, which became seals, walruses, and whales
			Regiones (alphanumeric; female titans); regions
Alpha	20–30S	0–10	First letter in the Greek alphabet
Asteria	18–30N	228–270	Greek titaness
Beta	20–35N	280–290	Second letter in the Greek alphabet
Eistla	15-30N	345-5	Norse giantess
Imdr	42S	211	Norse giantess
Metis	62N	255	Greek titaness
Phoebe	10N–20S	275–300	Greek titaness
Tellus	35N	80	Greek titaness
Tethus	55N	100	Greek titaness
Themis	37–40S	275–310	Greek titaness
Ulfran	28N-3S	220-230	Norse giantess
			Rupes (goddess of hearth, home); cliffs
Ut	48–53N	305–325	Turco-Tatar goddess of the hearth fire
Vesta	55–65N	295–335	Roman hearth goddess
			Terrae (goddess of love); continents
Aphrodite	10N–35S	60-140	Greek goddess of love
Ishtar	60–75N	315–60	Babylonian goddess of love

References

Campbell, D. B.; and Burns, B. A.: Earth-based Radar Imagery of Venus. J. Geophys. Res., vol. 85, no. A13, 1980, pp. 8271–8281.

Carpenter, R. L.; and Department of Astronomy, UCLA: Study of Venus by CW Radar—1964 Results. Astron. J., vol. 71 , no. 2, 1966, pp. 142–152.

Dyce, R. B.; Pettengill, G. H.; and Shapiro, I. I.: Radar Determination of the Rotations of Venus and Mercury. Astron. J., vol. 72, no. 3, 1967, pp. 351–359.

Goldstein, R. M.: Preliminary Venus Radar Results. J. Res., sec. D, Radio Science, vol. 69D, no. 12, Dec. 1965, pp. 1623–1625.

Jurgens, R. F.: Some Preliminary Results of the 70-cm Radar Studies of Venus. Radio Science, vol. 5, no. 2, 1970, pp. 435–442.

Masursky, Harold; et. al.: Pioneer Venus Radar Results: Geology from Images and Altimetry. J. Geophys. Res., vol. 85, no. A13, 1980, pp. 8232–8260.

Pettengill, G. H.; et. al.: Pioneer Venus Radar Results: Altimetry and Surface Properties. J. Geophys. Res., vol. 85, no. Al3, 1980, pp. 8261–8270.

Rogers, A. E. E.; and Ingalls, R. P.: Venus—Mapping the Surface Reflectivity by Radar Interferometry. Science, vol. 165, no. 3895, 1969, pp. 797–799.

Rogers, A. E. E.; Ingalls, R. P.; and Pettengill, G. H.: Radar Map of Venus at 3.8 cm Wavelength. Icarus, vol. 21, no 3, 1974, pp. 237–241.

U.S. Geological Survey: Altimetric and Shaded Relief Map of Venus: U.S. Geological Survey Miscellaneous Investigations Series Map I-1324, 1981.

U.S. Geological Survey and Massachusetts Institute of Technology: Globe of Venus. Replogie Globes, Inc., 1981.

Working Group for Planetary System Nomenclature. IAU Trans., vol. 16B, 1977, pp. 321–369.

Working Group for Planetary System Nomenclature. IAU Trans., vol. 17B, 1980, pp. 285–304.

Additional Reading

Aveni, A. F.: Venus and the Maya. American Scientist, vol. 67, no. 3, May–June 1979, pp. 274–285.

Azimov, I.: Venus, Near Neighbor of the Sun. Lothrop, Lee & Shepard, 1981.

Durant, W.: Our Oriental Heritage: Being a History of Civilization in Egypt and the Near East to the Death of Alexander, and in India, China and Japan from the Beginning to Our Own Day. Simon and Schuster, 1954.

Hawkes, J.: The Atlas of Early Man. St. Martin's Press, 1976.

Moore, P.: The Planet Venus. Second ed. Macmillan, 1959.

Wilson, C.: The Starseekers. First ed. Doubleday, 1980.

APPENDIX C

SCIENCE RULES AND WORKING GROUPS

A. Rules of the Road for Pioneer Venus Investigators

The Pioneer Venus Science Steering Group (PVSSG) developed a set of procedures and rules to assure an orderly and efficient analysis and interpretation of the mission's scientific results. These rules appear here for historical interest and for possible use in future projects.

1. Instrument Principal Investigators, Radio and Radar Science Team members, and Interdisciplinary Scientists (including the project scientist for the purpose of these rules) were designated Pioneer Venus Investigators (PVIs). Only PVIs could sponsor investigators (research projects involving unpublished Pioneer Venus data). (Later, the Principal Investigators were PIs, Radar and Radio Scientists were Radioscience Investigators (RIs), and Interdisciplinary Scientists were IDS. A planned Guest Investigator (GI) program began in 1981.)

2. Each instrument PVI was responsible for analysis and interpretation of data from his instrument. He and his co-investigators (COIs, originally Co-Is) were responsible for the initial analysis, interpretation, and publication of those data. The three months following the acquisition of any data by the PVI were important. During that time, he identified the investigation that he, his COIs, and associates expected to pursue with those data. (Associates were people such as graduate students or post-doctoral research fellows who were clearly identified with the PVI or his COIs. Normally, the criterion was funding for their salaries through PV data analysis contracts. They were specifically not senior, independent scientists who belonged to the same institution as the PVI or COI.)

3. PVIs and COIs had free access to all data acquired during the mission (and extended mission). They also had access to publications resulting from the use of those data. The normal vehicle for data dissemination was the Unified Abstract Data System.

4. If a PVI's unpublished data were used in an investigation, that PVI had the right to be included among the authors of any publication that resulted. During the formative stages of an investigation, it was the sponsoring investigator's responsibility to solicit the participation of the PVI whose data or results he was using. The PVI, whose cooperation the investigator solicited, could refuse coauthorship but not use of his data. He had, however, to provide information concerning the quality of the data in question. He also could require that the sponsoring investigator include suitable caveats regarding the data in publications.

5. The role of an Interdisciplinary Scientist (IDS) in this mission was to enhance the scientific output of the mission. He did this by promoting investigations that used data from more than one instrument. Mission personnel hoped that the IDS would be able to promote cooperation among other PIs. They also hoped that other PVIs would exploit any unusual insights the IDS had. In this way, they would enrich interpretation of data from specific instruments as well as from an ensemble of instruments. Thus, administration normally expected IDSs to participate in investigations that involved data from more than one instrument. This could occur either as a result of their proposing such investigations or by invitation from other PVIs to participate.

Suppose a group of PVIs proposed an investigation in an area in which an IDS was a specialist. Normal procedure was to invite him to participate. After the three-month period in Rule 2, an IDS could propose an investigation involving data that a single instrument produced. COIs of the PVI responsible for that instrument also had a right to participate in that investigation. They could ask their associates to participate, too.

6. PVIs or COIs could not preempt major science areas for themselves. They had to pursue an investigation promptly.

7. Scientific Working Groups normally provided the forum in which researchers discussed investigations. Researchers had to send titles and descriptions of proposed investigations to the Project Scientist. He served as the interface between investigators, project, and other PVIs. In particular, he informed all PVIs of proposed new investigations. Objections or comments by other PVIs went to the Science Steering Group's co-chairmen for settlement or other appropriate action.

8. PVIs could release their own data to whomever they wished, but not the data of other PVIs without their consent.

9. There was no Pioneer Venus mission policy for paper form or publication medium. The only exception was a possible agreement for publication of the mission's initial results.

10. Independent scientists who were not mission PVIs, COIs, or associates could participate in an investigation provided:

(a) A PVI sponsored them.

(b) They provided suitable correlative data for distribution to other PVIs through the sponsoring PVI.

(c) The rest of the PVIs gave their approval before the investigation started, and the SSG issued a letter of invitation and cooperation. Such scientists later formed the program's GIs. This group began in 1981. The objective was to involve new scientists in the program who would bring a fresh perspective to data analysis and interpretation.

B. Pioneer Venus Working Groups

The PVSSG developed a set of six Working Groups to address particular disciplines. These disciplines were Composition and Atmosphere Structure; Clouds; Dynamics; Thermal Balance; Solar Wind, Ionosphere, and Aeronomy; and, Surface and Interior. The Working Groups were very successful and wrote group papers synthesizing results from various experiments.

Composition/Atmosphere Structure

Primary

J. Hoffman (LNMS)—Chairman
A. Stewart (OUVS)
V. Oyama (LGC)
U. von Zahn (BNMS)
H. Niemann (ONMS)
A. Seiff (LAS/SAS)
D. Hunten (IS)
N. Spencer (IS)
T. Donahue (IS)
G. Keating (RADIO)
A. Kliore (RADIO)

Secondary

F. Taylor (OIR)
R. Knollenberg (LCPS)
H. Taylor (OIMS)
R. Goody (IS)
A. Nagy (IS)
J. Pollack (IS)
T. Croft (RADIO)

Cloud

Primary
R. Knollenberg (LCPS)—Chairman
R. Ragent (LN/SN)
F. Taylor (OIR)
J. Hansen (OCPP)

Secondary
A. Stewart (OUVS)
V. Oyama (LGC)
M. Tomasko (LSFR)
V. Suomi (SNFR)
D. Hunten (IS)
J. Pollack (IS)
T. Croft (RADIO)

Dynamics

Primary
G. Schubert (IS)—Chairman
C. Counselman (DLBI)
F. Taylor (OIR)
A. Seiff (LAS/SAS)
J. Hansen (OCPP)
R. Woo (RADIO)
T. Croft (RADIO)

Secondary
G. Pettengill (RADIO)
U. von Zahn (BNMS)
V. Suomi (SNFR)
R. Goody (IS)
J. Pollack (IS)

Thermal Balance

Primary
M. Tomasko (LSFR)—Chairman
F. Taylor (OIR)
R. Boese (LIR)
A. Seiff (LAS/SAS)
J. Hansen (OCPP)
V. Suomi (SNFR)
R. Goody (IS)
J. Pollack (IS)

Secondary
A. Stewart (OUVS)
V. Oyama (LGC)
D. Hunten (IS)

Solar Wind/Ionosphere Aeronomy

Primary
S. Bauer (IS) (later A. Nagy (IS)—Chairman
I. Stewart (OUVS)
F. Scarf (OEPD)
C. Russell (OMAG)
L. Brace (OETP)
H. Taylor (OIMS)
W. Knudsen (ORPA)
A. Barnes (OPA) formerly J. Wolfe (OPA)
N. Spencer (IS)
T. Donahue (IS)
T. Croft (RADIO)

Secondary
U. von Zahn (BNMS)
H. Niemann (ONMS)
D. Hunten (IS)
G. Keating (RADIO)
A. Kliore (RADIO)

Surface/Interior

Primary
H. Masursky (IS)—Chairman
C. Russell (OMAG)
G. Pettengill (ORAD)
W. Kaula (ORAD)
G. McGill (IS)
R. Phillips (RADIO)
I. Shapiro (RADIO)

Secondary
V. Oyama (LGC)
G. Schubert (IS)

C. Key Scientific Questions

Prior to the Pioneer spacecraft's launch, the six PVSSG Working Groups each developed a set of key scientific questions. These were questions that their members and the associated experiments could and would address during the mission.

Composition/Atmosphere Structure

Key Questions

- Present state of atmosphere

  Lower atmosphere composition
  - Apart from CO_2, what does the lower atmosphere consist of, and how are these constituents distributed?
  - What are the clouds made of?
  - What does the atmosphere tell us about the planet's surface and interior?

  Lower atmosphere structure
  - How do the state property profiles vary over the planet?
  - Why is the lower atmosphere so hot?
  - What role do phase changes play in the thermal structure?

  Upper atmosphere composition and structure
  - What are the composition and temperature profiles of the upper atmosphere and where is the homopause?
  - What are the spatial and temporal variations in Venus' upper atmosphere?
  - Is the stability of CO_2 due to global circulation or local turbulence?
  - How does the neutral composition influence the ionosphere and the thermal structure?
  - Does superrotation extend into the thermosphere?
  - How does the upper atmosphere respond to changes in solar extreme ultraviolet radiation and solar wind?

- Origin and evolution of Venus' atmosphere
  - Where did the atmosphere come from and where is it going?
  - Where is the water?
  - Why does Venus' atmosphere differ so much from Earth's?

Clouds

Key Questions

- What is the planetary cloud structure in altitude and horizontally?
- How deep do the sulfuric acid clouds extend?
- Do larger particles or denser clouds (higher concentration) exist at lower levels? What is their composition?
- Is the concentration of cloud particles proportional to gas pressure so that the scale heights of the particles and gas are identical?
- What substance is responsible for the ultraviolet absorption contrasts? Is the ultraviolet absorber well-mixed vertically and not horizontally?
- What is the structure and composition of the thin haze layers above the visible cloud deck (70-90 km (43-56 miles))? Do they correlate with the Mariner 10 radio-occultation inversions?
- What is the nature of the observed white polar caps?
- Is there aeolian transport of dust within 10 km (6 miles) of the surface?
- What are the couplings between cloud microphysics and Venusian dynamics?
- What are the cloud optical properties?

- What are the cloud formation and dissipation mechanisms? Coalescence? Coagulation? Condensation? Evaporation? Precipitation?

- Why is the cloud size spectrum so narrow?

Dynamics

Key Questions

- Upper atmosphere circulation

  Is the apparent four-day rotation an actual zonal motion of the atmosphere or is it a wave phenomenon?

  Do retrograde 100 m/sec upper atmosphere zonal winds flow all around the planet, even in the antisolar region?

  Is there a longitude-dependence of the zonal motion speed, especially with respect to the subsolar region?

  What is the latitude-dependence of the apparent zonal wind velocities?

  What is the altitude-dependence of zonal wind velocities? Is there essentially a decoupling of the upper atmosphere from the lower, with large zonal winds confined mainly to the upper atmosphere?

  What are the magnitudes of meridional motions?

  What mechanism drives the rapid zonal circulation of the upper atmosphere?

- Lower atmosphere circulation

  What is the nature of the lower atmosphere's circulation?

  Are the motions primarily zonal or meridional?

  What is the magnitude of the velocity?

  If the motions are meridional, do they represent a Hadley cell circulation?

  If the motions are zonal, is there an overall rotation of the lower atmosphere? Or is the circulation between subsolar and antisolar points?

  Are there unique motions (for example, small-scale convection) near the subsolar, antisolar, and polar regions in the deep atmosphere?

- Vertical flow and convection

  Are there strong upward and downward convective motions?

  What are the horizontal scales of convective cells?

  What are the magnitudes of vertical velocities?

- Waves and instabilities

  Are there any wave-like phenomena or instabilities that scientists can identify as occurring in the atmosphere?

- Distinctive features in the Mariner 10 imagery

  What atmospheric processes are responsible for circumequatorial belts, bow waves, spiral streaks, polar ring, and other distinctive features in Mariner 10 pictures?

- Turbulence and eddy diffusion

  What is the intensity of turbulence in the atmosphere? What are the altitudes of turbulent layers? What are their thicknesses? What are the turbulent eddy diffusion coefficients?

- Thermal contrast and energy deposition

  What are horizontal temperature contrasts that drive atmospheric motions? What is distribution of solar energy deposition in the atmosphere?

- Phase changes

  Do phase changes and associated latent heats of condensible species play an important role in the atmospheric dynamics?

- Nature of Ultraviolet Clouds

  What materials and physical processes are responsible for the ultraviolet albedo variations?

Thermal Balance

Key Questions

- What is the cause of the high surface temperature? If it is a greenhouse effect, what are the sources, other than CO_2, of the infrared opacity?

- Why are there small horizontal temperature contrasts near the cloud tops in the presence of strong apparent motions?

- Why are there small horizontal temperature gradients (both day-night and equator-pole) at the cloud tops and near the surface? (These occur despite an expected strong variation in the local deposition of solar energy over the illuminated hemisphere.)

- Why is the exospheric temperature so low?

- What are the roles of radiative and dynamical processes in maintaining the thermal balance of the atmosphere?

- What is the global (vertical and horizontal) temperature structure? How does dynamical heat transport determine it?

- Where are the sources and sinks of heating by solar and thermal radiation fields?

- What are the cloud optical properties?

- Do latent heat effects on convection produce subadiabatic regions in the generally adiabatic-looking vertical temperature profiles? Is the nearly adiabatic structure due to small-scale convection or planet-wide circulation?

Solar Wind/Ionosphere Aeronomy

Key Questions

- Venus ionosphere

  What is the ion composition, and what controls the plasma distribution of Venus' ionosphere?

  What is the plasma temperature of Venus' ionosphere and what controls its thermal structure?

  What are the mechanisms and the significance of mass, momentum, and energy transfer from solar wind to the upper atmosphere/ionosphere?

- Solar wind-Venus interaction

  Is there an intrinsic magnetic field?

How do ionospheric currents contribute to the deflection of solar wind?

How important are processes such as charge-exchange and mass-addition?

What is the source of the dayside ionosphere's variability?

How much solar wind does the ionosphere absorb?

Is there a magnetotail?

Is there a plasma sheet?

Are there substorms on Venus?

How does the plasma close behind the planet?

What maintains the nightside ionosphere?

What produces the two peaks in the electron density profile in the nightside ionosphere? What causes their variability?

What is the source of night-time airglow and the ashen light?

Is there a boundary layer or rarefaction region in the flow?

How does the Venus bow shock and upstream region differ from Earth's?

Surface/Interior

Key Questions

- What is the extent of endogenic activity leading to tectonics, crustal differentiation, and volcanism?
- What is the extent of exogenic processes such as impact cratering, weathering, and transportation and erosion of surface materials by winds and crustal recycling?
- What is Venus' gravity-field distribution? Is there evidence of density contrasts?
- Are tectonic features evident on the surface: arcuate mountain systems, strip-like faults of large displacement, rifts, volcanic craters, or chains of volcanic craters?
- Does Venus' interior consist of an iron core and a mantle of magnesium and iron silicates (like Earth)?
- What is, and what is the cause of, the offset of the center of mass from the center of figure?
- What is the subsurface temperature gradient? What has been Venus' thermal history?
- Can an exogenic effect (such as solar tidal torque or a planetesimal impact) explain Venus' slow retrograde spin?
- Does Venus possess an intrinsic magnetic field? How large is it?
- Is the surface in thermal and chemical equilibrium with the lower atmosphere?
- Is there a resonant lock between Venus' spin period and the relative orbital motions of Earth and Venus?
- Is Venus further along than Earth on the evolutionary path toward the end of complete compositional stratification and thermal quiescence?

GLOSSARY

A

ablative material: material that absorbs heat by converting state (i.e., solid to gas) and thereby carries the heat away.

adiabatic: without a loss or gain of heat from the surroundings.

aeronomy: study of the upper regions of the atmosphere where there is ionization, dissociation, and chemical reaction.

aeroshell: an insulating shell to protect a spacecraft from atmospheric heating during entry into an atmosphere.

airglow: a quasi-steady radiation from an atmosphere arising from collisions among molecules and atoms and distinct from aurora.

albedo: ratio of the amount of electromagnetic radiation reflected by a body to the amount incident upon it. For example, the albedo of Earth is 34%.

angstrom: unit of wavelength of light equal to 10^{-8} cm or 3.937×10^{-9} in.

anisotropic: exhibiting different properties when tested along different axes.

apoapsis: point in an elliptical orbit that is most distant from the center of attraction.

apsides: the two points in an orbit nearest and farthest from the center of attraction.

arcuate feature: a geological feature of bow shape.

ashen light: a glow from the dark side of Venus that some scientists claim is observable from Earth.

astronomical unit: the mean distance of Earth from the Sun, approximately 149,599,000 km (93,000,000 miles).

Atmosphere Explorer: an Earth-orbiting satellite in the NASA series used to explore the upper atmosphere.

B

ballistic trajectory: trajectory followed by a body moving solely under the influence of gravity.

bar: unit of pressure; 10^4 dyne/cm^2.

basalt: an igneous rock.

bifilar: consisting of two wires.

boundary layer: layer of fluid in immediate vicinity of a bounding surface.

bow shock: shock wave in front of a body at which the velocity changes abruptly.

caldera: a roughly circular volcanic depression whose diameter is many times that of the volcanic vent.

Carboniferous era: the fifth period of the Paleozoic Era.

Cassegrain telescope: a telescope using a primary and secondary mirror.

cold trap: a location in an atmosphere where gases can be trapped and prevented from rising higher in that atmosphere.

convective plume: a plume of hot magma rising from the interior of a planet toward its surface.

corona: the outer visible envelope of the Sun.

cryosphere: cold, upper atmospheric region of a planet.

deuterium: heavy isotope of hydrogen whose nucleus contains a neutron in addition to a proton.

differentiation: process in which light materials rise above heavier materials in a gravitational field.

Doppler shift: apparent change in frequency of a vibration such as sound or light or radio waves, resulting from relative movement between the observer and the source.

dropsonde: a capsule dropped from a spacecraft to investigate the atmosphere of a planet.

dynamo: a direct current generator that converts mechanical energy into electrical energy by motion of a conductor through magnetic field lines.

ecliptic plane: plane of Earth's orbit around the Sun. Ecliptic is the projection of this plane on the star sphere.

electron: subatomic particle that possesses the smallest possible negative electric charge.

electron density: a measure of the number of electrons per unit volume in an ionized gas.

eV: electron-volt; energy of an electron accelerated through a potential of one volt.

exosphere: outermost regions of an atmosphere where the molecules and atoms travel in ballistic paths and rarely collide with each other.

extreme ultraviolet radiation: ultraviolet radiation of very short wavelength.

F

Faraday cup: a device to measure plasma properties over a wide angular viewpoint.

flux rope: a unique magnetic structure consisting of a long, narrow region of strong magnetic field with field lines twisted like the threads of a rope.

G

gamma: a measurement of magnetic field intensity; 10^{-5} gauss.

gamma ray: electromagnetic radiation of very short wavelength beyond that of x-rays.

gamma burst: intense, short-lived pulse of gamma rays from deep space.

gravity wave: a wave disturbance in a fluid in which buoyancy, or reduced gravity, acts to restore displaced parts of the fluid back to hydrostatic equilibrium.

greenhouse effect: condition in which an atmosphere can absorb more radiation than it can emit back into space, thus causing a rise in temperature of that atmosphere.

H

Hadley cell: an atmospheric circulation pattern in which heated atmosphere rises at the equatorial region, travels at a high altitude toward the pole where it cools and descends and then travels back at a low altitude to the equatorial region.

heavy hydrogen: deuterium, an isotope of hydrogen whose nucleus contains a neutron in addition to a proton.

heliocentric: centered on the Sun.

high-gain antenna: an antenna that is designed to concentrate electromagnetic radiation into a tight beam.

hydrogen coma: a region around the head of a comet that hydrogen atoms occupy.

hydroxyl: a monovalent chemical group consisting of a hydrogen atom linked to an oxygen atom.

inertial space: a stationary frame of reference; a set of coordinates used for calculating trajectories.

I

inferior conjunction: position of a planet moving in an orbit within that of Earth when the planet is aligned between Earth and Sun.

ion: an atom or molecule that is positively or negatively charged.

ionopause: boundary between the shocked solar wind and the ionosphere of a planet.

ionosphere hole: a region of the ionosphere where the number of ions is severely depleted.

ionosphere: region of high atmosphere in which many of the molecules and atoms are ionized.

ionotail: an ionized region extending on the side of the planet away from the Sun.

isostatic: hydrostatic equilibrium maintained by flow of material from one part to another.

isothermal: thermodynamic change of state of a system that takes place at constant temperature.

isotope: atoms with the same chemical properties but with different atomic weights because of a different number of neutrons in the nucleus.

L

Langmuir probe: a device consisting of conductors inserted in a plasma to measure the plasma current.

line of apsides: the line connecting the closest and most distant points in an elliptical orbit from the center of attraction.

lithosphere: outer rocky shell of a planetary body.

Lyman alpha: radiation with a wavelength of 1216 angstroms in the extreme ultraviolet region of the spectrum; emitted by hydrogen atoms.

M

magma: hot volcanic rock.

magnetosphere: the volume around a planet affected by the planet's magnetic field.

magnetotail: magnetic field lines extending downstream from a planet away from the Sun.

mantle: the shell of a planetary body underlying the crust of that body.

mesosphere: atmospheric shell in which the temperature generally decreases with increasing height.

neutral atmosphere: atmosphere that is not ionized.

N

O

oblate: distorted from a sphere; equatorial diameter exceeds polar diameter.

occultation: hiding of one celestial body by another passing between that body and the observer.

orbital decay: loss of kinetic energy by an orbiting body so that it moves inward toward the center of attraction.

ozone hole: a region of the ozone layer in which the amount of ozone is depleted.

P

periapsis: point in an elliptical orbit that is closest to the center of attraction.

perihelion passage: usually refers to the closest point to the Sun in an orbit of a comet.

perturbation: disturbance of the orbit of one body orbiting another by the gravity of a third body.

photodissociation: breaking up of a molecule by radiation by the absorption of a photon.

photoelectrons: electrons released when a high-energy photon hits an atom.

planetary dynamo: circulation within a planet that produces a magnetic field.

plasma: an electrically conductive gas consisting of neutral particles, ionized particles, and free electrons.

plasma wave: a wave motion within a plasma.

plate tectonics: molding of a planetary surface by movement of plates of crust powered by forces acting from within the planet.

precession: change in the direction of the axis of a spinning body or the alignment of an orbit when acted upon by a torque.

prograde: motion in the usual direction of the bodies in a given system.

R

radio occultation: occultation of a radio source by a planetary body.

radionuclides: atoms that emit corpuscular or electromagnetic radiation.

redox reaction: chemical oxidation-reduction reaction.

regolith: surface material of a planet.

retrograde: motion opposite to the usual direction of the bodies in a given system.

runaway greenhouse: a condition in which the greenhouse effect continues to an extreme; for example, until all oceans boil and water is lost from a planet.

S

S-band: radio frequency band about 2.2 GHz allocated to space communications.

scale height: a measure of the relationship between density and temperature in an atmosphere.

sidereal day: a planet's period of rotation with respect to the stars.

solar cycle: the cycle of approximately 11 years over which solar activity varies in a repetitive fashion.

solar flare: a sudden outpouring of energy from the Sun.

solar panel: panel on a spacecraft that converts solar radiation into electrical energy.

solar wind: the blizzard of electrons and protons flowing from the Sun out across the Solar System.

spin axis: the axis on which a spacecraft spins to stabilize its orientation in space.

spreading center: location on a planet's surface from which crustal material emerges from within the planet to spread on its surface.

sub-spacecraft point: the point on the surface of a planet immediately beneath a spacecraft.

subsolar ionopause: the location in the ionopause immediately beneath the Sun.

sunspot number: a measure of sunspot activity based on the numbers of individual spots and groups of spots.

superior conjunction: conjunction of a planet and the Sun when the planet is on the far side of the Sun.

supernova: intense disruption of a star undergoing gravitational collapse with an enormous explosive production of energy and ejection into space of most of the star's mass.

superrotation: rotation of a planetary atmosphere faster than the rotation of the planet.

synodic period: period of rotation with respect to the Sun.

T

tectonics: molding of a planet's surface by forces arising from its interior.

telemetery: measurement at a distance.

terminator: the boundary on a planet between the sunlit and the dark hemispheres.

thermosphere: hot region of a planet's high atmosphere.

transfer ellipse: an ellipse connecting two planetary orbits.

transit: passage of one celestial body across the face of another as viewed from a third.

transponder: a combined receiver and transmitter designed to transmit signals automatically when interrogated.

troposphere: region of a planet's atmosphere where weather occurs.

U

ultraviolet: region of radiant energy beyond the visible region of the spectrum with wavelengths between 1000 and 3800 angstroms.

X-band: radio-frequency band allocated to space radio communications; about 8.5 GHz.

BIBLIOGRAPHY

SUGGESTED FURTHER READING

Over 500 journal articles, conference meeting papers, and reports on the Pioneer Venus Program have been published. A complete bibliography of these publications is available in two formats:

(1) a Macintosh disk "Pioneering Venus Exploration: A Bibliography"; and

(2) a printed document, NASA/Ames Research Center Pioneer Venus project document.

Contact the Pioneer Venus Mission's Science Chief Dr. Lawrence E. Lasher at Ames Research Center for instructions on how to obtain a copy of the disk or document.

The following bibliography contains major publications appropriate to the objectives of this NASA Special Publication. The references included provide suggestions for further reading that, except for the first three, should be readily available. The references are arranged chronologically in three groups: historical background; the Pioneer Venus mission and its results; and background on Pioneer Venus and other Venus missions.

Pioneer Venus Program: Historical

1970 Venus, Strategy for Exploration: Report of a Study by the Space Science Board. National Academy of Sciences, June 1970.

1972 Pioneer Venus: Report of a Study by the Science Steering Group, NASA TM-X-6237, June 1972.

1973 Schuyer, M.: Pioneer-Venus Orbiter: Assessment Report of Feasibility Studies by MBB, BAC, CIA. European Space Agency, 1973.

Pioneer Venus Mission and Results

1977 Colin, L. The Exploration of Venus—Pioneer Venus Program History. Space Science Reviews, vol. 20, May 1977, pp. 249–258.

Space Science Reviews, vol. 20, nos. 3–4, May–June 1977.

1979 Colin, L.: Encounter with Venus. Science, vol. 203, no. 4382, Feb. 23, 1979, pp. 743–745.

Science, vol. 203, no. 4382, Feb. 23, 1979.

Donahue, T. M.: Pioneer Venus Results: An Overview. Science, vol. 205, no. 4401, July 6, 1979, pp. 41–44.

Colin, L.: Encounter with Venus: An Update. Science, vol. 205, no. 4401, July 6, 1979, pp. 44–46.

Brodsky, R. F., ed.: Pioneer Venus Case Study in Spacecraft Design. American Institute of Aeronautics and Astronautics, Inc., 1979.

1980 Colin, L.: Pioneer Venus Overview. IEEE Trans. Geosci. Remote Sens., vol. GE-18, no. 1, Jan. 1980, pp. 3–4.

IEEE Transactions. Geoscience and Remote Sensing, vol. GE-18, no. 1, Jan. 1980.

Mutch, T. A.: Introduction to Pioneer Venus. Special Issue, J. Geophys. Res., vol. 85, no. A13, Dec. 30, 1980, p. 7573.

Colin, L.: The Pioneeer Venus Program. J. Geophys. Res., vol. 85, no. A13, Dec. 30, 1980, pp. 7575–7598.

Dorfman, S. D.; and Meredith, C. M.: The Pioneer Venus Spacecraft Program. Acta Astronaut., vol. 7, no. 6, 1980, pp. 773–795.

1981 Evans, W. D., et al.: Gamma-Burst Observations from the Pioneer Venus Orbiter. Astrophys. Space Sci., vol. 75, no. 1, Mar. 1981, pp. 35–46.

Head, J. W.; Yuter, S. E.; and Solomon, S. C.: Topography of Venus and Earth: A Test for the Presence of Plate Tectonics. Amer. Sci., vol. 69, no. 6, 1981, pp. 614–623.

Hunten, D. M.: The Upper Atmosphere of Venus: Implications for Aeronomy. Advances in Space Research, vol. 1, no. 9, 1981, pp. 3–4.

Limay, S. S.; and Suomi, V. E.: Cloud Motions on Venus: Global Structure and Organiza tion. J. Atmospheric Sci., vol. 38, no. 6, 1981, pp. 1220–1235.

Masursky, H.; Dial, A. L.; Schaber, G. G.; and Strobell, M. E.: Venus, a First Geologic Map Based on Radar Altimetric and Image Data. Proc. Lunar Sci. Conf., vol. XII, March 1981, pp. 661–663.

McGill, G. E.; Steenstrup, S. J.; Barton, C.; and Ford, P. G.: Continental Rifting and the Origin of Beta Regio, Venus. Geophys. Res. Lett., vol. 8, no. 7, July 1981, pp. 737–740.

Morrison, N. D.; and Morrison, D.: The Surface of Venus. Mercury, vol. 10, no. 1, Jan.–Feb., 1981, pp. 19–20.

Nozette, S.; and Ford, P.: A World Revealed: Venus by Radar. GAST, vol. 9, no. 3, 1981, pp. 6–15.

Phillips, R. J.; Kaula, W. M.; McGill, G. E.; and Malin, M. C.: Tectonics and Evolution of Venus. Science, vol. 212, no. 4497, 1981, pp. 879–887.

Reasenberg, R. D.; Goldberg, Z. M.; MacNeil, P. E.; and Shapiro, I.: Venus Gravity—A High-Resolution Map. J. Geophys. Res., vol. 86, no. 38, Aug. 10, 1981, pp. 7173–7179.

Russell, C. T.; Luhmann, J. G.; Elphie, R. C.; and Scarf, F. L.: The Distant Bow Shock and Magnetotail of Venus: Magnetic Field and Plasma Wave Observations. Geophysi. Res. Lett., vol. 8, no. 7, 1981, pp. 843–846.

Schubert, G.; and Covey, C.: The Atmosphere of Venus. Sci. Am., vol. 245, no. 1, 1981, pp. 66–74.

Taylor, F. W.; Elson, L. S.; McCleese, D. J.; and Diner, D. J.: Comparative Aspects of Venus and Terrestrial Meteorology. WTHRA, vol. 36, no. 2, 1981, pp. 34–40.

Watson, A. J.; Donahue, T. M.; and Walker, J. C. G.: The Dynamics of a Rapidly Escaping Atmosphere: Applications to the Evolution of Earth and Venus. Icarus, vol. 48, no. 2, Nov. 1981, pp. 150–166.

1982 Beatty, J. K.: Venus: The Mystery Continues. Sky and Telesc., vol. 63, no. 2, 1982, pp. 134–138.

Beatty, J. K.: Report from a Torrid Planet. Sky and Telesc., vol. 63, no. 5, 1982, pp. 452–453.

Campbell, P.: After Pioneer—Good Science, Bad News. Nature, vol. 296, no. 5852, 1982, pp. 13–14.

Donahue, T. M.; Hoffman, J. H.; Hodges, R. R., Jr.; and Watson, A. J.: Venus Was Wet: A Measurement of the Ratio of Deuterium to Hydrogen. Science, vol. 216, no. 4546, 1982, pp. 630–633.

Gold, M.: Through the Veil of Venus. Science82, vol. 3, no. 1, 1982, p. 12.

Kliore, A. J.; and Patel, I. R.: Thermal Structure of the Atmosphere of Venus from Pioneer Venus Radio Occultations. Icarus, vol. 52, no. 2, Nov. 1982, pp. 320–334.

Limaye, S. S.: Polarization Features on Venus. Bull. Am. Astron. Soc., vol. 14, 1982, pp. 740–741.

Nagy, A. F.; and Brace, L. H.: Structure and Dynamics of the Ionosphere. Nature, vol. 296, no. 5852, 1982, p. 19.

Seiff, A.; and Kirk, D. B.: Structure of the Venus Mesosphere and Lower Thermosphere from Measurements during Entry of the Pioneer Venus Probes. Icarus, vol. 49, no. 1, Jan. 1982, pp. 49–70.

Toon, O. B.; Turco, R. P.; and Pollack, J. B.: The Ultraviolet Absorber on Venus: Amorphous Sulfur. Icarus, vol. 51, no. 2, Aug. 1982, pp. 358–373.

1983 Phillips, R. J.: Plate Tectonics on Venus? Nature, vol. 302, no. 5910, 1983, pp. 655–656.

Schofield, J. T.; and Diner, D. J.: Rotation of Venus's Polar Dipole. Nature, vol. 305, no. 5930, Sept. 8, 1983, pp. 116–119.

Williams, B. G.; Mottinger, N. A.; and Panagiotacopulos, N. D.: Venus Gravity Field: Pioneer Venus Orbiter Navigation Results. Icarus, vol. 56, no. 3, Dec. 1983, pp. 578–589.

1987 McCormick, P. T.; Whitten, R. C.; and Knudsen, W. C.: Dynamics of the Venus Iono sphere Revisited. Icarus, vol. 70, no. 3, 1987, pp. 469–475.

Phillips, J. L.; and Russell, C. T.: Upper Limit on the Intrinsic Magnetic Field of Venus. J. Geophys. Res., vol. 92, no. A3, 1987, pp. 2253–2263.

Scarf, F. L.; Jordan, K. F.; and Russell, C. T.: Distribution of Whistler Mode Bursts at Venus. J. Geophys. Res., vol. 92, no. A11, 1987, pp. 12,407–12,411.

Stewart, A. I. F.: Pioneer Venus Measurements of H, O, and C Production in Comet P/ Halley near Perihelion. Astron. Astrophys., vol. 187, nos. 1–2, 1987, pp. 369–374.

Taylor, H. A., Jr.; Cloutier, P. A.; and Zheng, Z.: Venus "Lightning" Signals Reinterpreted as In Situ Plasma Noise. J. Geophys. Res., vol. 92, no. A9, 1987, pp. 9907–9919.

1988 Brace, L. H.; Theis, R. F.; Curtis, S. A.; and Parker, L. W.: A Precursor to the Venus Bow Shock. J. Geophys. Res., vol. 93, no. A11, 1988, pp. 12,735–12,749.

Fenimore, E. E.; et al.: Interpretations of Multiple Absorption Features in a Gamma-Ray Burst Spectrum, Part II. Astrophys. J., vol. 335, 1988, pp. L71–L74.

Kasting, J. F.; Toon, O. B.; and Pollack, J. B.: How Climate Evolved on the Terrestrial Planets. Sci. Amer., vol. 258, Feb. 1988, pp. 90–97.

Knudsen, W. C.: Solar Cycle Changes in the Morphology of the Venus Ionosphere. J. Geophys. Res., vol. 93, no. A8, 1988, pp. 8756–8762.

Mayr, H. G.; Harris, I.; and Kasprzak, W. T.: Gravity Waves in the Upper Atmosphere of Venus. J. Geophys. Res., vol. 93, no. A10, 1988, pp. 11,247–11,262.

Montoya, E. J.; and Fimmel, R. O.: Pioneers in Space: The Story of the Pioneer Missions (Part 1). Mercury, vol. 17, no. 2, Mar.–Apr. 1988, pp. 56–62.

Montoya, E. J.; and Fimmel, R. O.: Pioneers in Space: The Story of the Pioneer Missions (Part 2). Mercury, vol. 17, no. 3, May–June 1988, pp. 81–89.

Scarf, F. L.; and Russell, C. T.: Evidence of Lightning and Volcanic Activity on Venus; Pro and Con. Science, vol. 240, no. 4849, 1988, pp. 222–224.

1989 Kim, J.; Nagy, A. F.; Cravens, T. E.; and Kliore, A. J.: Solar Cycle Variations of the Elec tronic Densities Near the Ionospheric Peak of Venus. J. Geophys. Res., vol. 94, no. A9,1989, pp. 11,997–12,002.

Slavin, J. A.; Intriligator, D. S.; and Smith, E. J.: Pioneer Venus Orbiter Magnetic Field and Plasma Observations in the Venus Magnetotail. J. Geophys. Res., vol. 94, no. A3, 1989, pp. 2383–2398.

Woo, R.; et al.: Solar Wind Interaction with the Ionosphere of Venus Inferred from Radio Scintillation Measurements. J. Geophys. Res., vol. 94, no. 2, 1989, pp. 1473–1478.

Russell, C. T., ed.: Venus Aeronomy. Space Sci. Rev., vol. 55, nos. 1–4, Jan.–Feb. 1991, pp. 1–489.

Strangeway, R. J.: The Pioneer Venus Orbiter Entry Phase. Geophys. Res. Lett., vol. 20, no. 23, 1994, pp. 2715–2717.

Luhmann, J. G.; Pollack, J. B.; and Colin, L.: The Pioneer Mission to Venus. Sci. Amer., vol. 270, no. 4, April 1994, pp. 90–97.

Venus Exploration: Background and Other Missions

1963 Newlan, I.: First to Venus: The Story of Mariner II. McGraw-Hill, 1963.

1965 Mariner-Venus 1962: Final Project Report. NASA SP-59, 1965.

1971 Mariner-Venus 1967: Final Project Report. NASA SP-190, 1971.

1975 Hansen, J. E., ed.: The Atmosphere of Venus. NASA SP-382, 1975.

Conference on the Atmosphere of Venus. J. Atmos. Sci., vol. 32, no. 6, 1975, pp. 1005–1265.

1977 Murray, B.; and Burgess, E.: Flight to Mercury. Columbia Univ. Press, 1977.

1978 Dunne, J. A.; and Burgess, E.: The Voyage of Mariner 10: Missions to Venus and Mercury. NASA SP-424, 1978.

1979 Johnson, N. L.: Handbook of Soviet Lunar and Planetary Exploration. Science & Tech. Series, American Astronautical Soc., Univelt, Inc., vol. 47, 1979.

1981 Fichtel, C. E.; and Trombka, J. I.: Gamma Ray Astrophysics: New Insight Into the Universe. NASA SP-453, 1981.

1982 Burgess, E.: Venus: The Twin that Went Wrong. New Sci., vol. 94, no. 1310, June 17, 1982, pp. 786–789.

1983 Hunten, D. M.; et al., eds.: Venus. Univ. of Arizona Press, 1983.

Fimmel, R. O.; Colin, L.; and Burgess, E.: Pioneer Venus. NASA SP-461, 1983.

1985 Burgess, E.: Venus: An Errant Twin. Columbia Univ. Press, 1985.

1986 Byford, S.: Venus Unveiled by Vega Probes. SPFLA, vol. 28, no. 2, Feb. 1986, pp. 61–62.

1992 Robertson, D. F.: Venus—A Prime Soviet Objective. SPFLA, vol. 34, no. 5, May 1992, pp. 158–161.

Robertson, D. F.: Venus—A Prime Soviet Objective. SPFLA, vol. 34, no. 6, June 1992, pp. 202–205.

Saunders, R. S.; et al.: Magellan Mission Summary. J. Geophys. Res., vol. 97, no. E8, Aug. 25, 1992, pp. 13,067–13,090.

Magellan at Venus. J. Geophys. Res., vol. 97, no. E8, Aug. 25, 1992, pp. 13,062–13,689.

INDEX

A

B

C

D

E

F

G

H

I

J

K

L

M

N

P

Q

R

S

T

U

V

W

ACRONYMS

AO	Announcement of Opportunity
BIMS	Bus Ion Mass Spectrometer
BNMS	Bus Neutral Mass Spectrometer
CML	Ceramic microleak
COI	Co-investigator
CPAF	Cost plus award fee
DCE	Despin Control Electronics
DLBI	Differential Long Baseline Interferometry
DSN	Deep Space Network
DSU	Data Storage Unit
ESRO	European Space Research Organization
ETP	Electron Temperature Probe
GI	Guest Investigator
IAU	International Astronomical Union
ICBM	Intercontinental Ballistic Missile
ICE	International Cometary Explorer
IDS	Interdisciplinary Scientist
IMP	Interplanetary Monitoring Platform
IMS	Ion Mass Spectrometer
IR	Infrared Radiometer
IUE	Interplanetary Ultraviolet Explorer
JPL	Jet Propulsion Laboratory
LAS	Large Probe Atmospheric Structure Experiment
LCPS	Large Probe Cloud Particle Size Spectrometer
LED	Light-emitting diode
LGC	Large Probe Gas Chromatograph
LIR	Large Probe Infrared Radiometer
LN	Large Probe Nephelometer
LNMS	Large Probe Neutral Mass Spectrometer
LSFR	Large Probe Solar Flux Radiometer
NASCOM	NASA Communications System
NMS	Neutral Mass Spectrometer
OAD	Orbiter Atmospheric Drag Experiment
OCM	Orbiter Celestial Mechanics Experiment
OCPP	Orbiter Cloud Photopolarimeter
OEFD	Orbiter Electric Field Detector
OETP	Orbiter Electron Temperature Probe
OGBD	Orbiter Gamma Ray Burst Detector
OGPE	Orbiter Atmospheric Propagation Experiments
OIDD	Orbiter Internal Density Distribution Experiments
OIMS	Orbiter Ion Mass Spectrometer
OIR	Orbiter Infrared Radiometer
OMAG	Orbiter Magnetic Field Experiment
OMOP	Orbiter Mission Operations Planning
ONMS	Orbiter Neutral Mass Spectrometer
OPA	Orbiter Solar Wind Plasma Analyzer
OPTF	Operations Plan Task Force
ORAD	Orbiter Radar Mapping Instrument
ORO	Orbiter Dual Frequency Occultation Experiments
ORPA	Orbiter Charged Particle Retarding Potential Analyzer
OUVS	Orbiter Ultraviolet Spectrometer
PA	Plasma analyzer
PI	Principal Investigator
PAET	Planetary Atmosphere Experiments Test
PMCC	Pioneer Mission Computing Center
PMOC	Pioneer Mission Operations Center
PPO	Pioneer Project Office
PVI	Pioneer Venus Investigator
PVO	Pioneer Venus Orbiter
PVSSG	Pioneer Venus Science Steering Group
RFI	Radio frequency interference
RI	Radioscience Investigator
RPA	Retarding potential analyzer
SAS	Small Probe Atmospheric Structure Experiment
SN	Small Probe Nephelometer
SNFR	Small Probe Net Flux Radiometer
SSG	Science Steering Group
STDN	Spaceflight Tracking and Data Network
USGS	United States Geological Survey
UVS	Ultraviolet Spectrometer

THE AUTHORS

RICHARD O. FIMMEL

For 14 years until his retirement in 1994, Richard Fimmel was project manager for all the Pioneer missions at Ames Research Center. Earlier, he was responsible for encounter planning for the Pioneer missions to Jupiter and Saturn. As science chief, he planned for and operated the science instruments during their pioneering encounters with the two giant planets of the Solar System. Later he managed the extended missions of these spacecraft as they became the first explorers of the outer Solar System environment.

For Pioneer Venus he planned and directed the science-related activities of the Orbiter mission. He managed the 12-year Orbiter extended mission, during which a wealth of important information was gathered about Venus, its surrounding space environment, and its interaction with the solar wind. Fimmel was also an invited consultant to the German Space Agency for the HELIOS and AMPTE missions.

In his retirement, Fimmel continues to contribute technically to the NASA space program. He is an associate at Ames Research Center, applying his experience of over 50 years in research, design, development, and management of advanced instrumentation, computers, and data systems.

Fimmel received a BSEE in 1949 from Rutgers University followed by an MSEE in 1954. Co-author of several books on the Pioneer missions, he also has authored technical and scientific papers on spacecraft data management systems. Fimmel has participated on panels, working groups, and committees to develop design and systems criteria. NASA has twice awarded Fimmel the Exceptional Service Medal for his work on the Pioneer missions.

LAWRENCE COLIN

As project scientist for the Pioneer Venus program from its inception at Ames Research Center in late 1971 until 1994, Larry Colin participated in all mission phases—design, development, data acquisition, analysis and interpretation, and data dissemination. He was awarded the NASA Distinguished Service Medal for his participation and accomplishments.

Colin joined Ames Research Center as a research scientist in 1964 and, although he retired in 1988, he continued as the Pioneer Venus project scientist. Prior to his retirement Colin was the Chief of the Space Science Division. From 1962 to 1964 he was research scientist at the USAF Rome Air Development Center. Colin received a BEE from the Polytechnic Institute of Brooklyn in 1952, an MEE from Syracuse University in 1960, and a Ph.D. from Stanford University in 1964.

Since his retirement, Colin has devoted his time to teaching at Stanford University in space system engineering, astronomy, and the past, present, and future of space exploration.

ERIC BURGESS

Consultant, author, lecturer, and journalist, Eric Burgess has written about the Pioneer space missions since the first Air Force lunar probe in 1957. Author of many of the earliest books and technical and popular articles on upper atmosphere physics, guided missiles, rocket propulsion, and spaceflight, Burgess' by-line has appeared internationally in magazines, aerospace journals, and newspapers for over 50 years. He aided the formation of the British Interplanetary Society in the 1930s and the International Astronautical Federation in 1950. Burgess wrote the first detailed technical papers on unmanned communications satellites in geosynchronous orbit and the first technical paper on a planetary probe spacecraft. In the 1940s he researched and wrote the most comprehensive declassified survey of World War II rocket and missile development and had a Royal Air Force prize essay published on the future of large ballistic missiles.

Burgess is a Fellow of the Royal Astronomical Society, Fellow of the British Interplanetary Society, Associate Fellow of the American Institute of Aeronautics and Astronautics, and a Member of the National Association of Science Writers. He has held management, financial, and technical positions in companies ranging from textiles to advanced computing and electronic systems, and has acted as technical and storyboard consultant on major motion pictures. Most recently, Burgess concluded an eight-volume series of books with Columbia University Press on spacecraft exploration of the Moon and all the planets.

This book was produced by the Ames Technical Information Division, David A. Faust and Karen D. Thompson, designers. The text was set in 9 point type on a 13 point line, using Stone Serif typeface. Headlines and Subheads were set in Trajan bold typeface.